Progress in Physics
Volume 19

Series Editors
Anne Boutet de Monvel
Gerald Kaiser

Picture on the cover is a simplified version of a figure created by Oliver Conradt (Department of Physics and Astronomy, University of Basel, Klingelbergstrasse 82, CH-4056 Basel, Switzerland). It depicts a projective coordinate system in a planar field. For more information we refer to Conradt's paper *The Principle of Duality in Clifford Algebra and Projective Geometry* in Volume 1, Algebra and Physics, edited by Rafał Abłamowicz and Bertfried Fauser (*Progress in Physics*, **18**).

Clifford Algebras
and their Applications
in Mathematical Physics

Volume 2: Clifford Analysis

John Ryan
Wolfgang Sprößig
Editors

Springer Science+Business Media, LLC

John Ryan
Department of Mathematics
University of Arkansas
Fayetteville, AR 72701
U.S.A.

Wolfgang Sprößig
Faculty of Mathematics and Informatics
Freiberg University of Mining and Technology
09596 Freiberg
Germany

Library of Congress Cataloging-in-Publication Data

Clifford algebras and their applications in mathematical physics.
 p. cm. – (Progress in physics ; v. 18-19)
 Includes bibliographical references and indexes.
 Contents: v. 1. Algebra and physics / Rafał Abłamowicz, Bertfried Fauser, editors–v.
2. Clifford analysis / John Ryan, Wolfgang Sprößig, editors.
 ISBN 978-1-4612-7119-2   ISBN 978-1-4612-1374-1 (eBook)
 DOI 10.1007/978-1-4612-1374-1

 1. Clifford algebras. 2. Mathematical physics. I. Abłamowicz, Rafał. II. Progress in
physics (Boston, Mass.); v. 18-19.

QC20.7.C55 C55 2000
530.15'257–dc21   00-034310

AMS Subject Classifications: 15A66, 15A69, 30E20, 30E25, 30G35, 30J25, 31B05, 31C12, 31-XX,
32A26, 32-XX, 35A08, 35A22, 35G35, 35J70, 35-XX, 41A10, 42B05, 42B15, 42B20, 42B25, 42B30,
44A35, 47A40, 47A60, 53A04, 53A30, 53A55, 53C29, 53C20, 58A14, 58J05, 58J32, 78-XX, 81Q70,
81T13, 81T60

Printed on acid-free paper.
© 2000 Springer Science+Business Media New York
Originally published by Birkhäuser Boston in 2000
Softcover reprint of the hardcover 1st edition 2000

ISBN 978-1-4612-7119-2

Typeset by the editors in LᴬTₑX.

9 8 7 6 5 4 3 2 1

Contents

Preface to Volume 2

This volume of contributed papers arises from the section on Clifford analysis which took place as part of the "5th International Conference on Clifford Algebras and Their Applications in Mathematical Physics," held in Ixtapa, Mexico, June, 1999. Like in Volume 1, the majority of papers are an outgrowth and further development of the talks presented at the conference. All papers in this volume were refereed and were further developed after the conference. The editors are grateful to the referees for their extremely valuable assistance in creating this volume. The papers gathered here reflect some of the latest developments in the field of Clifford analysis and its applications. Topics range from the study of generalized Schwarzian derivatives to applications to boundary value problems and singular integrals. Topics covered also include links to a Möbius invariant function theory on hyperbolic space, analogues of Ahlfors-Beurling inequalities and their applications, differentiability properties of monogenic functions, links to supersymmetry, hyperbolic Dirac equations, and scattering theory. The papers appearing here can be broadly subdivided into the following categories.

Partial Differential Equations and Boundary Value Problems

As is well known, the complex Beltrami equation is of enormous importance in the general theory of elliptic equations and has many applications to other fields of analysis and geometry. Because of the many degrees of freedom in the combination of partial differential operators in higher dimensions, there is a large variety of generalized Beltrami equations. To motivate the treatment by Clifford analytic methods, U. Kähler shows some typical difficulties (integrability conditions, conditions on the Jacobian, etc.) in higher dimensions. The paper is a comprehensive survey of the study of Beltrami type equations in the three-dimensional setting. The results are based on integral operator methods and include detailed norm estimates for integral operators in different function spaces.

In her article, Xinhua Ji shows the relationship between Green's function, Möbius transformations and the Laplace-Beltrami operator acting on the one point compactification of $\mathbb{R}^n$. The author gives a complete description of all non-Euclidean translations of the corresponding Möbius group. With the aid of a geodesic distance, she obtains the fundamental solution of the degenerate Laplace-Beltrami equation. Finally, she solves the Dirichlet problem for the non-homogeneous Laplace-Beltrami equation. This paper contains a very nice example of a generalized Poisson kernel illustrating that the maximum principle is no longer valid for degenerate elliptic equations.

W. Sprößig considers a stationary problem from fluid mechanics using methods from Clifford analysis. Stationary Navier-Stokes equations are combined with field induction. Under certain conditions the solution can be obtained by an iteration procedure which converges rapidly in Sobolev spaces.

Singular Integral Operators

In their paper, Tao Qian, John Ryan and Xinhua Ji study Fourier multipliers and singular integral operators on Möbius images of Lipschitz graphs and starlike Lipschitz surfaces. All basic notation is explained. It is shown that the singular integral operators form an operator algebra. The main ideas are the application of a generalized Fourier transform and the use of Möbius transformations to pullback known results. One of the main results is the fact that, in the case of Lipschitz graphs, the operator algebra of singular integrals may be identified with the bounded holomorphic Fourier multipliers. The same problems are considered for starlike Lipschitz surfaces.

J. B. Reyes and R. A. Blaya study quaternionic Cauchy integrals on Ahlfors regular surfaces. In this very general setting Plemelj-Sokhotzkij formulae are deduced. These results can be applied to prove the solvability of a special case of Riemann's problem.

In M. Martin's article he proves some Hedberg type inequalities for the convolution operator associated with the Cauchy kernel in Euclidean space. Each of these inequalities involves a specific maximal operator and they all provide the best possible constants. Applications of Clifford analysis are also presented. Among them, one should single out a higher-dimensional generalization of a classical inequality in one-variable complex function theory due to Ahlfors and Beurling and some extensions of Alexander's inequality.

Applications in Geometry and Physics

S. Bernstein describes a new application for the Borel-Pompeiu formula in $\mathbb{C}^n$. It is obtained as a direct analogue of the Martinelli-Bochner formula. An extremely interesting feature here is her application of these results to scattering theory.

In G. Kaiser's contribution complex distance is applied to describe a useful extension of potential theory in $\mathbb{R}^n$ to $\mathbb{C}^n$. The resulting Newtonian potential is generated by an extended source distribution $\tilde{\delta}(z)$ in $\mathbb{C}^n$ whose restriction to $\mathbb{R}^n$ is the point source $\delta(x)$. This provides a possible model for extended particles in physics. In $\mathbb{C}^{n+1}$, interpreted as complex spacetime, $\tilde{\delta}$ acts as a propagator generating solutions of the wave equation from their initial values. This gives a new connection between elliptic and hyperbolic equations that does not assume analyticity of the Cauchy data. Generalized to Clifford analysis, it induces a similar connection between solutions of elliptic and hyperbolic Dirac equations, thereby extending earlier work of J. Ryan. There is a natural application to the time-dependent, inhomogeneous Dirac and Maxwell equations and the so-called electromagnetic wavelets.

The paper by J. Snygg deals with the use of Clifford algebras in differential geometry, especially involving the holonomy group. The Clifford algebra is used to express isometry operators.

In his paper F. Sommen presents an extension of Clifford analysis using both commuting as well as anti-commuting variables, thus, following the lines of thinking of supersymmetry. For abstract vector variables, the calculus remains the same. He illustrates that the radial algebra can be represented by both the use of commuting and anti-commuting variables. Another important fact is that the action of both the symplectic group and the rotation group are united in the super spin-group. This contribution suggests that the radial algebra is a natural background for supersymmetry because it is independent of dimension and invariant in the group theoretic sense.

J. Tolksdorf investigates bosonic and fermionic action in a Euclidean version of the standard model of particle physics in terms of elliptic Dirac operators on compact smooth even-dimensional manifolds without boundary. He explains the specific role of "standard Dirac operators" for fermionic action in contrast to the role of non-standard operators of Dirac type for bosonic action and he relates the Wodzicki residue (WR) to this action, thereby illuminating also an interesting interrelation with Connes' noncommutative geometry.

Möbius Transformations and Monogenic Functions

In their contribution M. Wada and O. Kobayashi define Schwarzian derivatives of transformations of $\mathbb{R}^n$ in terms of Schwarzian derivatives of regular curves and prove that a transformation of $\mathbb{R}^n$ is Möbius if and only if its Schwarzian derivatives are constantly zero. The framework then is modified to prove a result on immersions between Riemannian manifolds. The paper gives a new definition for Schwarzian derivative (cf. [1] and [3]) and shows accordingly the necessary and sufficient condition for a transformation being Möbius in terms of its Schwarzian derivatives.

In the last decades the construction of infinitesimal generators of the conformal group has played an important role in the study of monogenic functions. Several authors, including P. Lounesto, J. Ryan, F. Sommen and H. Leutwiler, have results on this topic. For instance F. Sommen [4] has obtained in his paper on monogenic operators a Taylor series formula using the conformal group. Here Y. Krasnov presents an explicit Taylor series formula by making use of the conformal group and its generators.

The paper of T. Hempfling is devoted to the hyperbolic modification of Clifford analysis. He considers the Cauchy-Riemann operator in $\mathbb{R}^{n+1}$. By splitting this operator into a radial and spherical part, he can formulate conditions for the radial part to vanish. This is especially true if the corresponding functions are hypermonogenic.

In 1978, A. Sudbery (cf. [5]) was the first to define in a quaternionic setting the derivative as differential coefficient between two forms of higher degree. Here H. Malonek succeeds in getting a similar result within the Clifford algebra $C\ell_{0,n}$ by considering differential forms of degree $n - 1$ and n as well as using the Hodge star operator; see also earlier results of J. Ryan cited in H. Malonek's paper. Here H. Malonek introduces the notion of the hypercomplex derivability of a function defined for paravectors with values in the n-dimensional real Clifford algebra. For instance, a real valued function f is called (left) derivable if and only if $A_{f,\ell}(z) \in C\ell_{0,n}$ exists and

$$d(d\tau f) = d\sigma A_{f,\ell}(z),$$

where $d\sigma$ denotes a hypercomplex differential of degree n and $d\tau$ a hypercomplex differential of degree $n - 1$. The main result is contained in a theorem that shows that a real differentiable function f is (left) derivable if and only if f is (left) monogenic.

H. Leutwiler shows in [2] and some other papers that the power functions are solutions of the modified Cauchy-Riemann system

$$x_n Df + (n - 1)f_n = 0.$$

This equation is closely related to the hyperbolic metric. Consequently, these functions are closely connected to the Laplace-Beltrami equation. H. Leutwiler and Sirkka-Liisa Eriksson-Bique introduce in their paper hypermonogenic functions as a generalization of classical complex functions. These hypermonogenic functions behave in relation to the Laplace-Beltrami operator like monogenic functions to the Laplacian. They introduce operators P and Q which can be understood as projections onto the real and "imaginary" parts of the Clifford algebra $C\ell_{n-1}$. The main theorems describe hypermonogenic functions, their relation to solutions of the hyperbolic Dirac operator and a representation formula.

Another paper devoted to modified Clifford analysis is presented by P. Cerejeiras. On the basis of results obtained by H. Leutwiler, J. Cnops, and J. Ryan, the existence of a Poisson-Szegö kernel for the Laplace-Beltrami equation associated with an n-dimensional hyperbolic space is proven. Poisson-Szegö kernels are explicitely constructed for a large class of orientable manifolds by using the initial kernel for the spherical model of hyperbolic spaces and the usual Möbius transform. These kernels solve a generalized Dirichlet problem.

The editors express their gratitude to Amy Knox for her very thorough proofreading of all papers.

John Ryan, Fayetteville, Arkansas, U.S.A.
Wolfgang Sprößig, Freiberg, Germany   April 1, 2000

REFERENCES

[1] L. V. Ahlfors, Cross-ratios and Schwarzian derivatives in $\mathbb{R}^n$, *Complex Analysis*, J. Hersch and A. Huber, eds., articles dedicated to Albert Pfluger on the occasion of his 80th birthday, Birkhäuser, 1988, 1–15.

[2] H. Leutwiler, Rudiments of a function theory in $\mathbb{R}^3$, *Forum Math.* **7** (1995), 279–305.

[3] J. Ryan, Generalized Schwarzian derivatives for generalized fractional linear transformations, *Ann Polon. Math.* **LVII** (1992), 29–44.

[4] F. Sommen, N. Van Acker, Monogenic differential operators, *Results in Math.* Vol. **22** (1992), 781–798.

[5] A. Sudbery, Quaternionic analysis, *Math. Proc. Cambr. Phil. Soc.* **85** (1979), 199–225.

Preface to Volume 1

The last conference on "Clifford Algebras and Their Applications in Mathematical Physics," the 5th of this well-known series, took place in Ixtapa-Zihuatanejo, Mexico, from June 27–July 4, 1999, in the beautiful surroundings of the Pacific Coast. The first conference of this series was organized in 1985 in Canterbury, United Kingdom, and was initiated by J.S.R. Chisholm at a time when Clifford algebras were just becoming recognized tools. Under the leadership of D. Hestenes, among others, Clifford algebras had not only entered various fields by providing an elegant and powerful tool for solving geometric problems, but, more importantly, Clifford algebras have provided a unique approach to reasoning in mathematics and physics. As a natural consequence of this development, the conferences of this series have had a large impact and have managed to form a "Clifford community."

The topics covered by the recent conference can be divided into two major parts: Clifford analysis and mathematical physics. This structure is reflected by the division of the presented contributions into two volumes: *Algebra and Physics* (Volume 1), and *Clifford Analysis* (Volume 2). The majority of papers are an outgrowth and further development of the talks given by the contributors at the conference. All papers in these two volumes have been refereed and were further developed after the conference.

There will also be a special issue of the *International Journal of Theoretical Physics*, edited by D. R. Finkelstein and Z. Oziewicz and containing invited papers. This too demonstrates the innovative and flourishing ideas in Clifford algebras.

During the main conference, two special sessions were organized. One was Applied Clifford Algebra in Cybernetics, Robotics, Image Processing, and Engineering (ACACSE), organized by E. J. Bayro-Corrochano and G. Sobczyk; the other was Global and Local Problems for Dirac Operators, organized by E. R. de Arellano, J. Ryan, and W. Spröβig. The goal was to gather people with interests in applications of Clifford algebras in engineering, robotics, computer vision, and symbolic computer algebra, or in the mathematics of Dirac Operators. The ACACSE activities will be presented elsewhere, while the Dirac operator contributions belong to *Clifford Analysis* (Volume 2).

The increasing interest in Clifford algebras has some deep foundations. Geometrical methods have seen a revival of making use of Clifford algebras which have been exactly designed to serve as geometrical algebras, combining the power of geometric intuition with the power of algebraic manipulations – which had been a dream of Leibniz followed by Graßmann, Peano, Hamilton, Cayley, Clifford, Boole, Rota, Hestenes, and others. It is remarkable to note the wealth of contributions to these volumes using conformal, projective, and hyperbolic geometries, not only in algebraic settings but also in Clifford analysis. The advent of these types of geometries during the 19th century gave birth to quaternions, Graßmann and Clifford algebras, and finally, to modern algebraic geometry.

Clifford algebras are used almost everywhere in mathematics and physics. Most problems can be encoded via a pair of a linear space and a quadratic form – some people even assign a value to such objects as line segments, areas, etc. However, this is already sufficient to construct a unique Clifford algebra. Multiplication makes the considered objects behave just like "numbers" and makes them easier to manipulate.

Clifford algebras have produced valuable applications. Besides their fundamental aspects, Clifford algebra, or quaternion, methods have produced well-recognized applications – even if sometimes disguised by matrix representations. Computer vision, robotics, navigation, space flight, and other areas also use these techniques.

Clifford algebras and their accompanying Graßmann-Cayley algebras broaden fields of thought. Automatic theorem proving, which might be important for autonomous robot systems, is based on this connection to Clifford algebras. Pioneered by Gian-Carlo Rota, the idea of connecting Hopf algebras with Clifford structures was also developed. Taking Clifford numbers as entities leads to new physical principles. Deformed Clifford algebras are used to solve problems in quantum field theory. Clifford algebraic computations provide a challenge for computer algebraic systems and open the new and fascinating area of experimental mathematics. Clifford algebras are bound to play a major role in quantum computing and the design of quantum computers.

Most of these currents in the Clifford community have found their way into these volumes. In this way, these books will contribute to the development of the field. The following outline will provide a subjective guide to the contributions (avoiding technical terms as much as possible) and will highlight their main features.

Dedication to Gian-Carlo Rota

Gian-Carlo Rota was invited to be a plenary speaker at this 5th International Conference on Clifford Algebras and Their Application in Mathematical Physics. His talk would have been a highlight of the conference. But he died suddenly, shortly before the conference. For this reason, the organizers of this conference arranged a special session dedicated to the memory of Professor Rota.

Several lecturers including David Ritz Finkelstein and Bernd Schmeikal, who knew Rota personally, took a chance to present some warm remarks about Professor Rota, showing him as a person, scientist, philosopher and a beloved man.

Physics – Applications and Models

Baylis: The linear space underlying a Clifford algebra is coming conventionally equipped with multi-vector gradings. The common use is that physically different entities are mapped onto the particular gradings. The advantage of this convention is that physical transformations become inner automorphisms while the disadvantage is the introduction of possibly non-physical or unnecessary variables. The present paper studies paravectors, which are sums of scalars and vectors and constitute themselves a graded space of a coarser grading than that of the multi-vector grading. Some mathematical outcomes are explained, and the usefulness of the paravector picture is demonstrated for plane waves and wave packets in electrodynamics. A paravector arises from a spacetime split, as discussed by Conradt.

Dray & Manogue: Two component Weyl spinors, also extensively used by Penrose, and four component Dirac spinors are widely known in mathematics and high energy physics. Usually the latter are constructed by the former in adding two unequivalent representations of Weyl spinors, differing by the outer automorphism of complex conjugation, in a direct sum. However, the authors present a generalization of Weyl spinors by extending the base field of complex numbers $\mathbb{C}$ to the quaternions $\mathbb{H}$, a skew field and a non-commutative division ring. As an advantage, one is able to describe massless and massive Dirac equations on the same footing by two-component quaternionic spinors. As a logical step, the two-component octonionic Dirac equation, which turns out to be 10-dimensional, is examined. Dimensional reduction takes place by singling out a unit – i.e., invertible element – in the octonions, or equivalently, a unique complex subalgebra. This allows one to find the variety of particles of a generation of leptons. The three distinct possibilities for singling out a sub-quaternion algebra are conjectured to carry the three families or generations. This mechanism should be compared with Schmeikal's finding several copies of $SU(3)$ as presented in his paper.

Just & Thevenot: As is well known, Dirac's theory contains the mass-term which comes with the time-like γ-matrix. In some phenomenological models, potentials are used which are not only vectorial in nature – having a γ-matrix – but involve scalar, vector, tensor, axial or pseudo vector, and pseudo scalar contributions. Especially the anomalous magnetic moment bears a tensorial character. Such terms are called Pauli terms. This paper considers the question of whether or not such general potential is possible and whether it is compatible with Dirac theory. As a surprising outcome, the authors present arguments against the presence of Pauli terms. One should note, however, that these terms are not necessary in the standard model and may be rendered superfluous.

Lewis, Lasenby, & Doran: Scattering experiments are still the common source of information in experimental high energy physics. S-matrix theory establishes the theoretical counterpart. An awkward and technical detail of this theory is the calculation of spin-sums of spinning particles. Using spacetime algebra (STA) – the Clifford algebra of Dirac theory – a spin direction can be introduced when the S-matrix is replaced by an STA operator. This operator can depend on the spin direction. In this way, a plain and straightened method is developed to work directly with the invariant spin direction avoiding spin sums and a choice of basis. However, some improvements have to be made for multi-particle spin states. Some achievements toward a proper formulation of multi-particle STA are made by the contributions of Fauser & Abłamowicz.

Physics – Structures

Bette's contribution reviews a twistor phase space picture which was developed in the past. After having introduced the twistor phase space, the first point is the introduction of shifted position coordinates which fulfill a non-trivial Poisson bracket and are, thereby, non-commuting. These positions are physically motivated by the requirement that the inner product of momentum and Pauli-Lubanski spin-part vanish. This requirement is motivated by Dirac quantization and conservation laws. In fact, since one wants to have momenta represented by derivatives, one looks for Noether currents. A classical spinning particle is given as an example in $\mathbf{Tp(2)}$, and the resulting equations of motion are shown to be different from Bargman-Michel-Telegdi equations.

Johnson: Clifford algebras have, as an interesting sub-structure, discrete groups sometimes called Dirac groups. In fact, one can construct Clifford algebras over the reals by a ring extension from the group algebras of these cyclic and dihedral discrete groups. Based on this idea, the author tensors two such structures to get a prototype of a fiber of a configuration space manifold. After defining a suitable subspace as the tangent space, he is able to incorporate the action in the complement which finally constitutes a constraint in the tensor space. A $1 + 1$ dimensional model is considered.

Pezzaglia criticizes that a mere reformulation of physics in new mathematical formalisms does not lead to new physics even if there might be a great achievement in straightening out the problem at hand. However, changing the perspective might be a key step in being able to generalize physical principles and to reach new physics. The author's key point is to look at a Clifford element as an entity even if it can be split into multi-vectors. Therefore, he interprets every multi-vector part of a Clifford number as a physical quantity with its own coordinate. Indeed, this is the situation in quaternion theory, where the "vector" part is given by the linear span of $\mathbf{i}, \mathbf{j}, \mathbf{k} = \mathbf{ij}$, and $\mathbf{k}$ is algebraically –but not linearly– dependent. As an example, special relativity is revisited, and a "polydimensional" (ungraded) formulation of physics is developed. Multi-vector valued, or "matrix," derivatives and the action principle are formulated. Papapetrou's equation and Crawford's hypergravity provide examples of the method in classical physics.

Piazzese: A real linear space can be equipped with different metrical forms, e.g., Euclidean or Minkowski. In this contribution, a relation is established for (time-like) vectors of a Minkowski space to such a Euclidean space without resorting to complex numbers. The remarkable fact is that one is able to transport transformation laws into the Euclidean picture. This allows the author to propose a quasi-classical description of a particle. Besides the classical energy, formed with use of the relativistic velocity – i.e., quasi classical – a second term arises in this description, which can be connected – via de Broglie's idea that every particle has an "internal clock," or an internal frequency – to an internal degree of freedom. This freedom might be of rotatory nature, and it might be connected to spin. The quasi-classical energy becomes frame independent.

Vargas & Torr provide a clear introduction to the unification program which had been developed by the authors in the past. The second section concentrates on notation not only for clarity but also to point out some difficulties, which are usually ignored, but which prove to be most important in the later development. A particular point presented in the paper is the distinction between the Dirac-Hestenes equation and Kähler theory of Dirac equations. Differential forms are introduced. Using the Hestenesian idea of mathematical viruses, two major diseases of common treatments, the "transmutation virus" and the "bachelor algebra virus" are described, and their prevention is discussed. A motivation to alter the Kähler approach by introducing a Kaluza-Klein type theory is given which is later on related to Finsler geometry. Connection is made with previous developments, pointing out a new way of deriving the results. Finally, the interior derivative in the particular case of the previously given Kaluza-Klein theory is introduced. The conclusion comes up with some very interesting outlooks about the further development of the theory. A full geometrical (invariant) Kähler equation for Clifford valued clifforms is given.

Geometry and Logic

Conradt: Duality is currently a well-recognized structure in physical theories. In string theory, duality connects strong and weakly interacting models which allows a perturbative approach to the former. Duality originated in projective geometry where it appeared as the striking fact that every projective theorem has a dual theorem if one interchanges several notions as point with plane, join with meet, etc. Since Clifford geometric algebras are known to describe metric and projective geometries, the paper calls for an implementation of this rather fundamental projective concept. Starting by defining a Clifford algebra, the author uses Poincaré duality of the underlying multi-vector space and asks the question if one could use $(n-1)$-vectors – the isomorphic picture of 1-vectors – to construct a dualized Clifford algebra. It turns out that the meet of the $(n-1)$-vectors can be seen to establish the dual outer product. In the same way, a dual Clifford product can be defined, and a dual Clifford algebra can be formulated in the same vector space as the original one, but with all r-multi-vectors mapped to $(n-r)$-multi-vectors of the Poincaré dual space. Such a duality for meet and join has been investigated by Rota and others. The resulting algebra is called double or Grassmann-Cayley algebra and constitutes a substructure of the Clifford approach presented here. The interpretation of such dual Clifford algebras in terms of projective geometry is given. Projective coordinate systems for points, lines, and planes are exemplified. The linear complex is introduced and a motivation provided as to where to use this structure in physics.

Li: Non-Euclidean geometry was one of the main working fields of W. K. Clifford, which might well have influenced his algebraic ideas. Furthermore, non-Euclidean geometries have already been described by Euclidean models by Felix Klein in the 19th century. This fact provides the basis of the study of hyperbolic geometries by Clifford algebraic techniques, singling out appropriate subspaces. Geometric facts can be encoded in the Grassmann-Cayley algebra, called "double algebra" by Rota. The main achievement of such an algebraization is that it opens an analytic and invariant approach to geometric problems. Automated geometric theorem proving is a possibility explored in this contribution. Not only machine based recalculations of theorems but the quite more interesting proving of new theorems, thus, becomes possible. Beside its beauty, this method is important also for applications in robotics and visualization.

Schmeikal: A ring of idempotent elements, or (alternatively) a ring over $\mathbb{Z}_2$, is a Boolean ring. Any assertions in such a ring can be interpreted as true or false, while the operations in this algebra become inference in a certain logic. Boole, de Morgan, Frege as McCulloch, Parry, or Peirce, or Zellweger have considered pictographical notations of logical conjunctions. Schmeikal adds in this contribution a description by the subring of idempotent elements of a Clifford algebra. The set of all idempotents constitutes a lattice, and, thereby, one can define "basic reflections," i.e., involutive automorphisms. Since representation spaces can be seen to be ideals generated by idempotents, such reflections connect different spaces. Following an idea of Chisholm, one can recover the fundamental $SU(3)$ representation by fixing one primitive idempotent out of four and then by looking at its stabilizer group. The octahedral symmetry of the idempotent lattice of the Dirac algebra is examined, and six copies of $SU(3)$ are detected. However, the picture gets more complicated by the introduction of generalized logic operators, which act on logic assertions. This tool, after being developed, is applied to establish the logic of Dirac spinors. Discrete symmetries are discussed as examples. Finally, it is outlined how this model can be enlarged to $Cl_{n,n}$. It is pointed out that using wave functions, and not $\mathbb{Z}_2$, provides an example of quantum logic where "tertium datur."

Mathematics – Deformations

Abłamowicz & Fauser: Indistinguishable particles are described by wave or partition functions invariant under the permutation group. This observation, made by physicists in thermodynamics and quantum mechanics, has influenced the theory of group representations to a large degree. Already Weyl showed that multi-particle states can be classified by a method developed by Young. Multi-particle Clifford algebras have, thus, to carry an action of the symmetric group. This article studies representations of the deformation of the group algebra of the symmetric group known as the Hecke algebra. Deformations have been proposed to serve as symmetries for composed entities. This provides the main motivation for this investigation and is in full accord with Fiore's contribution. Representations are constructed in ideals generated by q-idempotents in quantum Clifford algebras. These idempotents are Young operators. To get an intimate relation between both structures, reversion is taken to act as conjugation on Young operators, i.e., Young idempotents. Detailed computations are possible only by computer algebra. CLIFFORD, a Maple V package for (quantum) Clifford algebras – developed by one of the authors R.A. – has made algebraic computations possible. The representation theory in the 2- and 3-dimensional cases is developed in full detail.

Fiore: Any Clifford algebra naturally has a Lie algebra substructure. Therefore, it is interesting to ask if it is possible to q-deform this Lie algebra, or, more precisely, its enveloping algebra U_q. As a natural outcome of covariance under the action of such a quasi-triangular Hopf algebra, one obtains a q-deformed Clifford algebra. However, due to finite dimensionality in the orthogonal case, it is possible to express the deformed generators as polynomials in the undeformed ones. As a consequence, q-deformed creation and annihilation operators might be interpreted as creation and annihilation operators of effective, or compound, entities. A detailed analysis of the situation gives a connection between deformed and undeformed invariants.

It is shown under which conditions non-trivial deformations, having new invariants, actually occur.

Rosenbaum & Vergara point out that recently at two different places closely related Hopf algebra structures have popped up. The first one is the Connes-Moscovici Hopf algebra, which originates from non-commutative extensions of Riemannian geometry. The second is the Hopf algebra of rooted trees, which has been employed by Connes and Kreimer to produce the combinatorics of renormalization in perturbative quantum field theory. Decorated rooted trees are used to establish the forest theorems of renormalization, and the antipode action generates the counter terms in all orders. The aim of this work is to present both types of Hopf algebras in an invariant, coordinate-free language. Furthermore, it is shown that there might be a connection between the Dirac operator, spacetime at Planck scale, and the above two Hopf algebras, which may lead toward a finite quantum field theory. The connection to Schwinger-Dyson equations is discussed.

Vancliff: Neither physics nor mathematics is currently able to present a concise model of a non-commutative space as a coordinate space of a quantum group or other deformed algebras. Common methods fail to work since in such non-commutative spaces one cannot find a Poincaré-Birkhoff-Witt like basis. Following an idea of S.P. Smith, this work, being projective and geometric in nature, connects geometric data to deformed algebras. The Sklyanin algebra serves as a model to exemplify the problem at hand. The deep connection between projective point, line, etc., schemes and quantum spaces is explored. Roughly speaking, the quantum space of an algebra is a quotient category of graded modules. Certain modules play the role of points; others play the role of lines, and so forth. In the commutative setting, this idea can be traced back to J.-P. Serre. Quantum spaces are constructed via Poisson geometry; examples are provided.

Mathematics – Structures

Belinfante: Spinors have been discovered by Élie Cartan when classifying complex semi-simple Lie algebras. During further development done by Cartan, Freudenthal, Dynkin, Chevalley, and others, it became clear that one can construct spinor modules of **B** and **D** type Lie algebras over the integers. The integers play a fundamental role in the classification also. Spinors are connected, however, to Clifford algebras constituting their natural irreducible representation spaces. As a natural approach, spinor and semi-spinor modules are constructed for complex orthogonal Lie algebras – i.e., Lie algebras of types **B** and **D**. First and second Clifford algebras are introduced, and the mechanism using Dynkin diagrams to construct spinor modules is explicitly given. $\mathbf{B}_1$ and $\mathbf{B}_2$ provide examples; spin groups are discussed. MATHEMATICA code is provided which was used to check the results and invites the reader to redo the computations for gaining deeper insights into the theory.

Fauser & Abłamowicz: Clifford algebras over real and complex numbers are classified. As an outcome of this classification, every Clifford algebra can be decomposed into graded tensor products of "atomic," i.e., indecomposable, Clifford algebra factors. These factors are at most of algebraic dimension 4. This decomposition is the origin of periodicity theorems and vice versa. Albert Crumeyrolle stated that the decomposition properties of Clifford algebras provide the Mendeleïev periodic system of elementary particles. However, the situation can be much more complicated. Defining quantum Clifford algebras, i.e., Clifford algebras of an arbitrary bilinear form, it can be shown that common periodicity theorems in general fail to hold. A detailed introduction provides arguments that this is not the exception, but, instead, it should be seen as the rule. The Wick theorem of normal-ordering in quantum mechanics and quantum field theory establishes a quantum Clifford algebra structure. After the detailed and rigorous development of the theory, three examples are provided showing the theory at work.

Fernández, Moya & Rodrigues present in their paper the theory of covariant derivative operators on a Minkowski manifold. As a main tool of their investigation, the concept of multiform calculus is used. It is developed by passing directional covariant derivative operators and associated operators to covariant derivative operators which are compatible with a non-degenerate symmetric tensor in a Minkowski manifold. This seems to be one of the main goals of the paper because this result allows us to construct Riemann-Cartan geometries. As examples of applications of the presented theory, the Levi-Civita derivative and the Hestenes derivative are discussed in a detailed way, and they are interpreted in the framework of the developed theory. The paper presents an elegant way to handle covariant derivative operators, and it should initiate discussions about the different approaches.

Ławrynowicz & Suzuki: Twistors, introduced by Penrose, have been successfully used in gravity, the theory of non-linear differential equations, and representation theory of conformal groups. The "twistor program" has recently been geometrized by Ławrynowicz and Rembieliński. The geometric approach makes it possible to connect the Hurwitz problem of (de)composing quadratic forms with the so-called Hurwitz pairs, which constitute pseudo twistors. These pseudo twistors can be constructed for arbitrary signature and are not necessarily connected to conformal symmetry. This contribution deals with a special class of Hurwitz twistors, which includes signature $(3, 2)$ and its dual $(1, 4)$. Cohomological aspects and a generalization of Cartan's triality to a doubled triality as an atomization theorem and holomorphic embeddings are presented.

Oziewicz & Zeni: Symmetries can be used to reduce the order of differential equations via Lie's theorem. Sometimes this is not directly possible, but only after having prolonged the differential equation by a suitable integrating factor. The well-known multiplication of Newton's equations of motions by an $\dot{x}$ can be integrated yielding the energy conservation law. Lie had already shown under which conditions a "last multiplier" can be found, turning a differential equation into an exact one which can then be integrated. This article tackles the problem of finding last multiplier and the corresponding symmetries for ordinary differential equations in $n + 1$ dimensions. Differential forms are shown to be the natural language for the problem at hand, and a generalized Lie theorem is proved. Moreover, this method is constructive and allows one to find the integrating multiplier by direct calculations. The method is independent of a Riemannian or symplectic structure and does not rely on coordinate methods.

Tian: Besides their invariant coordinate-free character, Clifford algebras are still used by many physicists via matrix representations. Starting from a generator and relations approach to complex Clifford algebras, it is shown that there are only two distinct types of representations. One is for simple Clifford algebras, if the number of generators is even, and one is over a double field, in the semi-simple case when the dimension is odd. Complex similarity factorizations provide a one-to-one mapping of the abstract algebras into certain matrix algebras. In this way, each element a in the $\mathbb{C}$ algebra $C\ell_n$ could be regarded as an eigenvalue of its complex representation matrix $\phi_{n \times n}(a)$. This might be useful in application as calculating exponentials of Clifford numbers.

Acknowledgment: Since editing such a book cannot be done without valuable help of other people, we would like to express our gratitude to all contributors: Ann Kostant, Tom Grasso, and Caroline Graf of Birkhäuser, Elizabeth Loew of TEXniques for her help with TEXing, Amy Knox for her monumental proofreading, and all referees for their constructive criticism.

The first editor, R.A., thanks his wife Halina for her patience during this project. The second editor, B.F., would like to thank his wife Mechthild for her support during the laborious period when the book was completed.

Rafał Abłamowicz, Cookeville, Tennessee, U.S.A.
Bertfried Fauser, Konstanz, Germany   April 1, 2000

1.

PARTIAL DIFFERENTIAL EQUATIONS AND BOUNDARY VALUE PROBLEMS

On Quaternionic Beltrami Equations

Uwe Kähler

ABSTRACT One of the most interesting partial differential equations in complex analysis is the Beltrami equation. We will give an overview of possible generalizations of this equation in case of quaternions together with properties of these equations.

Keywords: Quaternions, Beltrami equation, singular integral operators.

1 Introduction

With the help of functional analytic methods, complex analysis is a powerful tool in treating linear and nonlinear first-order partial differential equations in the plane. Some of the most important of these partial differential equations are the Beltrami equations. This is due to the fact that the theory of Beltrami equations is related with many problems of geometry and analysis as we can see in the following list [Bo]:

1. The general theory of linear and quasilinear elliptic systems, for instance related with non-linear subsonic two-dimensional hydrodynamics,

2. Problems of conformal mappings of Riemannian manifolds,

3. Related problems of conformal and almost complex structure on general Riemannian surfaces – (introducing so-called isothermal coordinates),

4. The classical theory of uniformization and the theory of Teichmüller spaces,

5. Through the last theory, the Beltrami equations appear in various problems of deformation of complex structures on two-dimensional

This paper was written when the author was visiting the Universidade de Aveiro, Portugal, supported by a PRAXIS XXI-scholarship of the Fundação para a Ciência e a Tecnologia.

AMS Subject Classification: 30G35, 35A22.

surfaces and in the conformally invariant string theories in theoretical physics,

6. The theory appears in the so-called complex analytic dynamics or the study of iterations of one-dimensional complex rational mappings as well.

We illustrate this with a simple example. Let us consider the two-dimensional linear first-order partial differential equations

$$a_{11}\frac{\partial u}{\partial x} + a_{12}\frac{\partial u}{\partial y} + b_{11}\frac{\partial v}{\partial x} + b_{12}\frac{\partial v}{\partial y} + a_1 u + b_1 v = f_1,$$

$$a_{21}\frac{\partial u}{\partial x} + a_{22}\frac{\partial u}{\partial y} + b_{21}\frac{\partial v}{\partial x} + b_{22}\frac{\partial v}{\partial y} + a_2 u + b_2 v = f_2.$$

This system of partial differential equations can be transformed into a generalized Cauchy-Riemann equation

$$\frac{\partial w}{\partial \bar{\zeta}} = a\left(\zeta\right) w + b\left(\zeta\right) \bar{w},$$

whereby the necessary coordinate transform $\zeta = \zeta(z)$ has to satisfy the Beltrami equation

$$\frac{\partial \zeta}{\partial \bar{z}} = q(z)\frac{\partial \zeta}{\partial z}.$$

Here $q(z)$ depends on the coefficients a_{ij} and b_{ij}. In general the Beltrami equation

$$\frac{\partial w}{\partial \bar{z}} = q(z)\frac{\partial w}{\partial z} \tag{1.1}$$

is the complex form of the first order elliptic system

$$v_y = au_x + bu_y$$

$$-v_x = bu_x + du_y$$

where $w = u + iv$, $z = x + iy$, $ad - b^2 = 1$, $a > 0$ and $b > 0$. Using the *ansatz*

$$w(z) = z + T_\Omega h$$

with $T_\Omega h(z) = -\frac{1}{\pi}\int_\Omega \frac{h(\xi)}{\xi - z}d\xi_1 d\xi_2$, $\xi = \xi_1 + i\xi_2$, the Beltrami equation (1.1) can be transformed into the fixed-point equation

$$h = q(z)(1 + \Pi_\Omega h),$$

where $\Pi_\Omega h(z) = \frac{1}{2\pi i}\int_\Omega \frac{h(\xi)}{(\xi - z)^2}d\xi_1 d\xi_2$ is the complex Π-operator, defined as the complex partial derivative with respect to z of the T_Ω-operator. Obviously, the theory of Beltrami equations is strongly connected with the

Π-operator. This singular integral operator is immediately recognized as a two-dimensional Hilbert transform, known also under the name of integral operator with Beurling kernel, acting as an isometry from $L_2(\mathbb{C})$ onto $L_2(\mathbb{C})$.

This applicability and the general importance of the complex Beltrami equation were the main reason that, parallel to the investigations in the complex case by L. Ahlfors [Ahl], L. Bers [Ber], B. Bojarski [Bo], and I.N. Vekua [Vek], attempts were made to generalize these equations to the higher dimensional case. One of the first was by A. Newlander and L. Nirenberg [NeNi] in 1957. They considered the system

$$\overline{\partial}_{z_k} f(z) = \sum_{j=1}^{n} \partial_{z_j} f(z) \mu_{jk}(z), \qquad k = 1, \ldots, n$$

over a domain $U \subset \mathbb{C}^n$. Here, μ is a mapping from $U \subset \mathbb{C}^n$ into the set of complex $n \times n$ matrices. Of course, in the case of $n = 1$ this system coincides with the complex Beltrami equation. Newlander and Nirenberg showed that, in the case where the elements of the matrix μ are functions of class $\mathbb{C}^{2n}$, the integrability conditions

$$\overline{\partial}_{z_j} \mu_{lk} - \partial_{z_k} \mu_{lj} = \sum_{r=1}^{n} (\mu_{rj} \partial_{z_r} \mu_{lk} - \mu_{rk} \partial_{z_r} \mu_{lj}), \quad j, k, l = 1, \ldots, n,$$

are necessary and sufficient for the local existence of n independent solutions. Results connected with the Newlander-Nirenberg theorem later received further development by A. Nijenhuis and W.B. Woolf [NijWo], J.J. Kohn [Ko], L. Hörmander [H], B. Malgrange [M], and other mathematicians.

In 1992 I.V. Zhuravlev [Zh] showed the existence of a homeomorphic solution of the above multidimensional analogue of the complex Beltrami equation. But until now, it is not known whether this homeomorphic solution is also a quasi conformal solution, which is always true in the complex case. A quasi conformal solution is a homeomorphic mapping w which satisfies the condition

$$H_w(x) = \limsup_{r \to 0} \frac{\max\{|w(y) - w(x)| : |y - x| = r\}}{\min\{|w(y) - w(x)| : |y - x| = r\}} \le \varepsilon$$

for some $\varepsilon \ge 1$.

This was mainly the reason why T. Iwaniec and G. Martin [IMar] considered the n-dimensional system

$$D^T w(x) Dw(x) = J(x, w)^{\frac{2}{n}} G(x),$$

whereby Dw is the Jacobian matrix, $J(x, w)$ is the Jacobian determinant, and $G(x)$ is a symmetric matrix function, as a multidimensional generalization of the complex Beltrami equation. Obviously, this system can only be considered for mappings w with non-negative Jacobian. They could show that this system has a quasiregular solution, i.e., a solution which satisfies the dilatation condition, but does not need to be a homeomorphism.

One way to find generalizations of the complex Beltrami equation, which can provide quasiconformal solutions, is to look into the case of quaternionic or Clifford analysis. In this paper we will give an overview of possible generalizations in the quaternionic case and their properties.

2   Preliminaries

From now on, we will work in $\mathbb{H}$, the skew field of quaternions. This means we can write each element $x \in \mathbb{H}$ in the form

$$x = x_0 + x_1\mathbf{e}_1 + x_2\mathbf{e}_2 + x_3\mathbf{e}_3, \quad x_n \in \mathbb{R},$$

where $1, \mathbf{e}_1, \mathbf{e}_2, \mathbf{e}_3$ are the basis elements of $\mathbb{H}$. For these elements we have the multiplication rules $\mathbf{e}_1^2 = \mathbf{e}_2^2 = \mathbf{e}_3^2 = -1, \mathbf{e}_1\mathbf{e}_2 = -\mathbf{e}_2\mathbf{e}_1 = \mathbf{e}_3$. The conjugate element $\bar{x}$ is given by $\bar{x} = x_0 - x_1\mathbf{e}_1 - x_2\mathbf{e}_2 - x_3\mathbf{e}_3$ and we have the property $x\bar{x} = \bar{x}x = \|x\|^2 = x_0^2 + x_1^2 + x_2^2 + x_3^2$. Also, for the following, let $\Omega \subset \mathbb{H}$ be a bounded, simply connected domain with a sufficiently smooth boundary $\Gamma = \partial\Omega$. Moreover, we will consider functions f defined on Ω with values in $\mathbb{H}$.

We now define the generalized Cauchy-Riemann operator by

$$Df = \frac{\partial f}{\partial x_0} + \mathbf{e}_1\frac{\partial f}{\partial x_1} + \mathbf{e}_2\frac{\partial f}{\partial x_2} + \mathbf{e}_3\frac{\partial f}{\partial x_3}$$

and its conjugate operator by

$$\overline{D}f = \frac{\partial f}{\partial x_0} - \mathbf{e}_1\frac{\partial f}{\partial x_1} - \mathbf{e}_2\frac{\partial f}{\partial x_2} - \mathbf{e}_3\frac{\partial f}{\partial x_3}.$$

For this operator we have that

$$D\overline{D} = \overline{D}D = \Delta,$$

where Δ is the Laplacian. The Cauchy-Riemann operator has a right inverse of the form

$$Tf(x) = -\frac{1}{2\pi^2} \int_{\Omega} \frac{\overline{(y-x)}}{|y-x|^4} f(y)\, d\Omega, \quad x \in \Omega.$$

This so-called Teodorescu transform acts continuously from $\mathcal{W}_p^k(\Omega)$ into $\mathcal{W}_p^{k+1}(\Omega)$, $1 < p < \infty$, $k \in \mathbb{N} \cup \{0\}$ (see [GS2]). Moreover, we need the following Cauchy-type integral operator:

$$F_\Gamma f(x) = \frac{1}{2\pi^2} \int_{\Gamma} \frac{\overline{(y-x)}}{|y-x|^4} \alpha(y) f(y)\, d\Gamma, \quad x \in \Omega,$$

where $\alpha(y)$ is the outward pointing normal vector to Γ at the point y. This operator is a continuous mapping from $\mathcal{W}_p^{k+1/p}(\Gamma)$ into $\mathcal{W}_p^{k+1}(\Omega)$, $1 < p < \infty$, $k \in \mathbb{N} \cup \{0\}$ [GS2]. The above introduced operators are connected by the well-known Borel-Pompeiu formula

$$F_\Gamma f + TDf = f.$$

Taking the traces of $F_\Gamma f$ we introduce the projections

$$\begin{aligned}
(P_\Gamma f)(\tilde{x}) &= \lim_{x \to \tilde{x}, x \in \Omega, \ \tilde{x} \in \Gamma} (F_\Gamma f)(x), \\
(Q_\Gamma f)(\tilde{x}) &= \lim_{x \to \tilde{x}, x \in \mathbb{R}^4 \setminus \overline{\Omega}, \ \tilde{x} \in \Gamma} (-F_\Gamma f)(x).
\end{aligned}$$

P_Γ is the projection onto the space of all quaternion-valued functions which may be left-monogenic extended into the domain Ω. Q_Γ is the projection onto the space of all Quaternion-valued functions which may be left-monogenic extended into the domain $\mathbb{R}^4 \setminus \overline{\Omega}$.

For more information about these topics and general quaternionic analysis we refer to [BDS], [GS1] and [GS2].

3   Generalizations in the quaternionic case

In 1962 V.I. Shevchenko [Sh] introduced the first generalization of the complex Beltrami equation to the quaternions. He considered the system

$$Dw = q_1 D_1 w + q_2 D_2 w + q_3 D_3 w \qquad (3.1)$$

with measurable functions q_1, q_2, and q_3. Here D_i is the generalized Cauchy-Riemann operator where e_j is replaced by $\overline{e}_j$ for $j \neq i$. He could prove the existence of a global quasiconformal solution for this system, but was using for this a combined matrix-/quaternion-ansatz which severely handicap the applicability of these considerations. In fact, he considered w as a vector function, with D applying the partial derivatives to each of the components of the vector and multiplying the basis elements $1, e_1, e_2, e_3$ as 4×4-real matrices.

The fact that the results were published without proofs was mainly the reason for the article of M. Coroi-Nedelcu [C], where, for most of Shevchenko's results, proofs were given.

In 1979, W. Sprößig [Sp] proposed one of the most interesting generalizations. He considered the system

$$Dw(x) = q(x)\overline{D}w(x). \qquad (3.2)$$

This system was later investigated by Gürlebeck and Kähler [GK1], [K1] in detail. The most interesting observation about this system is the fact that the term $\overline{D}w$ can be considered as the derivative of a monogenic function [Sud], [GMal]. It plays an analogous role as the term $\partial_z w$ in the complex case.

In 1994, this system was also investigated by A. Yanushauskas [Y] in the language of $\mathbb{C}^2$ using the one-to-one correspondence between $\mathbb{C}^2$ and $\mathbb{H}$.

Moreover, in 1997, U. Kähler [K2] considered a more general system in the form of

$$^\psi Dw = q \, ^\varphi Dw$$

based on the idea of using different structural sets by M.V. Shapiro and I.N. Vasilevski [SV1], [SV2]. Here,

$$\psi := \{\psi_0, \psi_1, \psi_2, \psi_3\} \subset \mathbb{R}^4$$

and

$$\varphi := \{\varphi_0, \varphi_1, \varphi_2, \varphi_3\} \subset \mathbb{R}^4$$

are two given orthonormal basis in $\mathbb{R}^4$ which satisfy

$$\psi_i \cdot \overline{\psi_j} + \psi_j \cdot \overline{\psi_i} = 2\delta_{ij}$$

and

$$\varphi_i \cdot \overline{\varphi_j} + \varphi_j \cdot \overline{\varphi_i} = 2\delta_{ij}$$

for all $i, j \in \{0, 1, 2, 3\}$. These structural sets define the operators

$$^\psi Df = \sum_{k=0}^{3} \psi_k \frac{\partial f}{\partial x_k}$$

and

$$^\varphi Df = \sum_{k=0}^{3} \varphi_k \frac{\partial f}{\partial x_k}.$$

This system allows us to use structural sets ϕ and ψ, which are orthogonal to each other, i.e., $\sum \phi_k \overline{\psi}_k = 0$, in analogy to the complex sets $(1, i)$ and $(1, -i)$. We remark that this property does not hold in the case of the system (3.2).

Again, the idea of defining a derivative in Clifford analysis was also the basis for the generalization by H.R. Malonek and B. Müller [MalMü]. In the special case of quaternions we need the linear mappings $J_j : \mathbb{H} \mapsto \mathbb{H}$, $j = 1, \ldots, 3$, given by

$$J_j(\mathbf{e}_j) = \bar{\mathbf{e}}_j, \quad J_j(\mathbf{e}_k) = \mathbf{e}_k, \quad k, j = 1, \ldots, 3, \quad k \neq j.$$

In particular, we have the mapping

$$J_0 : \sum_{k=0}^{3} a_k \mathbf{e}_k \mapsto \sum_{k=0}^{3} a_k \bar{\mathbf{e}}_k.$$

Furthermore, suppose $A = \{\alpha_1, \ldots, \alpha_i\}$, $0 \leq \alpha_1 < \ldots < \alpha_i \leq 3$, and let $J_A = J_{\alpha_1} \ldots J_{\alpha_i}$ be a composition of mappings J_j and $J_\emptyset$ the identical mapping. Using these mappings J_A we can write every real-linear mapping $\mathcal{L} : \mathbb{H} \mapsto \mathbb{H}$ in the form

$$\mathcal{L}(x) = \sum_A c_A J_A(x)$$

with corresponding chosen coefficients $c_A \in \mathbb{H}$, $A \subseteq \{0, \ldots, 3\}$. Moreover, suppose we have the vector function

$$\vec{Q}_A = (Q_{1A}, Q_{2A}, Q_{3A}) : \mathbb{H} \mapsto \mathbb{H}^3$$

with $\vec{Q}_A \in \mathcal{L}_p(\Omega)$, $1 < p < \infty$, $A \subseteq \{0, \ldots, 3\}$. Then we can consider the Beltrami equation

$$Dw = \sum_A \prec \vec{Q}_A ; \nabla_{\bar{z}} J_A w \succ, \tag{3.3}$$

whereby $\prec \cdot ; \cdot \succ$ denotes the usual inner product in $\mathbb{H}^3$. It may be observed that also the term $\nabla_{\bar{z}}$ can be considered as a generalization of the complex derivative operator $\frac{\partial}{\partial z}$ for monogenic functions [Mal].

This system represents the basis for the transformation of a spatial partial differential system of first order to the hypercomplex form $Dw = \sum_A c_A J_A w$ of B. Goldschmidt [Gold].

These are the most important generalizations. Let us remark that the question of quasiconformality of the corresponding solutions has been an open question until now. A first answer was given by V.I. Shevchenko [Sh], but, as already mentioned, the applicability of his results were handicapped by the methods he used.

4 Existence theorems and some consequences

First we have to remark that all the above described quaternionic generalizations of the complex Beltrami equation are solved by making the *ansatz*

$$w = \Phi + Th,$$

where Φ is a monogenic function and T is the Teodorescu transform. Using this *ansatz* we can transform our Beltrami equation into a singular integral equation. Here, we have to observe that this *ansatz* is not artificial as we can see in the next theorem.

Theorem 1. *Suppose* $w \in \mathcal{W}_p^1(\Omega)$, $1 < p < \infty$; *then we can always find functions* $\Phi \in \ker D \cap \mathcal{W}_p^1(\Omega)$ *and* $h \in \mathcal{L}_p(\Omega)$ *such that*

$$w = \Phi + Th.$$

This theorem can be verified in an easy way by setting $h = Dw$ and $\Phi = w - Th$.

In general, applying this *ansatz* to our generalized Beltrami equation results in a transformation of the differential equation into a singular integral equation. As an example let us consider the case of equation (3.2). This differential equation will be transformed into the singular integral equation

$$h = q\overline{D}\Phi + q\Pi h,$$

where

$$\Pi h = -\frac{1}{\omega} \int_\Omega \frac{2 + 4\frac{\overline{(y-x)}^2}{|y-x|^2}}{|y-x|^4} f(y)\, d\Omega_y + \frac{1}{2} f(x)$$

is a generalization of the complex Π-operator. It may be observed that the integral-free term of the singular integral operator does not vanish in contrary to the complex case. Furthermore, the $\mathcal{L}_p(\Omega)$-norm of this operator is always larger than one, except in the case of $\mathcal{L}_2(\Omega)$ where it is one. This follows from the fact that this operator is an isometry over $\mathcal{L}_2(\Omega)$. Also, it may be observed that it follows from a theorem of M. Riesz [Ri] that $||\Pi||_{\mathcal{L}_p}^p$ is a logarithmic convex function of p. Therefore, for a ll $\epsilon > 0$ there exists a $\delta > 0$ such that $||\Pi||_{\mathcal{L}_p} - 1 < \epsilon$ if only $|p - 2| < \delta$. For more information about the properties of the Π-operator we refer to [GK1], [GK2].

If $q \in \mathcal{L}_\infty(\Omega), ||q\Pi||_{\mathcal{L}_p(\Omega)} \leq q_c < 1$, then we can solve the above singular integral equation by successive approximations. Applying Banach's fixed-point theorem we will establish:

Theorem 2. *Suppose* $q \in \mathcal{L}_\infty(\Omega), ||q\Pi||_{\mathcal{L}_p(\Omega)} \leq q_c < 1$; *then our singular integral equation has a unique solution* $h \in \mathcal{L}_p(\Omega)$. *Furthermore, our Beltrami equation (3.2) has also a solution given by* $w = \Phi + Th$.

Moreover, we can state a norm estimate of our solution:

Corollary 1. *Suppose* $\Omega \subset \mathbb{R}^4$ *is a bounded domain; then we have the norm estimate*

$$||w||_{W_2^1(\Omega)} \leq ||\Phi||_{W_2^1(\Omega)} + ||\overline{D}\Phi||_{\mathcal{L}_2(\Omega)} \frac{1}{1 - q_c}$$

$$\times \left(\left(\frac{1}{2\pi^2}\right)^{\frac{1}{4}} |\Omega|^{\frac{1}{4}} + 1 + 2\sqrt[4]{104}\sqrt{\frac{c_4}{\pi}} \right),$$

where

$$c_4 = \left(\int_S |\ln \frac{1}{|\cos\gamma|} + \frac{i\pi}{2} \operatorname{sign} \, \cos\gamma|^4 dS_{\theta'} \right)^{\frac{1}{2}}$$

is a constant and $|\Omega|$ *denotes the volume of the domain* Ω.

Proof. Obviously it holds that

$$||w||_{\mathcal{W}_2^1(\Omega)} \leq ||\Phi||_{\mathcal{W}_2^1(\Omega)} + ||Th||_{\mathcal{W}_2^1(\Omega)}.$$

It remains to estimate $||Th||_{\mathcal{W}_2^1(\Omega)} = ||T||_{[\mathcal{L}_2(\Omega), \mathcal{W}_2^1(\Omega)]} ||h||_{\mathcal{L}_2(\Omega)}$. Here, we have

$$||Th||_{\mathcal{W}_2^1(\Omega)} = ||Th||_{\mathcal{L}_2(\Omega)} + \sum_{i=0}^{3} ||\partial_i Th||_{\mathcal{L}_2(\Omega)}$$

$$= \left(||T||_{\mathcal{L}_2(\Omega)} + \sum_{i=0}^{3} ||\partial_i T||_{\mathcal{L}_2(\Omega)} \right) ||h||_{\mathcal{L}_2(\Omega)}.$$

In analogy to [Kip] it holds that

$$||T||_{\mathcal{L}_2(\Omega)} \leq \left(\frac{1}{2\pi^2} \right)^{\frac{1}{4}} |\Omega|^{\frac{1}{4}}.$$

Furthermore, according [MiP] for $i = 0, \ldots, 3$, we have

$$||\partial_i T||_{\mathcal{L}_2(\Omega)} = \operatorname{supess} |\Psi_i(\theta)|,$$

where $\Psi_i(\theta)$ is the symbol of the singular integral operator $\partial_i T$ and θ runs over the 4-dimensional unit sphere. Applying the representation of the symbol $\Psi_i(\theta) = \frac{1}{4} + \tilde{\Psi}_i(\theta)$ by the characteristic $\kappa_i(\theta)$, we get the estimate

$$
\begin{aligned}
|\tilde{\Psi}_i(\theta)|^2 &= \left| \int_S \kappa_i(\theta) \left[\ln \frac{1}{|\cos \gamma|} + \frac{i\pi}{2} \operatorname{sign} \, \cos \gamma \right] dS \right|^2 \\
&\leq \int_S |\kappa_i(\theta)|^2 \left| \ln \frac{1}{|\cos \gamma|} + \frac{i\pi}{2} \operatorname{sign} \cos \gamma \right|^2 dS \, \pi^2 \\
&\leq \left(\int_S |\kappa_i(\theta)|^4 dS \right)^{\frac{1}{2}} \left(\int_S \left| \ln \frac{1}{|\cos \gamma|} + \frac{i\pi}{2} \operatorname{sign} \, \cos \gamma \right|^4 dS \right)^{\frac{1}{2}} 2\pi^2 \\
&\leq \left(\int_S |\kappa_i(\theta)|^4 dS \right)^{\frac{1}{2}} c_4 2\pi^2,
\end{aligned}
$$

where $c_4 = \left(\int_S |\ln \frac{1}{|\cos \gamma|} + \frac{i\pi}{2} \operatorname{sign} \cos \gamma|^4 dS_{\theta'} \right)^{\frac{1}{2}}$ is a constant. The term $\left(\int_S |\kappa_i(\theta)|^4 dS \right)^{\frac{1}{2}}$ can be calculated exactly (c.f. [GK1]). In particular we obtain for $i = 0, \ldots, 3$,

$$|\tilde{\Psi}_i(\theta)|^2 \leq \sqrt{\frac{13}{2\pi^2}} c_4.$$

This results in

$$\|\partial_i T\|_{\mathcal{L}_2(\Omega)} \leq \frac{1}{4} + \sqrt{\sqrt{\frac{13}{2\pi^2}} c_4}.$$

Altogether, we derive the norm estimate

$$\|T\|_{\mathcal{W}_2^1(\Omega)} \leq \left(\frac{1}{2\pi^2}\right)^{\frac{1}{4}} |\Omega|^{\frac{1}{4}} + 1 + 2\sqrt[4]{104}\sqrt{\frac{c_4}{\pi}}.$$

From $h = (I - q\Pi)^{-1}\overline{D}\Phi$ it follows that

$$\|h\|_{\mathcal{L}_2(\Omega)} \leq \|(I - q\Pi)^{-1}\|_{\mathcal{L}_2(\Omega)}\|\overline{D}\Phi\|_{\mathcal{L}_2(\Omega)}$$

as well as

$$\|h\|_{\mathcal{L}_2(\Omega)} \leq \|(I - q\Pi)^{-1}\|_{\mathcal{L}_2(\Omega)}\|\overline{D}\Phi\|_{\mathcal{L}_2(\Omega)}.$$

Due to $\|(I - q\Pi)^{-1}\|_{\mathcal{L}_2(\Omega)} \leq \frac{1}{1-q_c}$ we have

$$\|h\|_{\mathcal{L}_2(\Omega)} \leq \frac{1}{1-q_c}\|\overline{D}\Phi\|_{\mathcal{L}_2(\Omega)}$$

and, therefore, our norm estimate for w. $\qquad\square$

Remark 1. *Similar norm estimates can also be derived in case of* $\mathbb{R}^4$, *where the main problem is finding appropriate estimates of* $\|T\|_{\mathcal{W}_2^1(\Omega)}$.

At this point we have to make a general remark about the solvability. Obviously, the Beltrami equation (3.2) itself is not uniquely solvable. If w is a solution of (3.2), then we can add any function $w_1 \in \ker D \cap \ker \overline{D}$ to get another solution. However, any solution has the form $w = \Phi + Th$. On the other hand, according to Banach's fixed-point theorem our system of singular integral equations is uniquely solvable. As a matter of fact by fixing our monogenic function Φ also h is uniquely determined. This is of special importance for boundary value problems of this Beltrami equation as we can see in the following theorem.

Theorem 3. *Suppose* $g \in \mathcal{W}_p^{k+1/p}(\Gamma) \cap \operatorname{im} P_\Gamma,\ 1 < p < \infty, k \geq 0,$ *and* $q \in \mathcal{W}_\infty^k(\Omega),\ \|q\Pi\|_{\mathcal{W}_p^k} \leq q_c < 1;$ *then the problem*

$$Dw = q\overline{D}w \quad \text{in } \Omega,$$
$$P_\Gamma w = g \qquad \text{on } \Gamma$$

has a unique solution $w \in \mathcal{W}_p^{k+1}(\Omega)$ *of the form*

$$w = F_\Gamma g + Th,$$

where h *is the solution of the corresponding singular integral equation*

$$h = q\overline{D}F_\Gamma g + q\Pi h.$$

The proof simply follows from the fact that $\Phi = F_\Gamma g$ is uniquely determined and $P_\Gamma \text{tr}\, Th = 0$. In closing, the above investigations of the Beltrami equation (3.2) should be considered as an example of the general treatment of these kinds of equations. We can also get similar results in the case of other quaternionic generalizations of the complex Beltrami equation.

REFERENCES

[Ahl]   L. Ahlfors, *Lectures on Quasi Conformal Mappings*, Van Nostrand, Princeton, 1966.

[Ber]   L. Bers, On a theorem of Mori and the definition of quasiconformality, *Trans. Am. Math. Soc.* **84** (1957), 78–84.

[BDS]   F. Brackx, R. Delanghe, and F. Sommen, Clifford analysis, *Research Notes in Mathematics* **76**, Pitman Advanced Publishing Program, London, 1982.

[Bo]   B. Bojarski, Old and new on Beltrami equations, in *Functional Analytic Methods in Complex Analysis and Applications to Partial Differential Equations, Proceedings of the ICTP*, Trieste, Italy, Feb. 8–19, 1988, A. S. A. Mshimba and W. Tutschke, eds., World Scientific, London, 1988, 173–188.

[C]   M. Coroi-Nedelcu, Asupra solutiei unui sistem de ecuatii cu derivate partiale de tip eliptic în E_m, analog ecuaţiei lui Beltrami în complex, *St. Cerc. Mat.* **17** 7 (1965), 1049–1058.

[Gold]   B. Goldschmidt, A theorem about the representation of linear combinations in Clifford algebras, *Beiträge zur Algebra und Geometrie* **13** (1982), 21–24.

[GK1]   K. Gürlebeck and U. Kähler, On a spatial generalization of the complex Π-operator, *J. Anal. Appl.* **15** (1996) 2, 283–297.

[GK2]   K. Gürlebeck and U. Kähler, On a boundary value problem of the biharmonic equation, *Math. Meth. in the Applied Sciences* **20** (1997), 867–883.

[GMal]   K. Gürlebeck and H. Malonek, A hypercomplex derivative of monogenic functions in $\mathbb{R}^{n+1}$ and its applications, *Complex Variables* **39** (1999), 1999–228.

[GS1]   K. Gürlebeck and W. Sprößig, *Quaternionic Analysis and Elliptic Boundary Value Problems*, ISNM 89, Birkhäuser, Basel, 1990.

[GS2]   K. Gürlebeck and W. Sprößig, *Quaternionic and Clifford Calculus for Engineers and Physicists*, John Wiley &. Sons, Chichester, 1997.

[H]   L. Hörmander, *An Introduction to Complex Analysis in Several Variables*, Van Nostrand, Princeton, 1966.

[IMar]   T. Iwaniec and G. Martin, Quasiregular mappings in even dimensions, *Acta Math.* **170** (1993), 29–81.

[K1]   U. Kähler, Application of hypercomplex Π-operators to the solution of Beltrami systems, in *Analytical and Numerical Methods in Quaternionic and Clifford Analysis*, W. Sprößig and K. Gürlebeck, eds., Freiberger Forschungshefte, Freiberg, 1997, 69–75.

[K2]   U. Kähler, On the solutions of higher-dimensional Beltrami-equations, *Digital Proc. of the IKM 97*, Weimar, 1997.

[Kip]   F. Kippig, Untersuchungen zu Randwert- und anfangswertaufgaben für partielle differentialgleichungen mit methoden der Clifford analysis, Thesis, TU Bergakademie Freiberg, 1997.

[Ko]   J. J. Kohn, Harmonic integrals on strongly pseudo-convex manifolds, *Ann. of Math.* **78** (1963), 112–148.

[KS]   V. V. Kravchenko and M. V. Shapiro, *Integral representations for spatial models of mathematical physics*, Pitman Research Notes in Mathematics Series 351, Pitman- Longman, 1996.

[M]   B. Malgrange, Sur l'intégrabilité des structures presque-complexes, *Symposia Math. (INDAM), Vol. 2*, Academic Press, 1969, 289–296.

[Mal]   H. Malonek, A new hypercomplex structure of the Euclidean space $\mathbb{R}^{m+1}$, *Complex Variables* **14** (1990), 25–33.

[MalMü]   H. Malonek and B. Müller, Definition and properties of a hypercomplex singular integral operator, *Results in Mathematics* **22** (1992), 713–724.

[MiP]   S. G. Michlin and S. Prößdorf, *Singuläre Integraloperatoren*, Akademie-Verlag, Berlin, 1980.

[MitS]   I. M. Mitelman and M. V. Shapiro, Differentiation of the Martinelli-Bochner integrals and the notion of hyperderivability, *Math. Nachr.* **172** (1995), 211–238.

[NeNi]   A. Newlander and L. Nirenberg, Complex analytic coordinates in almost complex manifolds, *Ann. of Math.* **65** (1957), 391–404.

[NijWo]   A. Nijenhuis and W. B. Woolf, Some integration problems in almost-complex and complex manifolds, *Ann. of Math.* **77** (1963), 424–489.

[Ri]   M. Riesz, Sur les maxima des formes billinéaires et sur les fonctionelles linéaires, *Acta Math.* **49** (1928), 465–497.

[SV1]   M. V. Shapiro and N. L. Vasilevski, On the Bergman kernel function in the Clifford analysis, in *Clifford Algebras and their Applications in Mathematical Physics*, F. Brackx, R. Delanghe, and H. Serras, eds., Kluwer, Dordrecht, 1993, 183–192.

[SV2]   M. V. Shapiro and N. L. Vasilevski, On the Bergman kernel function in hypercomplex analysis, *Acta Appl. Math.* **46** 1 (1997), 1–27.

[Sh]   Шевченко, В. И., *О локальном гомеоморфизме трехмерного пространства, осуществляемом решением некоторой еллиптической системы*, Доклады Академии Наук СССР **5** (1962), 1035–1038.

[Sp]   W. Sprößig, *Über eine mehrdimensionale operatorrechnung über beschränkten gebieten des $\mathbb{R}^n$*, Thesis, TH Karl-Marx-Stadt, 1979.

[Sud]   A. Sudbery, Quaternionic Analysis, *Math. Proc. Cambr. Phil. Soc.* **85** (1979), 199–225.

[Vek]   I. N. Vekua, *Generalized Analytic Functions*, Addison-Wesley, Reading, 1963.

[Y]   A. Yanushauskas, On multi-dimensional generalizations of the Cauchy-Riemann system, *Complex Variables* **26** (1994), 53–62.

[Zh]   I. V. Zhuravlev, On the existence of a homeomorphic solution for a multidimensional analogue of the Beltrami equation, *Russian Acad. Sci. Dokl. Math.* **45** (3) (1992), 649–652.

Uwe Kähler
Departamento de Matemática
Universidade de Aveiro
P-3810-193 Aveiro, Portugal
E-mail: uwek@mat.ua.pt

Received: September 30, 1999; Revised: January 5, 2000

The Möbius Transformation, Green Function and the Degenerate Elliptic Equation

Xinhua Ji

ABSTRACT In this article we show the relationship between Green function and the Möbius transformation corresponding to the Laplace-Beltrami operator acting on the compactification, $\mathbb{R}^n \cup \{\infty\}$, of $\mathbb{R}^n$. With the aid of Green function we solve the Dirichlet problem for the non-homogeneous Laplace-Beltrami equation. Then we show that for the Laplace-Beltrami equation (which is a degenerate elliptic type) there exists twice continuously differentiable solutions on the entire space including infinity.

Keywords: Degenerate elliptic equations, Möbius transformations, fundamental solutions.

1 Möbius group and space $\mathbb{R}^n \cup \{\infty\}$

Clifford analysis has its roots in various attempts to generalize one variable complex analysis. Many of the basic results can be found in the text of Brackx, Delanghe and Sommen ([6]). Also, quite considerable work has been done on the links between Clifford analysis and the conformal group (see, for instance, [5], [7], [16], [17]). This arose from the rediscovery and development by Ahlfors ([1, 2, 3, 4]) in the 1980's of long forgotten work from the turn of the century by Vahlen ([19]). In that work it is shown that arbitrary Möbius transformations over Euclidean space can be expressed in much the same form as Möbius transformations in the complex plane, now using 2×2 matrices with Clifford algebra valued coefficients, satisfying some constraints. The Möbius transformations used in this paper correspond to the ones described by Ahlfors using Clifford algebras.

In this article we show the relationship between the Möbius transformation and the Laplace-Beltrami operator acting on the compactification,

Research supported by the NSF of China.
AMS Subject Classification: 35J70, 35G35, 35A08.

$\mathbb{R}^n \cup \{\infty\}$, of $\mathbb{R}^n$. For the Laplace-Beltrami equation see [11, 12].

$$LU \equiv (1 - |x|^2) \sum_{i=1}^{n} \left[(1 - |x|^2) \frac{\partial^2 U}{\partial x_i^2} + 2(n-2)x_i \frac{\partial U}{\partial x_i} \right] = 0 \qquad (1.1)$$

is elliptic inside and outside the unit sphere, but has a degenerate surface $S^{n-1} = \{x \in \mathbb{R}^n : |x| = 1\}$ on which it degenerates into a partial differential equation of first order. Equation (1.1) has an important property, that is, it is invariant under the action of the Möbius group defined as follows.

Definition 1. *By the Möbius group we mean the group of orientation preserving transformations acting on the compactification, $\mathbb{R}^n \cup \{\infty\}$, of $\mathbb{R}^n$, generated by rigid body motions: non-euclidean translations, rotations and reflections with respect to the unit sphere where $\{\infty\}$ is defined by*

$$\{\infty\} \overset{\text{def}}{=} \left\{ y = \frac{x}{|x|^2} \Big|_{x = col(0, \cdots, 0)} \right\}.$$

Let us denote the transpose of a matrix by $'$. Thus, the operation $'$. transposes a row vector $x \in \mathbb{R}^n$ into a column vector $x' = col(x_1, \ldots, x_n)$. So, we have $xx' = |x|^2$.

Definition 2. *All rotations of Möbius group from $\mathbb{R}^n \cup \{\infty\}$ to $\mathbb{R}^n \cup \{\infty\}$ are of the form*

$$y = x\Gamma \quad \text{with} \quad \Gamma\Gamma' = I_{n \times n}$$

where Γ is an orthogonal matrix of order n and $I_{n \times n}$ is the identity matrix.

Definition 3. *All reflections (with respect to the unit sphere) of Möbius group from $\mathbb{R}^n \cup \{\infty\}$ to $\mathbb{R}^n \cup \{\infty\}$ are of the form*

$$y = \frac{x}{xx'}.$$

Definition 4. *All non-euclidean translations of Möbius group from $\mathbb{R}^n \cup \{\infty\}$ to $\mathbb{R}^n \cup \{\infty\}$ are of the form*

$$y = \frac{(1 - aa')(x-a) - (x-a)(x-a)'a}{1 - 2ax' + aa'xx'} \qquad (1.2)$$

where $x, y, a \in \mathbb{R}^n$.

We explain that translation (1.2) is an extension of the usual Möbius transformation in complex plane. To see this, for $n = 2$ we change vectors $y = (y_1, y_2), x = (x_1, x_2), a = (a_1, a_2)$ into the form of complex numbers with $w = y_1 + iy_2$, $z = x_1 + ix_2$, $b = a_1 + ia_2$. Transformation (1.2) then reduces to

$$w = \frac{z - b - bz\bar{z} + b^2\bar{z}}{1 - \bar{b}z - b\bar{z} + b\bar{b}z\bar{z}} = \frac{z - b}{1 - \bar{b}z}$$

which is just a Möbius transformation in the complex plane.

Proposition 1. *Translation (1.2) of the Möbius group is a one-to-one mapping. The inverse of (1.2) is given by*

$$x = \frac{y + a + ayy' + y(2a'a - aa'I)}{1 + 2ay' + aa'yy'}$$

whose only singular point is the point $y = \frac{-a}{aa'}$.

Substituting translations (1.2) (or rotations, or reflections with respect to the unit sphere) of the Möbius group into equation (1.1) we get

$$(1 - |y|^2) \sum_{i=1}^{n} \left[(1 - |y|^2) \frac{\partial^2 U}{\partial y_i^2} + 2(n-2)y_i \frac{\partial U}{\partial y_i} \right] =$$

$$(1 - |x|^2) \sum_{i=1}^{n} \left[(1 - |x|^2) \frac{\partial^2 U}{\partial x_i^2} + 2(n-2)x_i \frac{\partial U}{\partial x_i} \right].$$

Therefore, we have

Proposition 2. *Laplace-Beltrami equation (1.1) is invariant under the action of the Möbius group.*

Differentiating both sides of (1.2) we get

$$\frac{dydy'}{(1 - yy')^2} = \frac{dxdx'}{(1 - xx')^2}.$$

Substituting transformations of rotations, or reflections (with respect to the unit sphere) of the Möbius group into $\frac{dydy'}{(1-yy')^2}$, we obtain $\frac{dydy'}{(1-yy')^2} = \frac{dxdx'}{(1-xx')^2}$ as well. So, we have

Proposition 3. *Corresponding to the Laplace-Beltrami equation (1.1) the Poincaré metric $\frac{dxdx'}{(1-xx')^2}$ is invariant under transforms from the Möbius group.*

For convenience in later applications we put

$$dx \, G(x) \, dx' = \frac{dxdx'}{(1 - xx')^2} \quad \text{where} \quad G = (g_{ij})_{n \times n},$$

$$g = \det G, \quad G^{-1} = (g^{ij})_{n \times n} \quad \text{with} \quad G^{-1}G = GG^{-1} = I. \qquad (1.3)$$

From (1.2) we get more formulae:

$$yy' = \frac{(x - a)(x - a)'}{1 - 2ax' + aa'xx'} \qquad (1.4)$$

and

$$1 - yy' = \frac{(1 - aa')(1 - xx')}{1 - 2xa' + aa'xx'}. \qquad (1.5)$$

Proposition 4. *Translation (1.2) of the Möbius group has the following properties:*

(i) *It translates any point $a \in \{|x| < 1\}$ to the origin.*

(ii) *It maps the unit sphere S^{n-1} onto itself.*

(iii) *When the point a is inside the unit sphere, the map (1.2) maps the unit ball, $\{|x| < 1\}$, onto itself $\{|y| < 1\}$; and it maps the domain outside of the unit sphere $\{|x| > 1\}$ onto itself $\{|y| > 1\}$.*

(iv) *When $a \in \{|x| > 1\}$, it maps the unit ball $\{|x| < 1\}$ onto the domain outside $S^{n-1}, \{|y| > 1\}$, and it maps the domain outside $S^{n-1}, \{|x| > 1\}$ onto the unit ball $\{|y| < 1\}$.*

Proof. From equations (1.2) and (1.5), obviously we have (i) and (ii). To get (iii) and (iv), from (1.5) we have only to see

$$\left. \begin{array}{c} aa' < 1 \\ xx' < 1 \end{array} \right\} \quad or \quad \left. \begin{array}{c} aa' > 1 \\ xx' > 1 \end{array} \right\} \quad \Leftrightarrow \quad yy' < 1,$$

$$\left. \begin{array}{c} aa' < 1 \\ xx' > 1 \end{array} \right\} \quad or \quad \left. \begin{array}{c} aa' > 1 \\ xx' < 1 \end{array} \right\} \quad \Leftrightarrow \quad yy' > 1. \qquad \square$$

To define the fundamental solution we need the notion of geodesic distance. Suggested by equation (1.4) we define by $\rho(0, y) = \sqrt{yy'}$ the geodesic distance between 0 and y.

Definition 5. *Geodesic distance $\rho(x, a)$ between x and a is defined by*

$$\rho(x, a) \overset{\text{def}}{=} \sqrt{\frac{(x - a)(x - a)'}{1 - 2ax' + aa'xx'}}. \tag{1.6}$$

The definition of geodesic distance comes from formula (1.4) and the fact

$$1 - 2ax' + aa'xx' \geq 0.$$

In fact if we suppose the lengths of vectors a and x are R and r, and θ is the angle between a and x, then we have

$$1 - 2ax' + aa'xx' = (1 - ax')^2 + aa'xx' - (ax')^2$$
$$= (1 - Rr \cos \theta)^2 + R^2 r^2 (1 - \cos^2 \theta) \geq 0.$$

Proposition 5. *In space $\mathbb{R}^n \cup \{\infty\}$,*

1. *the meaning of zero distance is the same as Euclidean distance;*

2. *the infinite distance is between points a and $\frac{a}{aa'}$.*

Proof. From (1.5) we see, if at least one of two points is on the unit sphere, the distance between these points cannot be zero. From (1.4) we know that when both x and a are not on the unit sphere,

$$\rho(x, a) = 0$$

holds if and only if $x = a$ holds (in this case $1 - 2ax' + aa'xx' = (1 - aa')^2 > 0$). So, the meaning of zero distance is the same as in Euclidean space.

Since $x = \frac{a}{aa'}$ is equivalent to

$$1 - 2ax' + aa'xx' = 0 \tag{1.7}$$

and in this case

$$(x - a)(x - a)' \neq 0 \quad \text{for} \quad aa' \neq 1, \tag{1.8}$$

(1.7) and (1.8) imply

$$\rho(x, a) = \infty \quad \text{for} \quad x = \frac{a}{aa'}.$$

$\square$

Definition 6. *A sphere with center at a and with radius r is represented by*

$$\rho(x, a) = r$$

where $r(> 0)$ is a constant.

Proposition 6. *There are two meanings for the sphere $\rho(x, a) = r$ in $\mathbb{R}^n$:*

1. If $r^2 = \frac{1}{aa'}$, the sphere $\rho(x, a) = r$ represents a plane

$$-ax'(1 - r^2) + aa' - r^2 = 0 \quad in \quad \mathbb{R}^n.$$

2. If $r^2 \neq \frac{1}{aa'}$, the sphere $\rho(x, a) = r$ represents a sphere

$$\left(x - \frac{(1 - r^2)a}{1 - aa'r^2}\right)\left(x - \frac{(1 - r^2)a}{1 - aa'r^2}\right)' = \left(\frac{r(1 - aa')}{1 - aa'r^2}\right)^2 \quad in \quad \mathbb{R}^n,$$

i.e., the sphere with center at $\frac{(1-r^2)a}{1-aa'r^2}$ and with radius $\frac{r(1-aa')}{1-aa'r^2}$.

2   Green function and the fundamental solution

We rewrite the Laplace-Beltrami equation (1.1) in the following form

$$LU \equiv \frac{1}{\sqrt{g}} \sum_{i,j=1}^{n} \frac{\partial}{\partial x_i} \left(\sqrt{g} g^{ij} \frac{\partial U}{\partial x_j}\right) = 0$$

where g and g^{ij} are given by (1.3).

Definition 7. *Let D be a domain, let point a be fixed $a, x \in D$, and let $\rho(x,a)$ be the geodesic distance between x and a. The function $V(\rho(x,a))$ is said to be a fundamental solution with singularity $a \in D$ for equation (1.1) if*

(a) $V(\rho(x,a))$ *is twice continuously differentiable for $x \neq a$.*

(b) $V(\rho(x,a))$ *satisfies (1.1), or*

$$L_x V(\rho(x,a)) = 0 \quad for \quad x \neq a.$$

(c) *Let $S_\epsilon = \{x : \rho(x,a) = \epsilon, \ \epsilon > 0\}$ be a generalized small sphere with radius ϵ and the center at a. Then, as $\epsilon \to 0$ we have*

$$\lim_{\epsilon \to 0} \oint_{S_\epsilon} \sum_{i,j=1}^{n} \sqrt{g}g^{ij} \frac{\partial V(\rho(x,a))}{\partial x_j} n_i(x) d\sigma(x) = -1$$

where $\bar{n}(x) = (n_1(x), \cdots, n_n(x))$ is the outward normal vector at points of S_ϵ, and $d\sigma(x)$ is the area element of S_ϵ.

(d) $V(\rho(x,a)) = O(\rho^{2-n}(x,a))$.

Theorem 1. *The function $V(\rho(x,a))$*

$$V(\rho(x,a))$$

$$= \frac{(-1)^{\frac{n+1}{2}} 2^{n-2}}{\omega_n} \sum_{k=0}^{\frac{n-3}{2}} \binom{\frac{n-3}{2}}{k} \frac{(-1)^k}{2k+1} \left[\frac{\rho(x,a) + \rho^{-1}(x,a)}{2} \right]^{2k+1}$$

for n odd, and

$$V(\rho(x,a))$$

$$= \frac{-1}{\omega_n} \left(\sum_{\substack{k=0 \\ k \neq \frac{n-2}{2}}}^{n-2} \binom{n-2}{k} \frac{\rho^{2k-n+2}(x,a)}{2k-n+2} + (-1)^{\frac{n-2}{2}} \binom{n-2}{\frac{n-2}{2}} \log \rho(x,a) \right),$$

for n even, is a fundamental solution (with singularity $a \in D$) for equation (1.1), where $\rho(x,a)$ is given by (1.6) and ω_n is the area of the unit sphere S^{n-1} in $\mathbb{R}^n$.

Proof. From the expression of $V(\rho(x,a))$ it is easy to see that function $V(\rho(x,a))$ satisfies conditions (a) and (d) obviously. So, to prove Theorem 1 we have only to show (b) and (c).

To prove (b), notice that $V(\rho(x,a))$ satisfies equation (1.1) for $x \neq a$. By differentiating the function $V(\rho(x,a))$ we get

$$\frac{dV}{d\rho} = \frac{-1}{\omega_n} \frac{(1-\rho^2)^{n-2}}{\rho^{n-1}}. \tag{2.1}$$

Thus, we have

$$L_y V(\rho(y,0)) = \frac{(1-\rho^2)^n}{\rho^{n-1}} \frac{d}{d\rho} \left[\frac{\rho^{n-1}}{(1-\rho^2)^{n-2}} \frac{dV}{d\rho} \right] = 0 \quad \text{for} \quad y \neq 0.$$

Then, with the aid of Proposition 2 and Proposition 4(i) we obtain

$$L_x V(\rho(x,a)) = L_y V(\rho(y,0)) = 0 \quad \text{for} \quad x \neq a, \ y \ ne0$$

which establishes (b).

To prove (c) we apply Proposition 4(i) to get

$$\oint_{S_\epsilon} \sum_{i,j=1}^n \sqrt{g(x)} \, g^{ij}(x) \frac{\partial V(\rho(x,a))}{\partial x_j} n_i(x) d\sigma(x)$$

$$= \oint_{yy'=\epsilon^2} \sum_{i,j=1}^n \sqrt{g(y)} \, g^{ij}(y) \frac{\partial V(\rho(y,0))}{\partial y_j} n_i(y) d\sigma(y).$$

In the spherical coordinate system for equation (1.1) we have

$$g^{ij} = \frac{\delta^{ij}}{h_i}, \quad \delta_{ij} = \begin{cases} 0, & \text{for } i \neq j, \\ 1, & \text{for } i = j, \end{cases}.$$

$$h_1 = \frac{1}{(1-\rho^2)^2}, \quad h_2 = \frac{\rho^2}{(1-\rho^2)^2}, \quad h_3 = \frac{\rho^2 \sin^2 \theta_1}{(1-\rho^2)^2}, \quad \cdots$$

$$\cdots, \quad h_n = \frac{\rho^2 \sin^2 \theta_1 \cdots \sin^2 \theta_{n-2}}{(1-\rho^2)^2}, \quad g = h_1 \cdots h_n.$$

Therefore,

$$\oint_{yy'=\epsilon^2} \sum_{i,j=1}^n \sqrt{g(y)} \, g^{ij}(y) \frac{\partial V(\rho(y,0))}{\partial y_j} n_i(y) d\sigma(y)$$

$$= \oint_{\rho=\epsilon} g^{11} \frac{\partial V}{\partial \rho} \sqrt{g(\rho, \theta_1, \cdots, \theta_{n-1})} \, d\theta_1 \cdots d\theta_{n-1}$$

$$= \oint_{\rho=\epsilon} (1-\rho^2)^2 \left(\frac{-(1-\rho^2)^{n-2}}{\omega_n \, \rho^{n-1}} \right) \times$$

$$\left[\frac{\rho^{n-1}}{(1-\rho^2)^n} \sin^{n-2} \theta_1 \sin^{n-3} \theta_2 \cdots \sin \theta_{n-2} \right] d\theta_1 \cdots d\theta_{n-1}$$

$$= -\frac{1}{\omega_n} \oint_{S^{n-1}} \sin^{n-2} \theta_1 \sin^{n-3} \theta_2 \cdots \sin \theta_{n-2} d\theta_1 \cdots d\theta_{n-1}$$

$$= -1$$

which proves (c). □

Definition 8. *Assume $D \subset \mathbb{R}^n \cup \{\infty\}$ is a domain. The function $G(x, a)$ is said to be a Green function for the domain D if and only if*

1. $G(x, a)$ *is a fundamental solution; and*

2. $G(x, a)$ *is vanishing on the boundary of D.*

Theorem 2. *Let $D \subset \mathbb{R}^n \cup \{\infty\}$ be a domain bounded by ∂D and let function $V(\rho(x, a))$ be the fundamental solution given by Theorem 1. Then function $G(x, a)$ defined by*

$$G(x, a) \overset{\text{def}}{=} V(\rho(x, a)) - V(1)$$

is Green function for equation (1.1) in the domain

$$D = \{x : \rho(x, a) \le 1\}. \tag{2.2}$$

Proof. Because of Theorem 1 and because the boundary of D is

$$\partial D = \{x : \rho(x, a) = 1\}$$

which implies

$$G(x, a)|_{x \in \partial D} = 0,$$

by the definition of Green function the theorem is obvious. $\qquad\square$

Proposition 7. *For the domain D given by (2.2), when point a is inside the unit sphere, the domain D is the unit ball; when point a is outside the unit sphere, the domain D is the entire domain outside of the unit sphere.*

Proof. From (1.5) and (1.6) we have

$$1 - \rho^2(x, a) = \frac{(1 - aa')(1 - xx')}{1 - 2xa' + aa'xx'}.$$

Noting

$$D = \{x : \rho(x, a) \le 1\} = \begin{cases} \{x : xx' \le 1\}, & \text{for } a \text{ inside } S^{n-1}, \\ \{x : xx' \ge 1\}, & \text{for } a \text{ outside } S^{n-1}, \end{cases}$$

we complete the proof. $\qquad\square$

3   Solution to non-homogeneous equation

Let D be a domain with smooth boundary ∂D, and let functions $u(x), v(x)$ be twice continuously differentiable functions on $D \cup \partial D$. Then, the Laplace-Beltrami operator has the following Green integral formula:

$$\int_D (v\, Lu - u\, Lv)\sqrt{g}\, dx_1 \cdots dx_n$$

$$= \oint_{\partial D} \sum_{i,j=1}^n g^{ij}\left(v\, \frac{\partial u}{\partial x_j} - u\, \frac{\partial v}{\partial x_j} \right) \sqrt{g}\, n_i(x) d\sigma(x), \tag{3.1}$$

where g^{ij}, g are given by (1.3), $\bar{n}(x) = (n_1(x), \ldots, n_n(x))$ is an outward normal vector at points of ∂D, and $d\sigma(x)$ is the area element of ∂D.

3.1 Dirichlet problem for non-homogeneous equation

Books and papers such as [12, 13, 15] study the Dirichlet problem for the Laplace-Beltrami equation (1.1)

$$LU \equiv (1 - |x|^2) \sum_{i=1}^{n} \left[(1 - |x|^2) \frac{\partial^2 U}{\partial x_i^2} + 2(n-2)x_i \frac{\partial U}{\partial x_i} \right] = 0$$

which is homogeneous. In this section, we study the Dirichlet problem for the corresponding non-homogeneous equation , i.e.,

$$Lu(x) = f(x) \tag{3.2}$$

where the continuous function $f(x)$ is satisfying the condition

$$\lim_{|x| \to 1} \frac{f(x)}{1 - |x|^2} = p(x) \quad \text{with} \quad |p(x)| < \infty. \tag{3.3}$$

Let $D \subset \mathbb{R}^n \cup \{\infty\}$ be a domain bounded by ∂D. We consider Dirichlet problem in D. We explain that, for this purpose, we have only to find out any differentiable solution v for equation (3.2) on $\bar{D} = D \cup \partial D$ without considering boundary value function.

In fact, suppose v is such a solution; then let

$$u = v + w$$

where w is the solution for equation (1.1) with the following boundary condition

$$Lw = 0, \quad w|_{\partial D} = u|_{\partial D} - v|_{\partial D}.$$

Then, function u will be the solution we are looking for, that is, u is the solution of Dirichlet problem for equation (3.2). Thus, the proof of existence and uniqueness for Dirichlet problem to non-homogeneous equation (3.2) is reduced to the same problem to Laplace-Beltrami equation (1.1) that we studied before.

In the remainder of this section we assume domain D given by (2.2).

3.2 Solution to the non-homogeneous equation

Theorem 3. *Let $G(x, a)$ be Green function with singularity $a \in D$ and domain D given by (2.2). Then, function*

$$u(a) = - \int_D G(x, a) \, f(x) \, \sqrt{g(x)} \, dx_1 \cdots dx_n$$

satisfies equation

$$L_a u(a) = f(a).$$

Proof. First we show the following fact, that is,

$$\int_B L_a G(x,a)\, dx = \begin{cases} 0, & a \bar{\in} B, \\ -1, & a \in B, \end{cases} \tag{3.4}$$

where B is any domain, $dx = \sqrt{g(x)}dx_1 \cdots dx_n$, and L_a means the operator L acting on the variable a. We obviously have

$$\int_B L_a G(x,a)\, dx = 0 \quad \text{for} \quad a \bar{\in} B$$

because a is not in B and x in B imply $L_a G(x,a) = 0$.

In order to prove

$$\int_B L_a G(x,a)\, dx = -1 \quad \text{for} \quad a \in B,$$

we employ

$$y = -\frac{(1 - xx')(a - x) - (a - x)(a - x)'x}{1 - 2ax' + aa'xx'} \tag{3.5}$$

which is a transformation composed of translation and rotation from the Möbius group, and it maps a to y. Applying (3.5) and Proposition 2 we get

$$L_a G(x,a) = L_y G(y,0)$$

and

$$dx = \sqrt{g(x)}dx_1 \cdots dx_n = \sqrt{g(y)}dy_1 \cdots dy_n = dy.$$

Thus, for $a \in B$

$$\int_B L_a G(x,a)\, dx$$

$$= \int_{\rho(y,0)\leq\delta} L_y G(y,0)\, dy$$

$$= \int_{\rho\leq\delta} \frac{(1 - \rho^2)^n}{\rho^{n-1}} \left(\frac{d}{d\rho}\left[\frac{\rho^{n-1}}{(1-\rho^2)^{n-2}} \frac{d\, G(y,0)}{d\rho}\right]\right) \frac{\rho^{n-1}}{(1-\rho^2)^n}\, d\rho\, d\sigma(\theta)$$

$$= \int_{\rho\leq\delta} \frac{d}{d\rho}\left[\frac{\rho^{n-1}}{(1-\rho^2)^{n-2}} \frac{d\, V(\rho)}{d\rho}\right] d\rho\, d\sigma(\theta) \tag{3.6}$$

where

$$d\sigma(\theta) = \sin^{n-2}\theta_1 \sin^{n-3}\theta_2 \cdots \sin\theta_{n-2}\, d\theta_1 \cdots d\theta_{n-1}$$

is the area element on S^{n-1}. Substituting (2.1) into (3.6) we have

$$\int_{\rho \leq \delta} \frac{d}{d\rho} \left[\frac{\rho^{n-1}}{(1-\rho^2)^{n-2}} \frac{d\,V(\rho)}{d\rho} \right] d\rho \, d\sigma(\theta) = -1.$$

Therefore (3.4) holds true.

With the aid of (3.4) we may do the following calculation

$$Lu(a)$$
$$= -\int_D L_a G(x,a) f(x) \, dx = -\int_{\rho(x,a) \leq \delta} L_a G(x,a) f(x) \, dx$$
$$= -f(a) \int_{\rho(x,a) \leq \delta} L_a G(x,a) dx - \int_{\rho(x,a) \leq \delta} L_a G(x,a)[f(x) - f(a)] \, dx.$$

Since $\delta > 0$ is an arbitrary number and $f(x)$ is a continuous function, by (3.4) we immediately obtain $Lu(a) = f(a)$. In the above proof we have assumed $aa' \neq 1$.

In the case $aa' = 1$ we let $V(\rho(x,a)) = c_o$ with c_o constant. Then we get

$$L_a G(x,a)|_{aa'=1} = 0 \quad \text{and} \quad f(a) = 0.$$

So, in case of $aa' = 1$ we also have $Lu(a) = f(a)$. Therefore Theorem 3 has been proven. $\square$

Theorem 4. *Assume $G(x,a)$ is Green function in the domain D with singularity $a \in D$. Then for the Dirichlet problem for equation (3.2) there exists a unique solution (with a as its variable) explicitly represented by*

$$u(a) = -\int_D G(x,a) f(x) \sqrt{g(x)} \, dx_1 \cdots dx_n$$
$$- \oint_{\partial D} \sum_{i,j=1}^{n} u(x) \frac{\partial G(x,a)}{\partial x_j} \, g^{ij}(x) \, \sqrt{g(x)} \, n_i(x) d\sigma(x) \quad (3.7)$$

where $\bar{n}(x) = (n_1(x), \dots, n_n(x))$ is an outward normal vector at points of ∂D, and $d\sigma(x)$ is the area element of ∂D.

Sufficiency. We consider the two integrals on the right-hand side of (3.7) term by term, and let (3.7) be expressed by $u(a) = u_1(a) + u_2(a)$ where $u_1(a)$ is the integral over domain D and $u_2(a)$ is the integral over ∂D.

From Theorem 3 we know that the first term on the right-hand side of (3.7) is a solution of equation (3.2), i.e., $Lu_1(a) = f(a)$. As to the second term, we see that a is in the interior of the domain D which implies that a is not on ∂D and obviously $a \neq x$ (because x is on the boundary ∂D). Thus, when the operator L_a acts on the second term of the right-hand side of (3.7), the operator L_a can be commuted with the integral operator and the partial differential operator there, and L_a may act on Green function $G(x,a)$ directly. Since $L_a G(x,a) = 0$ for $x \neq a$, we have that $L_a u_2(a) = 0$. Therefore, $Lu(a) = Lu_1(a) + Lu_2(a) = f(a)$ and we obtain the sufficiency.

Necessity. We use Green integral formula (3.1) and place $v(x) = G(x,a)$ (in Theorem 2) with singularity a in the interior of D and domain D given by (2.2). Let $S_\epsilon = \{x : \rho(x,a) = \epsilon, \ \epsilon > 0\}$ be a generalized small sphere with radius ϵ and the center at a. Let $\epsilon > 0$ be sufficiently small so that S_ϵ is contained in the interior of D, and let $D_\epsilon = D - S_\epsilon$. Applying Green integral formula (3.1) over the domain D_ϵ, we have

$$\int_{D_\epsilon} \{G\,Lu - u\,LG\}\sqrt{g(x)}\,dx_1\cdots dx_n$$
$$= \left(\oint_{\partial D} - \oint_{S_\epsilon}\right) \sum_{i,j=1}^{n} g^{ij}\left(G\frac{\partial u}{\partial x_j} - u\frac{\partial G}{\partial x_j}\right)\sqrt{g(x)}\,n_i(x)d\sigma(x) \quad (3.8)$$

where g^{ij}, g are given by (1.3), $\bar{n}(x) = (n_1(x),\ldots,n_n(x))$ is the outward normal vector at points of ∂D or S_ϵ and $d\sigma(x)$ is the area element of ∂D or S_ϵ. Since $LG(x,a) = 0$ for $x \in D_\epsilon$ from (3.8), we have

$$\int_{D_\epsilon} G\,Lu\,\sqrt{g(x)}\,dx_1\cdots dx_n$$
$$= \left(\oint_{\partial D} - \oint_{S_\epsilon}\right) \sum_{i,j=1}^{n} g^{ij}\left(G\frac{\partial u}{\partial x_j} - u\frac{\partial G}{\partial x_j}\right)\sqrt{g(x)}\,n_i(x)d\sigma(x). \quad (3.9)$$

And since

$$\frac{\partial G(x,a)}{\partial x_i} = \frac{\partial V(\rho(x,a))}{\partial x_i}$$

and applying Definition 7 (definition of the fundamental solution) condition (c)

$$\lim_{\epsilon \to 0} \oint_{S_\epsilon} \sum_{i,j=1}^{n} \sqrt{g}g^{ij}\frac{\partial V(\rho(x,a))}{\partial x_j}n_i(x)d\sigma(x) = -1,$$

it is easy to know that as $\epsilon \to 0$ we have

$$\lim_{\epsilon \to 0} \oint_{S_\epsilon} u(x) \sum_{i,j=1}^{n} g^{ij}(x)\frac{\partial G(x,a)}{\partial x_j}\sqrt{g(x)}n_i(x)d\sigma(x)$$

$$= \lim_{\epsilon \to 0}\left(u(a) \oint_{S_\epsilon} \sum_{i,j=1}^{n} g^{ij}(x)\frac{\partial V(\rho(x,a))}{\partial x_j}\sqrt{g(x)}n_i(x)d\sigma(x)\right.$$

$$\left. + \oint_{S_\epsilon} [u(x) - u(a)] \sum_{i,j=1}^{n} g^{ij}\frac{\partial G}{\partial x_j}\sqrt{g(x)}\,n_i(x)d\sigma(x)\right)$$

$$= -u(a). \quad (3.10)$$

Moreover, applying the inverse mapping of translation (1.2)

$$x = \frac{y + a + ayy' + y(2a'a - aa'I)}{1 + 2ay' + aa'yy'}, \quad yy' = \rho^2,$$

we have

$$\oint_{S_\epsilon} G(x,a) \sum_{i,j=1}^{n} g^{ij}(x) \frac{\partial u}{\partial x_j} \sqrt{g(x)}\, n_i(x)\, d\sigma(x)$$

$$= \oint_{yy'=\epsilon^2} G(y,0) \sum_{i,j=1}^{n} g^{ij}(y) \frac{\partial u}{\partial y_j} \sqrt{g(y)}\, n_i(y)\, d\sigma(y)$$

$$= \oint_{\rho=\epsilon} [V(\rho) - V(1)]\, g^{11} \frac{\partial u}{\partial \rho} \sqrt{g}\, d\theta_1 \cdots d\theta_{n-1}$$

$$= \oint_{\rho=\epsilon} [V(\rho) - V(1)](1 - \rho^2)^2 \frac{\partial u}{\partial \rho} \left(\frac{\rho^{n-1} \sin^{n-2}\theta_1 \cdots \sin\theta_{n-2}}{(1 - \rho^2)^n} \right)$$

$$\times\, d\theta_1 \cdots d\theta_{n-1}.$$

From Definition 7 condition (d), we know that as $\epsilon \to 0$

$$V(\rho) - V(1) = O(\rho^{2-n});$$

and since $u \in C^2(D)$, there exists a positive number $M > 0$ such that

$$\left| \frac{\partial u}{\partial \rho} \right| < M,$$

and we obtain

$$\lim_{\epsilon \to 0} \oint_{S_\epsilon} G(x,a) \sum_{i,j=1}^{n} g^{ij}(x) \frac{\partial u}{\partial x_j} \sqrt{g(x)}\, n_i(x)\, d\sigma(x) = 0. \tag{3.11}$$

Substituting (3.10) and (3.11) into (3.9) and letting ϵ approach zero on both sides of (3.9), we have

$$u(a) = -\int_D G(x,a) Lu(x) \sqrt{g(x)}\, dx_1 \cdots dx_n$$

$$+ \oint_{\partial D} \sum_{i,j=1}^{n} g^{ij}(x) \left(G\frac{\partial u}{\partial x_j} - u\frac{\partial G}{\partial x_j} \right) \sqrt{g(x)}\, n_i(x)\, d\sigma(x).$$

But

$$G(x,a)|_{x \in \partial D} = 0.$$

Thus, for any twice continuously differentiable function $u(x)$, if $u(x)$ satisfies equation (3.2), then it must have the following representation

$$u(a) = - \int_D G(x, a) f(x) \sqrt{g(x)} \, dx_1 \cdots dx_n$$

$$- \oint_{\partial D} \sum_{i,j=1}^n u(x) \frac{\partial G(x, a)}{\partial x_j} \, g^{ij}(x) \sqrt{g(x)} \, n_i(x) \, d\sigma(x)$$

which is just formula (3.7). Therefore, we have shown that (3.7) is the unique solution of Dirichlet problem for equation (3.2). The proof of Theorem 4 is complete.   □

4   The Poisson formula

In this section we consider the Laplace-Beltrami equation (1.1) as a special case of the non-homogeneous degenerate elliptic equation (3.2) with $f = 0$. Let D be the unit ball. Then by Theorem 4 the unique solution of Dirichlet problem for the corresponding homogeneous equation (1.1) in the unit ball is represented by

$$U(a) = - \oint_{\partial D} \sum_{i,j=1}^n U(x) \frac{\partial G(x, a)}{\partial x_j} \, g^{ij}(x) \, \sqrt{g(x)} \, n_i(x) d\sigma(x).$$

Applying the above representation, we shall obtain the Poisson formula for equation (1.1) in the unit ball. For this purpose we use the spherical coordinate system and suppose the vectors $x = Rv$ and $a = ru$ with $|x| = R$, $|a| = r$, and $u, v \in S^{n-1}$, i.e.,

$$v = (\cos\theta_1, \ \sin\theta_1 \cos\theta_2, \ \sin\theta_1 \sin\theta_2 \cos\theta_3, \ \cdots,$$
$$\sin\theta_1 \cdots \sin\theta_{n-2} \cos\theta_{n-1}, \ \sin\theta_1 \sin\theta_2 \cdots \sin\theta_{n-1}).$$

Noting (1.3) and substituting

$$1 - \rho^2(x, a) = 1 - \rho^2(Rv, ru) = \frac{(1 - r^2)(1 - R^2)}{1 - 2rRuv' + r^2 R^2}$$

and

$$\left. \frac{\partial \rho}{\partial R} \right|_{R=1} = \frac{1 - r^2}{1 - 2ruv' + r^2}$$

into the above representation of the solution of the Dirichlet problem to equation (1.1) –and since $\partial D = \{x : |x| = R = 1\}$ implies $\rho = 1$ on $\partial D-$

we obtain

$$U(a) = -\oint_{\partial D} \sum_{i,j=1}^{n} U(x) \frac{\partial G(x,a)}{\partial x_j} \, g^{ij}(x) \, \sqrt{g(x)} \, n_i(x) d\sigma(x)$$

$$= -\oint_{R=1} U(Rv) \, g^{11} \, \frac{\partial V(\rho(x,a))}{\partial R} \, \sqrt{g} \, d\theta_1 \cdots d\theta_{n-1}$$

$$= -\oint_{R=1} U(v) \, (1-R^2)^2 \left[-\frac{1}{\omega_n} \frac{(1-\rho^2)^{n-2}}{\rho^{n-1}} \frac{\partial \rho}{\partial R} \right]$$

$$\times \left[\frac{R^{n-1}}{(1-R^2)^n} \, \sin^{n-2}\theta_1 \cdots \sin\theta_{n-2} \right] d\theta_1 \cdots d\theta_{n-1}$$

$$= \frac{1}{\omega_n} \oint_{R=1} U(v) \, \frac{R^{n-1}}{(1-R^2)^{n-2}} \frac{1}{\rho^{n-1}} \left[\frac{(1-r^2)(1-R^2)}{1-2rR\,uv'+r^2R^2} \right]^{n-2}$$

$$\times \left(\frac{1-r^2}{1-2ruv'+r^2} \right) d\sigma(v)$$

$$= \frac{1}{\omega_n} \oint_{S^{n-1}} U(v) \left(\frac{1-r^2}{1-2ruv'+r^2} \right)^{n-1} d\sigma(v),$$

where

$$d\sigma(v) = \sin^{n-2}\theta_1 \cdots \sin\theta_{n-2} d\theta_1 \cdots d\theta_{n-1}$$

is the spherical element on S^{n-1}. Therefore, we obtain the following Poisson formula for Laplace-Beltrami equation (1.1)

$$U(x) = \frac{1}{\omega_n} \oint_{S^{n-1}} U(v) \left(\frac{1-|x|^2}{1-2xv'+|x|^2} \right)^{n-1} d\sigma(v). \qquad (4.1)$$

Proposition 8. *The Poisson kernel*

$$P(ru,v) = \left(\frac{1-r^2}{1-2ruv'+r^2} \right)^{n-1} \qquad (4.2)$$

is a counterexample showing that the well-known maximum principle for a strictly elliptic equation is not valid for degenerate elliptic equations.

A well-known maximum principle for elliptic partial differential equations of second order says that for a strictly elliptic equation a non-constant solution of its Dirichlet problem can attain its non-negative maximum (non-positive minimum) only on the boundary of a bounded domain. By offering the Poisson kernel $P(ru,v) = P(x,v)$ given by (4.2) as a counterexample, we explain that the above maximum principle is not true for degenerate elliptic equations. It is easy to see that in the case where n is an odd number, the Poisson kernel (4.2) is positive everywhere except that the value of $P(ru, v)$ is zero on the unit sphere when $u \neq v$. Therefore, if we take a bounded domain Ω containing a piece of the unit sphere in its interior but excluding $x = v$, then there is a function $P(x, v)$ satisfying equation (1.1), that is, $L_x P(x, v) = 0$ in Ω, but $P(x, v)$ attains its minimum, zero, on the degenerate surface, the unit sphere S^{n-1}, included in the interior of the bounded domain Ω.

We call the solutions of equation (1.1) harmonic functions.

Theorem 5. *Assume that $f \in C^2(S^{n-1})$ and $n \geq 3$. Then, there exists a harmonic function on the entire space (including ∞) given by*

$$U(ru) = \begin{cases} \dfrac{1}{\omega_n} \oint_{S^{n-1}} \left(\dfrac{1-r^2}{1-2ruv'+r^2} \right)^{n-1} f(v)\, d\sigma(v), & r < 1, \\[4mm] \dfrac{1}{\omega_n} \oint_{S^{n-1}} \left(\dfrac{r^2-1}{r^2-2ruv'+1} \right)^{n-1} f(v)\, d\sigma(v), & r > 1, \end{cases} \tag{4.3}$$

such that $U(ru)$ is the unique solution of equation (1.1) satisfying

$$\lim_{r \to 1+0} U(ru) = \lim_{r \to 1-0} U(ru) = f(u) \quad for \quad u \in S^{n-1}$$

and $U(ru) = U(x) \in C^2(\mathbb{R}^n \cup \{\infty\})$ where $\{\infty\}$ was defined in Definition 1.

Proof. Proposition 2 showing that equation (1.1) is invariant under reflections (with respect to the unit sphere) of the Möbius group, the function $U(x)$ obviously satisfies equation (1.1) inside and outside S^{n-1}. The only point we need to prove is if $U(x) \in C^2(S^{n-1})$. For this purpose we expand (4.3) as follows:

$$U(ru) = \frac{1}{\omega_n} \oint_{S^{n-1}} \left(\frac{1-r^2}{1-2ruv'+r^2} \right)^{n-1} f(v)\, d\sigma(v)$$

$$= \frac{1}{\omega_n} \sum_{k=0}^{\infty} \frac{(2k+n-2)\, \Gamma(\frac{n}{2})\, \Gamma(k+n-1)}{(n-2)\, \Gamma(n-1)\, \Gamma(k+\frac{n}{2})}$$

$$\times F(k, 1-\frac{n}{2}, k+\frac{n}{2};\, r^2)\, r^k \oint_{S^{n-1}} P_k^{(\frac{n}{2}-1)}(uv') f(v)\, d\sigma(v)$$

for $r < 1$ and

$$U(ru) = \frac{1}{\omega_n} \oint_{S^{n-1}} \left(\frac{r^2-1}{r^2-2ruv'+1} \right)^{n-1} f(v)\, d\sigma(v)$$

$$= \frac{1}{\omega_n} \sum_{k=0}^{\infty} \frac{(2k+n-2)\, \Gamma(\frac{n}{2})\, \Gamma(k+n-1)}{(n-2)\, \Gamma(n-1)\, \Gamma(k+\frac{n}{2})}$$

$$\times F(k, 1-\frac{n}{2}, k+\frac{n}{2};\, r^{-2})\, r^{-k} \oint_{S^{n-1}} P_k^{(\frac{n}{2}-1)}(uv') f(v)\, d\sigma(v),$$

for $r > 1$, where $\Gamma(z)$ is a Gamma function, $F(a, b, c; z)$ is a hypergeometric function, and

$$P_k^{(\lambda)}(\xi) = \sum_{m=0}^{[\frac{k}{2}]} (-1)^m \frac{\Gamma(k - m + \lambda)}{\Gamma(\lambda)\, m!\, (k - 2m)!} (2\xi)^{k-2m} \quad \text{for} \ -1 \le \xi \le 1$$

is a Gegenbauer polynomial (where k is integer, $\lambda > 0$, and $[\frac{k}{2}]$ is the greatest integer less than or equal to $\frac{k}{2}$).

Applying properties of hypergeometric functions, we get

$$\lim_{r \to 1-0} \frac{d}{dr}\left[r^k\, F\left(k,\, 1 - \frac{n}{2},\, k + \frac{n}{2};\, r^2\right)\right] = 0,$$

$$\lim_{r \to 1+0} \frac{d}{dr}\left[r^{-k}\, F\left(k,\, 1 - \frac{n}{2},\, k + \frac{n}{2};\, r^{-2}\right)\right] = 0$$

which shows $U(ru) = U(x) \in C^1(S^{n-1})$. Moreover, we have

$$\frac{\partial^2 U(r^{-1}, u)}{\partial r^2} = 2\, r^{-3}\left[\frac{\partial U(r, u)}{\partial r}\right]_{r=r^{-1}} + r^{-4}\left[\frac{\partial^2 U(r, u)}{\partial r^2}\right]_{r=r^{-1}}$$

that implies (noting $\left.\frac{\partial U(r,u)}{\partial r}\right|_{r=1} = 0$)

$$\left.\frac{\partial^2 U(r^{-1}, u)}{\partial r^2}\right|_{r=1} = \left.\frac{\partial^2 U(r, u)}{\partial r^2}\right|_{r=1},$$

i.e., $U(x) \in C^2(S^{n-1})$. Therefore, the proof is complete. $\square$

Laplace-Beltrami equation (1.1) is a degenerate elliptic equation. It has a degenerate surface S^{n-1} on which it degenerates into a partial differential equation of first order. We showed that because the degeneracy equation (1.1) is quite different from the strictly elliptic equations, e.g., Laplace's equation. We showed that for the Laplace-Beltrami equation (1.1) the well-known maximum principle for elliptic equations is not valid by offering a counterexample. Then we conclude that there exists a non-constant twice continuously differentiable solution of Dirichlet problem for equation (1.1) on the entire space including infinity. Moreover, it is easy to verify that for equation (1.1) the solution $U(x)$ given by (4.3) is not in $C^3(S^{n-1})$. This is another point showing the difference between degenerate elliptic equations and strictly elliptic equations.

5   Upper half space

There is another degenerate elliptic equation in upper half space. The following propositions (see [12]) suggest that the results obtained for the

Laplace-Beltrami equation (1.1), such as the Möbius group, Green function, the Poisson formula and Dirichlet problem for the non-homogeneous equation, etc., will be able to transfer to the degenerate elliptic equation degenerating on the boundary of upper half space.

Proposition 9. *The transformation*

$$y = \frac{(\ |x|^2 - 1\ ,\ 2x_2\ ,\ \ldots\ ,\ 2x_n\ )}{(x_1 + 1)^2 + x_2^2 + \cdots + x_n^2} \tag{5.1}$$

transforms the upper half space $x_1 > 0$ in to the unit ball $\{|y|^2 < 1\}$ of n dimensions, and transforms the equation (1.1) into

$$x_1^2 \sum_{i=1}^{n} \frac{\partial^2 \Phi}{\partial x_i^2} + (2 - n)x_1 \frac{\partial \Phi}{\partial x_1} = 0. \tag{5.2}$$

Similarly, we can also consider Dirichlet problem for equation (5.2) in a domain D including a piece of the degenerate hyperplane $x_1 = 0$ in its interior.

Proposition 10. *From (5.1) and (4.1) it is easy to deduce that in upper half space the Poisson formula for equation (5.2) is*

$$\Phi(x_1, \cdots, x_n) = \frac{1}{\omega_n} \int_{-\infty}^{\infty} \cdots \int_{-\infty}^{\infty} \frac{(2x_1)^{n-1}\Phi(0, \xi_2, \cdots, \xi_n)d\xi_2 \cdots d\xi_n}{[x_1^2 + (\xi_2 + x_2)^2 + \cdots + (\xi_n + x_n)^2]^{n-1}}.$$

REFERENCES

[1] L. Ahlfors, Old and new in Möbius groups, *Ann. Acad. Sci. Fenn. Ser. A I Math.* **9** (1984), 93–105.

[2] L. Ahlfors, Möbius transforms and Clifford numbers, *Differential Geometry and Complex Analysis*, Isaac Chavel and Hershel M. Farkas, eds., Springer-Verlag, Berlin, 1985, 65–73.

[3] L. Ahlfors, Clifford-numbers and Möbius transformations in $\mathbb{R}^n$, *Clifford Algebras and Their Role in Mathematics and Physics*, J. R. Chisholm and A. K. Common, eds., Canterbury, 1985, Reidel, Dordrecht, 1986, 167–175.

[4] L. Ahlfors, Möbius transformations in $\mathbb{R}^d$ expressed through 2×2 Clifford matrices, *Complex Variables Theory Appl.* **5** (1986), 215–224.

[5] B. Bojarski, Conformally covariant differential operators, *Proceedings, XXth Iranian Math. Congress*, Teheran, 1989.

[6] F. Brackx, R. Delanghe, and F. Sommen, Clifford analysis, *Pitman Res. Notes Math. Ser.*, Vol. **76**, 1982.

[7] J. Cnops, Hurwitz pairs and applications of Möbius transformations, Ph.D. Dissertation, University of Gent, Belgium, 1994.

[8] R. Courant and D. Hilbert, *Method of Mathematical Physics, Vol. II*, Interscience, New York, 1962.

[9] A. Erdélyi, W. Magnus, F. Oberhettinger, and F. G. Tricomi, *Higher Transcendental Functions, Vols. I, II, III*, McGraw-Hill, New York, 1954.

[10] D. Gilbarg and N. Trudinger, *Elliptic Partial Differential Equations of Second Order*, Springer, Berlin, 1983.

[11] Loo-Keng Hua, *Harmonic Analysis of Functions of Several Complex Variables in the Classical Domains*, Science Press, Beijing, China, 1965.

[12] Loo-Keng Hua, *Starting with the Unit Circle*, Science Press, Beijing, China, 1977.

[13] Xinhua Ji and Dequan Chen, A partial differential equation with a degenerate surface in n-dimensional extended space, *Chinese Annals of Mathematics* **1:3** (4) (1980), 375–386.

[14] Xinhua Ji and Dequan Chen, Fundamental properties of solutions of a degenerate elliptic equation in extended space, *Journal of Differential Equations, Vol. 58*, No. **2** (1985), 192–211.

[15] Xinhua Ji and Dequan Chen, Applications of expansion of Poisson kernel to the Dirichlet Problem for a degenerate elliptic equation in extended space, *Geometry, Analysis, and Mechanics*, John Rassias, ed., World Scientific Publ. Co., 1994, 219–226.

[16] J. Peetre and T. Qian, Möbius covariance of iterated Dirac operators, *J. Austral. Math. Soc. Ser. A* **56** (1994), 403–414.

[17] T. Qian and J. Ryan, Conformal transformation and Hardy spaces arising in Clifford analysis, *J.Operator Theory* **35** (1996), 349–372.

[18] E. M. Stein and G. Weiss, *Introduction to Fourier Analysis on Euclidean Space*, Princeton University Press, 1971.

[19] K. Th. Vahlen, Uber Bewegungen und Complexe Zahlen, *Math. Ann.* **55** (1902), 585–593.

Xinhua Ji
Institute of Mathematics
Academia Sinica
Beijing, 100080 China
E-mail: xhji@math03.math.ac.cn

Received: October 1, 1999; Revised: February 20, 2000

Quaternionic Analysis in Fluid Mechanics

Wolfgang Sprößig

ABSTRACT We give a survey of problems in fluid mechanics which could be considered successfully by methods from quaternionic analysis. In particular we study a special problem where stationary Navier–Stokes equations are combined with field induction.
Keywords: Fluid mechanics, quaternionic analysis, Navier–Stokes equations.

1 Introduction

Quaternionic and Clifford analysis have in recent years become increasingly important tools in the analysis of partial differential equations and their application in mathematical physics and engineering. Above all, this paper reflects the importance of quaternionic methods for the treatment of boundary value problems of stationary equations in fluid mechanics. There exists already a well-developed mathematical theory of viscous fluids. Well-known mathematicians in this field like O.A. Ladyzenskaja [11], V.A. Solonnikov [12], W. Borchers [2], G.P. Galdi [4] and W. Varnhorn [15] achieved essential results. In this paper we will present an alternative approach to the treatment of fluid problems. Some of our results were already presented at the Conference on Computational Fluid Mechanics and 3-dimensional Complex Flows held in Lausanne in 1995. Our aim is to give a survey of problems which could be successfully dealt with by methods from Clifford analysis. The scope of these problems reaches from linear and non-linear Stokes equations via versions of Navier–Stokes equations involving viscous fluids under the influence of temperature or field induction. The most important theorems will be formulated.

In the second part we will show the efficiency of these methods with help of an example where Navier–Stokes equations are combined with field induction. We restrict our analysis to problems which are posed in bounded domains with piecewise smooth boundaries. In papers by S. Bernstein [1], U. Kähler [9],[10] and P. Cerejeiras and U. Kähler [3], problems of Navier–Stokes type in unbounded domains with boundaries satisfying very weak

---

AMS Subject Classification: 30G35, 30J25.

conditions are studies using methods from Clifford analysis. We will also note here that in [5] and [6] at least for the case of the stationary Navier–Stokes equations a discrete analogue is formulated. Relations between the continuous and discrete models are worked out there in detail. For mesh width tending to zero, the difference between the solutions of both the continuous problem and the discrete problem may be estimated under weak conditions.

2   Preliminaries

Let e_1, e_2, e_3 be unit vectors in $\mathbb{R}^3$ and $e_0 = 1$. Assume that these basis vectors fulfill the relations $e_i e_j + e_j e_i = -2\delta_{ij}$ for $i, j = 1, 2, 3$. Furthermore, let $e_1 e_2 = e_3$. We consider all elements of the form $u = u_0 + \mathbf{u}$ where $\mathbf{u} := u_1 e_1 + u_2 e_2 + u_3 e_3$. The set of all these elements supplied with the above mentioned multiplication rules is called skew-field of **real quaternions**. $u_0 =: Sc(u)$ is called **scalar part** and $\mathbf{u} =: Vec(u)$ is called **a vector part** of the quaternion u. Furthermore, $\overline{u} =: u_0 - \mathbf{u}$ is called **a conjugated quaternion**. Obviously, $u\overline{u} = \overline{u}u =: |u|^2$ is the norm of u. Let u, v be quaternions. Then it follows

$$uv = u_0 v_0 + \mathbf{u} \cdot \mathbf{v} + u_0 \mathbf{v} + v_0 \mathbf{u} + \mathbf{u} \times \mathbf{v}.$$

Here $\mathbf{u} \cdot \mathbf{v}$ denotes the scalar product in $\mathbb{R}^3$ and $\mathbf{u} \times \mathbf{v}$ denotes Gibbs' cross product in $\mathbb{R}^3$. After identifying vectors as vector parts of quaternions we also will use the outer product, e.g., $u \wedge v$. The spaces

$$L_p(G, \mathbb{H}), \quad W_p^k(G, \mathbb{H}), \quad W_p^r(G, \mathbb{H}), \quad C^{0,\alpha}(G, \mathbb{H}),$$

$r \in \mathbb{R}$, $p \geq 1$, $k \in N$, $0 < \alpha \leq 1$, are defined componentwise. These are Lebesgue spaces, Sobolev spaces, Sobolev–Slobodetzkij spaces and Hölder spaces, respectively. We introduce the following abbreviations:

$$\|u\|_p := \left(\int_G |u(x)|^p dx \right)^{\frac{1}{p}}, \quad \|u\|_{2,1} := \left(\|u\|_2^2 + \sum_{i=1}^3 \|\partial_i u\|_2^2 \right)^{\frac{1}{2}},$$

$$(u, v)_2 := \int_G \overline{u}v \, dx \in \mathbb{H}.$$

In this way $\|u\|_p, \|u\|_{2,1}$ are the norms in the spaces $L_p(G, \mathbb{H}), W_2^2(G, \mathbb{H})$, respectively. The item $(u, v)_2$ is the scalar product in $L_2(G, \mathbb{H})$. $\overset{\circ}{W}_2^1(G, \mathbb{H})$ denotes the subspace of all functions of $W_2^1(G, \mathbb{H})$ whose trace on $\partial G = \Gamma$ is identically zero.

Now let us introduce the so-called **Dirac type operator** $D = \sum_{i=1}^{3} \partial_i e_i$, which acts in the following way: Let $u \in C^1(G)$. Then

$$Du = \sum_{j=0}^{3} \sum_{i=1}^{3} \partial_i u_j e_i e_j = \text{grad } u_0 + \text{rot } \mathbf{u} - \text{div } \mathbf{u} \; ,$$

$$uD = \sum_{j=0}^{3} \sum_{i=1}^{3} \partial_i u_j e_j e_i = \text{grad } u_0 - \text{rot } \mathbf{u} - \text{div } \mathbf{u} \; .$$

All functions which belong to $\ker D = \{u : Du = 0\}$ are called **left-$\mathbb{H}$-regular**. Functions which fulfill $(uD) = 0$ we call **right-$\mathbb{H}$-regular**. In the following when confusions can be excluded, we will name left-$\mathbb{H}$-regular functions simply $\mathbb{H}$-regular functions. The function

$$\mathbf{e}(x) = -\frac{1}{4\pi} \frac{x}{|x|^3}$$

is both left- and right-$\mathbb{H}$-regular. $\mathbf{e}(x)$ is a fundamental function for D. We have $D^2 = -\Delta = -(\partial_1^2 + \partial_2^2 + \partial_3^2)$.

Now we introduce the following integral operators, namely the so-called **Teodorescu transform**

$$(T_G u)(x) := -\int_G \mathbf{e}(x - y)u(y)dy, \quad x \in \mathbb{R}^3,$$

the **Cauchy-type operator**

$$(F_\Gamma u)(x) := \int_\Gamma \mathbf{e}(x - y)\alpha(y)u(y)d\Gamma_y, \quad x \notin \Gamma,$$

and a multidimensional singular integral operator

$$(S_\Gamma u)(x) := 2\int_\Gamma \mathbf{e}(x - y)\alpha(y)u(y)d\Gamma_y, \quad x \in \Gamma,$$

where $\alpha(y) = \sum_{i=1}^{3} e_i \alpha_i(y)$ is the unit vector of the outer normal at the point y. The integral which defines the operator S_Γ is understood in the sense of Cauchy's principal value. This operator is sometimes called the **Cauchy–Bitzadse operator**. The operator $P_\Gamma := \frac{1}{2}(I + S_\Gamma^\delta)$ denotes the **Pompeiu projection** onto the space of all $\mathbb{H}$-valued functions which may be $\mathbb{H}$-regular extended into the domain G. $Q_\Gamma := \frac{1}{2}(I - S_\Gamma^\delta)$ denotes the **Pompeiu projection** onto the space of all $\mathbb{H}$-valued functions which can be $\mathbb{H}$-regular extended into the domain $\mathbb{R}^3 \setminus \overline{G}$ and vanish at infinity. The operator S_Γ^δ is defined by

$$S_\Gamma^\delta u := \frac{4\pi - 2\delta(x)}{4\pi} u(x) + (S_\Gamma u)(x), \quad 0 < \delta \leq 4\pi,$$

where δ is the space angle taken from outside at the point x. It is easy to see that $(S_\Gamma^\delta)^2 = I$.

Note that Pompeiu projections coincide in case of the unit ball in $\mathbb{R}^3$ with the well-known Szegö projections. We have the following statements:

Proposition 1. [5] *Let $u \in C^1(G, \mathbb{H}) \cap C(\overline{G}, \mathbb{H})$. Then we have the following formulas:*

(i) $(F_\Gamma u)(x) + T_G Du(x) = \begin{cases} u(x), & \text{if } x \in G, \\ 0, & \text{if } x \in \mathbb{R}^3 \setminus \overline{G}, \end{cases}$

   (Borel-Pompeiu Formula)

(ii) $(DT_G u)(x) = \begin{cases} u(x), & \text{if } x \in G, \\ 0, & \text{if } x \in \mathbb{R}^3 \setminus \overline{G}, \end{cases}$

(iii) $(DF_\Gamma u(x) = 0 \quad \text{in} \quad G \cup (\mathbb{R}^3 \setminus \overline{G}).$

Remark 1. *Most of the statements which are known in the classical function theory of one complex variable and follow from the Borel-Pompeiu formula have analogue also in Clifford analysis (cf. [5], [6]).*

Theorem 1. [5] **(Plemelj-Sokhotzkij's Formulas)** *Let $u \in C^{0,\alpha}(G, \mathbb{H})$, $0 < \alpha < 1$. Then we have*

(i) $\lim\limits_{\substack{x \to \xi \in \Gamma \\ x \in G}} (F_\Gamma u)(x) = (P_\Gamma u)(\xi),$

(ii) $\lim\limits_{\substack{x \to \xi \in \Gamma \\ x \in \mathbb{R}^n \setminus \overline{G}}} (F_\Gamma u)(x) = (-Q_\Gamma u)(\xi),$

for any $\xi \in \Gamma$.

Corollary 1. [5] *Let $u \in C^{0,\alpha}(\Gamma, \mathbb{H})$. Then the relations*

   (i) $(S_\Gamma^2 u)(\xi) = u(\xi),$ (ii) $(F_\Gamma P_\Gamma u)(\xi) = F_\Gamma u(\xi),$

   (iii) $(P_\Gamma^2 u)(\xi) = (P_\Gamma u)(\xi),$ (iv) $(Q_\Gamma^2 u)(\xi) = (Q_\Gamma u)(\xi)$

are valid for any $\xi \in \Gamma$.

Let us note that the operators F_Γ, S_Γ, P_Γ and Q_Γ admit extensions to $L_2(G, \mathbb{H})$. The restriction of an $\mathbb{H}$-function u to a function defined on the boundary Γ is expressed by $tr_\Gamma u$.

Proposition 2. [14] *Let $G \subset \mathbb{R}^3$ be a bounded domain with a piecewise smooth boundary. Then the right-linear set*

$$\mathcal{L} := L_2(G, \mathbb{H}) \cap \ker D$$

is a subspace in $L_2(G, \mathbb{H})$.

It is important for our calculus to have the following decomposition of the Hilbert space $L_2(G, \mathbb{H})$, namely,

$$L_2(G, \mathbb{H}) = \ker D \cap L_2(G, \mathbb{H}) \oplus D \overset{\circ}{W}_2^1 (G, \mathbb{H}).$$

The corresponding projections $\mathcal{P}$ (onto $\ker D \cap L_2(G, \mathbb{H})$) and $\mathcal{Q}$ onto $D \overset{\circ}{W}_2^1 (G, \mathbb{H})$ are orthoprojections and $\mathcal{P}$ is just the well-known **Bergman projection**. In [13] it is proved that

$$f \in \operatorname{im} \mathcal{Q} \cap L_2(G, \mathbb{H}) \iff tr_\Gamma T_G f = 0,$$

where tr_Γ denotes the restriction of $T_G f$ onto the boundary Γ. It is shown (cf. [6]) that the Bergman projection $\mathcal{P}$ permits the representation

$$\mathcal{P} = F_\Gamma (tr_\Gamma V)^{-1} tr_\Gamma T_G,$$

where $(Vu)(x) = (T_G F_\Gamma u)(x) = \frac{1}{4\pi} \int_\Gamma \frac{\alpha}{|x-y|} u(y) d\Gamma_y$, $(u \in \ker D)$. We can see that the operator V is of a single layer potential type.

3   A review on fluid mechanics problems treated with quaternionic analysis

We will assume for simplicity that the domain G is bounded by a piecewise smooth bounded Liapunov surface. In the last years these assumptions have been considerably weakened. We will shortly discuss these generalizations.

3.1   Linear equations of Stokes' type

$$-\Delta u + \frac{1}{\eta} \nabla p = \frac{\rho}{\eta} f \quad \text{in} \quad G, \quad \operatorname{div} u = f_0 \quad \text{in} \quad G, \quad u = g \quad \text{on} \quad \Gamma.$$

Here η is the viscosity and ρ the density of the fluid. We have to look for the velocity u and the hydrostatic pressure p. Between f_0 and g the relation:

$$\int_G f_0 dx = \int_\Gamma \alpha g d\Gamma$$

has to be fulfilled. For $g = 0$ the measure of the compressibility f_0 satisfies the identity

$$\int_G f_0 dx = 0.$$

For all such real functions f_0 the unique solution (p is unique up to a real constant) can be represented as follows:

Theorem 2. [6] *Let* $f := f_0 + \mathbf{f} \in W_p^k(G, \mathbb{H})$ $(k \geq 0, 1 < p < \infty)$. *Then we have*

$$\mathbf{u} = \frac{\rho}{\eta} T_G VecT_G \mathbf{f} - \frac{\rho}{\eta} T_G VecF_\Gamma (tr_\Gamma T_G VecF_\Gamma)^{-1} tr_\Gamma T_G VecT_G \mathbf{f} - T_G f_0,$$

$$p = \rho Sc T_G \mathbf{f} - \rho Sc F_\Gamma (tr_\Gamma T_G VecF_\Gamma)^{-1} tr_\Gamma T_G VecT_G \mathbf{f} + \eta f_0.$$

In that way, we strongly separated velocity and pressure.

3.2   Nonlinear equations of Stokes' type

Now we assume that the compressibility depends on the velocity and the nonlinear outer forces. The equations describing this state are the following:

$$-\Delta \mathbf{u} + \frac{1}{\eta} \nabla p = \Lambda f(u) \quad \text{in } G,$$
$$\text{div } (\eta^{-1} \mathbf{u}) = 0 \quad \text{in } G, \tag{3.1}$$
$$\mathbf{u} = 0 \quad \text{on } \Gamma.$$

The parameter of viscosity $\eta > 0$ depends on the position. The main result is given by:

Theorem 3. [8]

1. *Let* $f \in L_2(G, \mathbb{H}), p \in W_2^1(G), \eta \in C^\infty(G)$. *Then every solution of the system (3.1) permits the operator representation*

$$u = \Lambda RBf - RBDp$$
$$0 = Sc\Lambda QT_G Bf - ScQT_G BDp. \tag{3.2}$$

Here B is the operator denoting multiplication by η^{-1} *and* $R :=$ $T_G QT_G$.

2. *If the operator function* $f(u)$ *satisfies the estimates*

   (i) $\|f(u) - f(v)\|_2 \leq L\|u - v\|_{2,1}$ *for* $\|u\|_{2,1}, \|v\|_{2,1} \leq 1$,

   (ii) $\|B\|_2 \leq K$ *for positive constants* K, L,

   (iii) $\Lambda < \{\|T\|_{im\ Q \cap L_2, L_2} \|T\|_{L_2, L_2} KL\}^{-1}$

   the iteration process

$$u_n = \Lambda RBf(u_{n-1}) - RBDp_n \Lambda Sc DBRf(u_{n-1})$$
$$= Sc DBRDp_n \|u_0\|_{2,1} \leq 1$$

for $u_0 \in \overset{\circ}{W}{}_2^1 (G, \mathbb{H})$ *converges to a unique solution* $\{u, p\} \in \overset{\circ}{W}{}_2^1$ $(G, \mathbb{H}) \cap \ker (\text{div}B) \times L_2(G)$ *of (3.2), where p is unique up to a real constant.*

3.3   Problems of Navier–Stokes type

In the stationary case, Navier–Stokes equations are described in the following way:

$$-\Delta u + \frac{\rho}{\eta}(u \cdot \nabla)u + \frac{1}{\eta}\nabla p = \frac{\rho}{\eta}f \quad \text{in} \quad G,$$

$$\text{div } u = 0 \quad \text{in} \quad G, \tag{3.3}$$

$$u = 0 \quad \text{on} \quad \Gamma.$$

We will abbreviate $M^*(u) - f$, where $M^*(u) = (u \cdot \text{grad})u$, as $M(u)$. The main result is now the following:

Theorem 4. [5]

1. *Let $f \in L_2(G, \mathbb{H})$, $p \in W_2^1(G)$. Every solution of (3.3) permits the operator representation*

$$u = -\frac{\rho}{\eta}RM(u) - \frac{1}{\eta}T_G Q p,$$

$$\frac{\rho}{\eta}ScQT_G M(u) - \frac{1}{\eta}ScQp = 0. \tag{3.4}$$

2. *The system (3.4) has a unique solution*

$$\{u, p\} \in \overset{\circ}{W}_2^1 (G, \mathbb{H}) \cap \ker(\text{div}B) \times L_2(G),$$

where p is unique up to a real constant, if

(i) $\|f\| \leq (18K^2 C_1)^{-1}$, $K := \frac{\rho}{\eta}\|T_G\|_{[L_2 \cap \text{ im } Q, W_2^1]}\|T\|_{[L_p, L_2]}$,

(ii) $u_0 \in \overset{\circ}{W}_2^1 (G, \mathbb{H}) \cap \ker(\text{div}B)$, $\|u_0\|_{2,1} \leq \min(V, \frac{1}{4KC_1} + W)$,

holds. Here $V := (2KC_1)^{-1}$, $W := [(4KC_1)^{-2} - \frac{\rho\|f\|_p}{\eta C_1}]^{\frac{1}{2}}$ and $C_1 := 9^{\frac{1}{p}}C$, where C is the embedding constant from W_2^1 in L_2. The iteration process (starting with u_0)

$$u_n = \frac{\rho}{\eta}RM(u_{n-1}) - \frac{1}{\eta}RDp_n \frac{\rho}{\eta}ScQT_G M(u_{n-1})$$

$$= -\frac{1}{\eta}ScQp_n, \quad u_0 \in \overset{\circ}{W}_2^1 (G, \mathbb{H}) \cap \ker \text{div})$$

converges in $W_2^1(G, \mathbb{H}) \times L_2(G)$.

3.4   Navier–Stokes equations with heat conduction

We will now consider the flow of a viscous fluid under the influence of temperature. The corresponding equations read as follows:

$$-\Delta \mathbf{u} + \frac{\rho}{\eta}(u \cdot \nabla)\mathbf{u} + \frac{1}{\eta}\nabla p + \frac{\gamma}{\eta}gw = f \quad \text{in} \quad G,$$

$$-\nabla w + \frac{m}{\kappa}(\mathbf{u} \cdot \nabla)w = \frac{1}{\kappa}h \quad \text{in} \quad G,$$

$$\text{div } \mathbf{u} = 0 \quad \text{in} \quad G, \qquad (3.5)$$

$$\mathbf{u} = 0 \quad \text{on} \quad \Gamma,$$

$$w = 0 \quad \text{on} \quad \Gamma.$$

We denote by ρ the density of the fluid, by η the viscosity, by γ the Grashof number, by κ the temperature conductivity number and by m the Prandtl number. As usual $\mathbf{u}$ stands for the velocity, w for the temperature and p for the hydrostatic pressure. Here we will only formulate the main result. It can be shown that the solutions of this system fulfill the following system of operator integral equations, where boundary conditions are satisfied automatically. Here we have this system:

$$u = -R[M(u) - \frac{\gamma}{\eta}e_3 w] - \frac{1}{\eta}T_G Qp,$$

$$0 = ScDR[M(u) - \frac{\gamma}{\eta}e_3 w] - \frac{1}{\eta}Qp, \qquad (3.6)$$

$$w = -\frac{m}{\kappa}RSc(uD)w + Rg,$$

where $M(u) := \frac{\rho}{\eta}(\mathbf{u} \cdot \text{grad})u + f(u) - F$.

Theorem 5. [7]

1. *We consider the following iteration procedure:*

$$u_n = -R[M(u_{n-1}) - \frac{\gamma}{\eta}e_3 w_{n-1}] - \frac{1}{\eta}T_G Qp_n,$$

$$0 = ScDR[M(u_{n-1}) - \frac{\gamma}{\eta}e_3 w_{n-1}] - \frac{1}{\eta}Qp_n, \qquad (3.7)$$

$$w_n = -\frac{m}{\kappa}RSc(u_n D)w_n + Rg.$$

The computation of w_n will be done by the inner iteration

$$w_n^j = \frac{m}{\kappa}RSc(u_n D)w_n^{j-1} + Rg.$$

2. *Let $u_n \in \overset{\circ}{W}_2^1$. Further, let $m \neq 4\kappa$ and $\|u_n\| < \frac{\kappa}{mKC}$. Then, the sequence $\{w_n^{\{j\}}\}_{j \in N}$ converges in $W_2^1(G)$.*

3. *Let $F \in L_2(G, \mathbb{H}), g \in L_2(G), f : W_2^1(G, \mathbb{H}) \to L_2(G, \mathbb{H})$ with*

$$\|f(u) - f(v)\|_2 \le L\|u - v\|_{2,1}$$

and $f(0) = 0$. Under the additional smallness conditions

(i) $\frac{\rho}{\eta}\|F\|_2 + \frac{\gamma}{\eta}K|d|^{-1}\|g\|_2 < \frac{1}{16K^2C}$, $\left(d := (4 - \frac{m}{\kappa})\kappa\right)$,

(ii) $\|g\|_2 < (1 - \frac{1}{\sqrt{2}})\frac{\eta d^2}{32K^3Cm}$,

(iii) $m < 4\kappa$,

the sequence $\{u_n, w_n, p_n\}_{n \in \mathbb{N}}$ converges in

$$W_2^1(G, \mathbb{H}) \times W_2^1(G, \mathbb{H}) \times L_2(G)$$

to the unique solution $(u, w, p) \in \overset{\circ}{W_2^1}(G, \mathbb{H}) \times \overset{\circ}{W_2^1}(G, \mathbb{H}) \times L_2(G)$ of the original boundary value problem, where p is unique up to a real constant.

Remark 2. *We note that conditions (i) and (ii) can always be realized for fluids with big enough viscosity number.*

4   Treatment of a problem in magneto-hydromechanics

4.1   Different formulations of the problem

In this section we will observe in more detail a special problem for viscous flows under the influence of a magnetic field **B**. Let $G \subset \mathbb{R}^3$ be a bounded domain with a sufficiently smooth boundary Γ. We consider now the following hydromechanical problem with unknown velocity **u** and field induction B.

$$
\begin{aligned}
-\Delta \mathbf{u} + \frac{\rho}{\eta}(\mathbf{u} \cdot \nabla)\mathbf{u} + \frac{1}{\eta}\nabla p - \frac{1}{\mu\eta}\mathbf{B} \times \operatorname{rot} \mathbf{B} &= \frac{\rho}{\eta}f \quad \text{in}\;\; G, \\
\Delta \mathbf{B} + \mu\sigma \operatorname{rot}(\mathbf{B} \times \mathbf{u}) &= \mathbf{g} \quad \text{in}\;\; G, \\
\operatorname{div} \mathbf{B} &= 0 \quad \text{in}\;\; G, \\
\operatorname{div} \mathbf{u} &= 0 \quad \text{in}\;\; G, \\
\mathbf{B} = \mathbf{u} &= 0 \quad \text{on}\;\; \Gamma,
\end{aligned}
$$
(4.1)

where μ is the permeability and σ is a measure of the electric charge.

Now we have to give these equations a quaternionic formulation. This reformulation reads as follows:

$$DD\mathbf{u} + \frac{1}{\eta}D\,p + M(\mathbf{u}) = \frac{1}{\mu\eta}\mathbf{B} \times (D \times \mathbf{B}) \quad \text{in} \quad G,$$
$$DD\mathbf{B} + \mu\sigma[\mathbf{u}, \mathbf{B}] = \mathbf{g} \qquad\qquad\qquad \text{in} \quad G,$$
$$Sc\,D\,\mathbf{u} = 0 \qquad\qquad\qquad\qquad \text{in} \quad G, \qquad (4.2)$$
$$Sc\,D\mathbf{B} = 0 \qquad\qquad\qquad\qquad \text{in} \quad G,$$
$$\mathbf{B} = \mathbf{u} = 0 \qquad\qquad\qquad\qquad \text{on} \quad \Gamma.$$

Only for simplicity will we add the following trivial problems:

$$\Delta u_0 = 0, \quad \Delta B_0 = 0 \quad \text{in} \quad G,$$
$$u_0 = 0, \qquad B_0 = 0 \quad \text{on} \quad \Gamma.$$

Now we will embed $\mathbb{R}^3$ into $\mathbb{H}$ and identify quaternions with scalar part zero with vectors in $\mathbb{H}$. Furthermore, we abbreviate

$$M(u) := \frac{\rho}{\eta}[M^*(u) - f], \ M^*(u) := (\mathbf{u} \cdot D)\mathbf{u}$$
$$[u, B] := D \wedge (B \wedge u) = (u \cdot D)B - (B \cdot D)\mathbf{u}.$$

Notice that $[u, B]$ is just a Lie bracket, $f := \mathbf{f}$ and $g := \mathbf{g}$. Finally, we get

$$DD\mathbf{u} + \frac{1}{\eta}\,Dp + M(u) = \frac{1}{\mu\eta}\,B \wedge (D \wedge B) \quad \text{in} \quad G,$$
$$DD\,B + \mu\sigma[u, B] = g \qquad\qquad\qquad \text{in} \quad G,$$
$$Sc\,Du = 0 \qquad\qquad\qquad\qquad \text{in} \quad G, \qquad (4.3)$$
$$Sc\,DB = 0 \qquad\qquad\qquad\qquad \text{in} \quad G,$$
$$u = 0 \qquad\qquad\qquad\qquad\qquad \text{on} \quad \Gamma.$$
$$B = 0 \qquad\qquad\qquad\qquad\qquad \text{on} \quad \Gamma.$$

After transforming into an operator integral equation, which can be done by the help of the preliminary results, we obtain

$$u = -\frac{\rho}{\eta}\,T_G Q T_G\,M(u) - \frac{1}{\eta}\,T_G Qp + \frac{1}{\mu\eta}\,T_G Q T_G(B \wedge (D \wedge B)) \quad (4.4a)$$
$$B = T_G Q T_G\,g - \mu\sigma\,T_G Q T_G[u, B] \qquad\qquad\qquad\qquad\qquad (4.4b)$$
$$0 = Sc\,\rho\,Q T_G M(u) + Sc\,Qp - \frac{1}{\mu}\,Sc\,Q T_G(B \wedge (D \wedge B)) \qquad (4.4c)$$
$$0 = Sc\,Q T_G g - \mu\sigma\,Sc\,Q T_G[u, B]. \qquad\qquad\qquad\qquad\qquad (4.4d)$$

Note again that the boundary conditions are satisfied automatically. It is sufficient to show this for the velocity. Indeed, we know that any element

Qu of im$Q$ has the representation $Qu = Dv$ with $v \in \overset{\circ}{W}{}_2^1 (G, \mathbb{H})$. Then it follows

$$tr_\Gamma T_G Qu = tr_\Gamma T_G Dv = tr_\Gamma v - tr_\Gamma F_\Gamma \, tr \, v = 0.$$

Remark 3. *Let $g = 0$. The solution $\{u, p, 0\}$ of (4.4) solves the Navier-Stokes equations. We look for solutions with $B \not\equiv 0$.*

4.2   An iteration procedure

We will now consider the following iteration method. For $n = 0, 1, 2, \ldots$ we get

$$u_n = -T_G Q T_G [\frac{\rho}{\eta} M(u_{n-1}) - \frac{1}{\mu\eta} B_{n-1} \wedge (D \wedge B_{n-1})] - \frac{1}{\eta} T_G Q p_n$$

$$0 = Sc[\rho Q T_G M(u_{n-1}) + Q p_n - \frac{1}{\mu} Q T_G [B_{n-1} \wedge (D \wedge B_{n-1})]$$

$$B_n^{(j)} = -\mu\sigma T_G Q T_G [u_n, B_n^{(j-1)}] + T_G Q T_G g \qquad (j = 1, 2, \ldots)$$

$$0 = -\mu\sigma Sc \, Q T_G [u_n, B_n^{(j-1)}] + Sc \, Q T_G g.$$

Here B_n will be computed by use of "inner" iteration.

Remark 4. *In this way in each step one has to solve only a linear Stokes problem.*

Let $u, B \in \overset{\circ}{W}{}_2^1 (G, \mathbb{H})$ and $1 < p < \frac{3}{2}$. Then we obtain

$$\|[u, B]\|_p^p \le 2 \sum_{j,i=1}^3 C^p \|u_i\|_{1,2}^p \|B_j\|_{1,2}^p \le 18 C^p \|u\|_{1,2}^p \|B\|_{1,2}^p.$$

The embedding constant C can be calculated by the estimate

$$\|u\|_q = \|T_G Du\|_q \le \|T_G\|_{[L_2, L_q]} \|Du\|_2 \le \|T_G\|_{[L_2, L_q]} \|u\|_{2,1}$$

also $\|T_G\|_{[L_2, L_q]} = C$, $q < 6$, $p = \frac{2q}{2+q}$. Set now: $C_1 = 9^{\frac{1}{p}} C$. So we get

$$\|[u, B]\|_p^p \le 2 C_1 \|u\|_{1,2} \|B\|_{2,1}.$$

4.3   Convergence of the inner iteration

Next we consider the inner iteration

$$B_n^{(j)} = \mu\sigma \, T_G Q T_G [(B_n^{(j-1)} \cdot D) u_n - B_n^{(j-1)} (u_n \cdot D)] + T_G Q T_G \, g$$

$$B_n^{(0)} = 0.$$

We will abbreviate

$$K := \|T_G\|_{[L_2 \cap \text{ im } Q, W_2^1]} \|T_G\|_{[L_p, L_2]}.$$

It follows that

$$\|B_n^{(j)}\|_{2,1} \le \mu\sigma K 2 C_1 \|u_n\|_{2,1} \|B_n^{(j-1)}\|_{2,1} + K\|g\|_p$$

Because of

$$\|u_n\|_{2,1} < \frac{1}{4C_1 \mu\sigma K},$$

we have

$$\|B_n^{(j)}\|_{2,1} \le 2K\|g\|_p.$$

Hence, $(B_n^{(j)})$ is bounded in $W_2^1(G, \mathbb{H})$ and it exists as a weakly convergent subsequence. On the other hand, we have

$$\|B_n^{(j)} - B_n^{(j-1)}\|_{2,1} = 2C_1 \mu\sigma \|T_G Q T_G\,[u_n, B_n^{(j-1)} - B_n^{(j-2)}]\|_{2,1}$$

$$\le 2C_1 K \mu\sigma \|u_n\|_{2,1} \|B_n^{(j-1)} - B_n^{(j-2)}\|_{2,1}$$

$$< \|B_n^{(j-1)} - B_n^{(j-2)}\|_{2,1}.$$

From these relations, we get the strong convergence of $(B_n^{(j)})$ in $W_2^1(G, \mathbb{H})$ to the limit function B_n which satisfies the estimate

$$\|B_n\|_{2,1} \le 2K\|g\|_p.$$

4.4 On some a priori estimates

It is easy to obtain

$$\|Du\|_2 \ge \frac{1}{\|T_G\|_{[\text{im }\varrho \cap L_2, W_2^1]}} \|u\|_{2,1}.$$

Using the a priori estimate (cf. [5])

$$\|Du\|_2 + \frac{1}{\eta}\|Qp\|_2 \le 2^{1/2} \left\| Q T_G \left[\frac{\varrho}{\eta} M^*(u) - \frac{\varrho}{\eta} f - \frac{1}{\mu\eta} B \wedge (D \wedge B) \right] \right\|,$$

we get

$$\frac{1}{\|T_G\|_{[\text{im }Q, W_2^1]}} \|u\|_{2,1} +$$

$$\frac{1}{\eta}\|Qp\|_2 \le 2^{1/2}\|T_G\|_{[L_p, L_2]} \left[\frac{\varrho}{\eta}\|M^*(u)\|_p + \frac{\varrho}{\eta}\|f\|_p + \frac{2}{\mu\eta}\|B\|_{2,1}^2 C_1 \right]$$

and so

$$\frac{1}{\|T_G\|_{[\text{im }Q, W_2^1]}} \|u\|_{2,1} \le$$

$$2\|T_G\|_{[L_p, L_2]} \left[\frac{\varrho}{\eta} C_1 \|u\|_{2,1}^2 + \frac{\varrho}{\eta}\|f\|_p + \frac{2}{\mu\eta}\|B\|_{2,1}^2 C_1 \right].$$

Hence,

$$\|u\|_{2,1} \leq 2K\frac{\rho}{\eta}\,C_1\|u\|_{2,1}^2 + 2K\frac{\rho}{\eta}\,\|f\|_p + \frac{4K}{\mu\eta}\,C_1\|B\|_{2,1}^2.$$

We have the following quadratic inequality for $\|u\|_{2,1}$:

$$2K\frac{\rho}{\eta}\,C_1\|u\|_{2,1}^2 - \|u\|_{2,1} + 2K\frac{\rho}{\eta}\|f\|_p + \frac{4K}{\mu\eta}\,C_1\|B\|_{2,1}^2 \geq 0$$

$$\|u\|_{2,1}^2 - \frac{\eta}{2K\rho C_1}\,\|u\|_{2,1} + \frac{1}{C_1}\,\|f\|_p + \frac{2}{\rho\mu C_1}\,\|B\|_{2,1}^2 = 0$$

such that

$$\frac{\eta}{4K\rho C_1} - \sqrt{\frac{\eta^2}{16K^2\rho^2 C_1^2} - \frac{1}{C_1}\,\|f\|_p - \frac{2}{\rho\mu}\,\|B\|_{2,1}^2}$$

$$\leq \|u\|_{2,1} \leq \frac{\eta}{4K\rho C_1} + \sqrt{\frac{\eta^2}{16K^2\rho^2 C_1^2} - \frac{1}{C_1}\,\|f\|_p - \frac{2}{\rho\mu}\,\|B\|_{2,1}^2}.$$

There arises the necessary condition

$$\frac{1}{C_1}\|f\|_p + \frac{2}{\rho\mu}\|B\|_{2,1}^2 < \frac{\eta^2}{16K^2\rho^2 C_1^2}. \tag{4.5}$$

On the other hand we have to realize for any u_n the condition

$$\|u_n\|_{2,1} \leq \frac{1}{4C_1\mu\sigma K}.$$

4.5   Convergence of the velocity sequence

Now we are able to estimate

$$\|u_n - u_{n-1}\|_{2,1} \leq \|T_G Q T_G \frac{\rho}{\eta}\,[M(u_{n-1}) - M(u_{n-2})]\|_{2,1}$$

$$+ \|T_G Q T_G \frac{1}{\mu\eta}[B_{n-1} \wedge (D \wedge B_{n-1}) - B_{n-2} \wedge (D \wedge B_{n-2})]\|_{2,1}$$

$$+ \frac{1}{\eta}\,\|T_G Q(p_n - p_{n-1})\|_{2,1}.$$

From the a priori estimate (cf. [5])

$$\sqrt{2}\|T_G\|_{[\mathrm{im}\,Q, W_2^1]}\|T_G N(u,B)\|_2 \geq \|u\|_{2,1} + \frac{1}{\eta}\,\|T_G\|_{[\mathrm{im}\,Q, W_2^1]}\|Qp\|$$

with

$$N(u,B) = \frac{\rho}{\eta}\,f + \frac{1}{\mu\eta}\,M^*(B) + \frac{1}{2\mu\eta}\,D|B|^2 - \frac{\rho}{\eta}\,M^*(u)$$

$$= \frac{1}{\eta}\left[\rho f + \frac{1}{\mu}\,M^*(B) - \rho\,M^*(u) - \frac{1}{2\mu}\,D|B|^2\right].$$

Because
$$B_1 \wedge (D \wedge B_2) = (B_1 \cdot D)B_B - D(B_1 \cdot B_2),$$
holds for some vector fields B_1 and B_2,
$$\|(B_1 \cdot D)B_2\|_p \leq C_1 \|B_1\|_{2,1} \|B_2\|_{1,2}$$
and
$$\|D B_1 \cdot B_2\|_p \leq C_1 \|B_1\|_{1,2} \|B_2\|_{1,2}.$$

We obtain
$$\|u_n - u_{n-1}\|_{2,1}$$
$$\leq 3\|T_G Q T_G\| \left\|\frac{\rho}{\eta}[M^*(u_{n-1}) - M^*(u_{n-2})]\right\|_{2,1}$$
$$+ 2\left\|T_G Q T_G \frac{1}{\mu\eta}\left[B_{n-1} \wedge (D \wedge (B_{n-2} + B_{n-1}))\right.\right.$$
$$\left.\left. + (B_{n-1} - B_{n-2}) \wedge (D \wedge B_{n-2})\right]\right\|_{2,1}$$
$$\leq 3\|T_G\|_{[\operatorname{im} Q, W_2^1]}\|T_G\|_{[L_p, L_2]}\| \left\|\frac{\rho}{\eta}[M^*(u_{n-1}) - M^*(u_{n-2})]\right\|_{2,1}$$
$$+ 3\|T_G\|_{[\operatorname{im} Q, W_2^1]}\|T_G\|_{[L_p, L_2]}\| \frac{1}{\mu\eta}\left\|\left[B_{n-1} \wedge (D \wedge (B_{n-2} + B_{n-1}))\right.\right.$$
$$\left.\left. + (B_{n-1} - B_{n-2}) \wedge (D \wedge B_{n-2})\right]\right\|_{2,1}$$
$$\leq \frac{3KC_1}{\eta}\left\{\rho\|u_{n-1} - u_{n-2}\|_{2,1}\left[\|u_{n-1}\|_{2,1} + \|u_{n-2}\|_{2,1}\right]\right.$$
$$\left. + \frac{4}{\mu}\|B_{n-1} - B_{n-2}\|_{2,1}\left[\|B_{n-1}\|_{2,1} + \|B_{n-2}\|_{2,1}\right]\right\}.$$

On the other hand
$$\|B_n - B_{n-1}\|_{2,1} \leq 8\mu\sigma K C_1 \|g\|_p \|u_n - u_{n-1}\|_{2,1},$$
so we obtain for
$$\frac{\eta}{\rho} < \frac{1}{\mu\sigma}$$
and
$$\|u_n\| \leq \frac{1}{12\mu\sigma K C_1}$$
$$\|u_n - u_{n-1}\|_{2,1} \leq \left(384 K^3 C_1^2 \|g\|_{2p} + \frac{1}{2}\right)\|u_{n-1} - u_{n-2}\|_{2,1}.$$

A sufficient condition should be
$$\|f\|_p + \frac{8K^2}{\rho\mu}\|g\|_p^2 < \frac{63}{64}\frac{\eta^2}{K^2\rho^2 C_1}. \tag{4.6}$$

By straightforward calculation one will find that the sequence $\{\|u_n\|_{2,1}\}$ is monotonic decreasing and separated from zero. We have proved the following theorem:

Theorem 6. *Let $f, g \in L_p(G, \mathbb{H})$. Further we assume*

(i) $\frac{\mu\sigma\eta}{\rho} < 1$,

(ii) *Condition (4.6),*

(iii) $\|g\|_p < \frac{\eta}{768K^3C_1^2\sigma}$,

with

$$K := \|T_G\|_{[L_2 \cap \operatorname{im} Q, W_2^1]} \|T_G\|_{[L_p, L_2]}$$

and

$$C_1 := 9^{\frac{1}{p}} \|T\|_{[L_2, L_q]}, \quad q < 6, \ p = \frac{2q}{2+q},$$

then the operator integral equation (4.4) has the unique solution

$$\{u, B, p\} \in \overset{\circ}{W}{}_2^1 (G, \mathbb{H}) \times \overset{\circ}{W}{}_2^1 (G, \mathbb{H}) \times L_2(G, \mathbb{R})$$

where u, B are uniquely defined and p is unique up to an additive real constant. Our iteration method converges in

$$\overset{\circ}{W}{}_2^1 (G, \mathbb{H}) \times \overset{\circ}{W}{}_2^1 (G, \mathbb{H}) \times L_2(G, \mathbb{R})$$

to this solution if $B_0, u_0 \in \overset{\circ}{W}{}_2^1 (G, \mathbb{H}) \cap \ker \operatorname{div}$ are sufficiently small.

5   Some remarks to a further interesting problem

We will now formulate the so-called **Poisson–Stokes problem**. It reads as

$$-\mu_1 \Delta \mathbf{u} - (\mu_1 + \mu_2)\nabla \operatorname{div} \underline{u} + \nabla\rho = \rho\mathbf{B} - \rho(\mathbf{u} \cdot \nabla)\mathbf{u} \quad \text{in} \quad G,$$

$$\operatorname{div} \rho\mathbf{u} = 0 \qquad\qquad \text{in} \quad G,$$

$$\mu_1 > 0, \ \mu_2 \geq -\frac{2}{3}\mu_1, \quad \mathbf{u} = 0 \qquad\qquad \text{on} \quad \Gamma,$$

where μ_1, μ_2 are Lame constants. We have to look for the unknown density ρ and the unknown velocity $\mathbf{u}$.

It is easy to find the corresponding quaternionic formulation by addition of a suitable trivial problem for u_0 :

$$DMDu - \frac{1}{\mu_1}D\rho = -\frac{\rho}{\mu_1}\mathbf{B} + \frac{\rho}{\mu_1}(\mathbf{u} \cdot D)u,$$

$$\operatorname{Sc}(D \operatorname{Vec}(u)) = -\frac{\nabla\rho}{\rho}u,$$

$$u = 0.$$

The operator M is given by

$$Mu = \frac{m_1 + 2\mu_2}{2(\mu_1 + \mu_2) + 1} u_0 + \mathbf{u}.$$

REFERENCES

[1] S. Bernstein, Operator calculus for elliptic boundary value problems in unbounded domains, *ZAA* **10** (4) (1993), 447–460.

[2] W. Borchers and T. Miyakawa, Algebraic L^2-decay for Navier–Stokes flow in exterior domains, *Acta Math.* **165** (1990), 189–227.

[3] P. Cerejeiras and U. Kähler, Elliptic boundary value problems of fluid dynamics over unbounded domains, *Math. Meth. Appl. Sci.* **23** (2000), 81–101.

[4] G. P. Galdi, An introduction to the mathematical theory of the Navier–Stokes equations, Vol. 1, *Linearized Problems, Springer Tracts in Natural Philosophy*, Springer, New York, 1994.

[5] K. Gürlebeck and W. Sprößig, *Quaternionic Analysis and Elliptic Boundary Value Problems*, Birkhäuser, Basel, 1990.

[6] K. Gürlebeck and W. Sprößig, *Quaternionic Calculus for Physicists and Engineers*, John Wiley & Sons, Chichester, New York, Brisbane, Weinheim, Singapore, Toronto. Series *Mathematical Problems in Practice* Vol. 1, 1998.

[7] K. Gürlebeck and W. Sprößig, Methods of quaternionic analysis for the treatment of non-linear boundary value problems, *Contemporary Mathematics (AMS)*, accepted for publication, 1999.

[8] K. Gürlebeck, W. Sprößig, and U. Wimmer, Hypercomplex function theory for consideration of non-linear Stokes problems with variable viscosity, *Complex Variables*, Vol. **22**, 1993, 195–202.

[9] U. Kähler, Clifford analysis and elliptic boundary value problems in unbounded domains, in *Clifford Algebras and Their Applications in Mathematical Physics*, V. Dietrich, K. Habetha, and G. Jank, eds., Kluwer Academic Publishers, Dordrecht, 1998, 145–160.

[10] U. Kähler, Die Anwendung der hyperkomplexen funktionentheorie auf die Lsung partieller differentialgleichungen, Thesis, TU Chemnitz, 1998, 108 pp.

[11] O. A. Ladyzenskaja, *The Mathematical Theory of Viscous Incompressible Fluid*, Gordon & Breach, New York, 1966.

[12] V. A. Solonnikov and V. E. Scadilov, On the boundary value problem for a stationary system of Navier–Stokes equations, *Proc. Steklov Inst. Math.* **125** (1973), 186–199.

[13] W. Sprößig, On decompositions of the Clifford valued Hilbert space and their applications to boundary value problems, *Advances in Applied Clifford Algebras* **5** No. **2** (1995), 167–186.

[14] W. Sprößig and K. Gürlebeck, On the treatment of fluid problems by methods of Clifford analysis, *Mathematics and Computers in Simulation* **44** (1997), 401–413.

[15] W. Varnhorn, *The Stokes Equation*, Akademie-Verlag, Berlin, 1994.

Wolfgang Sprößig
Freiberg University of Mining and Technology
Faculty of Mathematics and Informatics
Bernhard-von-Cotta-Str. 2
D-09596 Freiberg, Germany
E-mail: sproessig@math.tu-freiberg.de

Received: October 19, 1999; Revised: March 1, 2000

2.

SINGULAR INTEGRAL OPERATORS

Fourier Theory
Under Möbius Transformations

Xinhua Ji, Tao Qian, and John Ryan

ABSTRACT We study Fourier multipliers and singular integrals on curves and surfaces that are Möbius transformation images of Lipschitz graphs and starlike Lipschitz surfaces. We show that the singular integrals in each case form an operator algebra identical to the bounded holomorphic Fourier multipliers and the Cauchy-Dunford bounded holomorphic functional calculus of the associated Dirac operator in the context considered here.
Keywords: Functional calculus, Dirac operator, Fourier multiplier, singular integral, Lipschitz surfaces, Möbius transformation.

1 Introduction

Fourier multipliers and singular integrals that form operator algebras have been studied on two types of curves and surfaces: One is the graphs of Lipschitz functions defined in one- and higher-dimensional Euclidean spaces ([McQ1-2], [LMcQ], [LMcS], [GLQ], [T], [Mc2]); the other is starlike Lipschitz curves and surfaces ([Q1], [Q3-Q6]). The purpose of the theory in each case is to establish the identical relationship between the singular integral forms and the Fourier multiplier forms, as well as the forms of Cauchy-Dunford functional calculus of the associated Dirac operator in the context.

We also prove boundedness of the operators under consideration. We establish Fourier transform theory in each context. The operators under study in each case are induced by bounded holomorphic functions defined on sectors on the complex plane containing the spectrum of the Dirac operator on the smooth curve or surface of which the non-smooth one is a Lipschitz perturbation of the line. For more information on the established theory, we refer the reader to the introduction section and the references of [Q5]. In the present work we point out that the same theory is available for the Möbius transformation of images (abbreviated as "Möbius images" below) of the two types of curves and surfaces already studied. In most cases we only indicate the right forms of the relations under study, and only outline the proofs. We leave the detailed proofs to the interested reader.

AMS Subject Classification: 30G35, 42B15, 42B20, 47A60, 42B05, 42B30.

The work is a continuation of the study initialized by [QR] (also see [Ry1-3]). The writing plan is as follows. Section 2 contains preliminary knowledge of Clifford analysis that will be used throughout the paper. Section 3 contains an account on Lipschitz surfaces in relation to Möbius transformations. Section 4 is devoted to singular integral and Fourier transform theory on the Möbius images of Lipschitz graphs. Section 5 deals with singular integral and Fourier series theory on the Möbius images of closed and starlike Lipschitz surfaces.

Sincere thanks are due to Alan McIntosh who has shown a continuous interest in this project and, in several discussions, contributed fundamental ideas to the work. In particular, he pointed out that on Lipschitz perturbations of smooth curves and surfaces one should use the same Dirac operator as used on the smooth curves and surfaces when studying Fourier theory.

The third author wishes to acknowledge the support of the New England University Research Grant and that of the United Nations Development Program under the scheme: Transfer Knowledge Through Expatriate Nationals. The author, in particular, thanks the Institute of Mathematics of Academia Sinica for its invitation and hospitality where the first draft of the paper was written.

2   Preliminaries

We shall be working in the real Clifford algebra $\mathbb{R}^{(n)}$ generated by $\mathbf{e}_1, \dots, \mathbf{e}_n$, called *basic vectors*, over the real number field. Denote by $\mathbb{R}^n$ the linear subspace of $\mathbb{R}^{(n)}$ spanned by $\mathbf{e}_1, \dots, \mathbf{e}_n$. Denote by $\mathbf{e}_0$ the algebraic unit element, i.e., $\mathbf{e}_0 = 1$. A typical element of $\mathbb{R}^n$ is denoted by $\underline{x} = x_1 \mathbf{e}_1 + \cdots + x_n \mathbf{e}_n$. The basic vectors satisfy the conditions $\mathbf{e}_i \mathbf{e}_j + \mathbf{e}_j \mathbf{e}_i = -2\delta_{ij}, 1 \leq i, j \leq n$. Elements of $\mathbb{R}^{(n)}$ are denoted by $x, y, \dots$ and called *Clifford numbers*. Denote $\mathbb{R}_1^n = \{x = x_0 + \underline{x}, x_0 \in \mathbb{R}, \underline{x} \in \mathbb{R}^n\}$. An element in $\mathbb{R}_1^n$ is called a *vector*. Define two operations on the basis elements: $(\mathbf{e}_{i_1} \cdots \mathbf{e}_{i_l})^* = \mathbf{e}_{i_l} \cdots \mathbf{e}_{i_1}$ and $(\mathbf{e}_{i_1} \cdots \mathbf{e}_{i_l})' = (\mathbf{e}_{i_1})' \cdots (\mathbf{e}_{i_l})'$, where $(\mathbf{e}_0)' = \mathbf{e}_0, (\mathbf{e}_j)' = -\mathbf{e}_j, j = 1, \dots, n$, and extend them by linearity to $\mathbb{R}^{(n)}$, and hence, to $\mathbb{R}^n$. By combining them we define a third operation $^-$ by $\overline{x} = (x^*)'$. If x and y are two Clifford numbers in $\mathbb{R}^{(n)}$, then we have $\overline{xy} = \overline{y}\,\overline{x}$. If $x \in \mathbb{R}_1^n$ and $x \neq 0$, then its inverse x^{-1} exists: $x^{-1} = \frac{\overline{x}}{|x|^2}$, and $x^{-1}x = xx^{-1} = 1$. We also use the complex Clifford algebra $\mathbb{C}^{(n)}$ generated by $\mathbf{e}_1, \dots, \mathbf{e}_n$ over the complex number field. The complex imaginary element $\mathbf{i}$ commutes with all the $\mathbf{e}_j$'s, $j = 0, 1, \dots, n$ and $\mathbf{i}' = -\mathbf{i}$. So we extend the definitions of * and $'$ and, therefore, $^-$ to $\mathbb{C}^{(n)}$. The natural inner product between x and y in $\mathbb{C}^{(n)}$, denoted by $< x, y >$, is the complex number $\sum_S x_S \overline{y_S}$, where $x = \sum_S x_S \mathbf{e}_S, y = \sum_S y_S \mathbf{e}_S$, S runs over all the subsets $(i_1, i_2, \cdots i_l)$ of the set $\{1, 2, \dots, n\}$, where $i_1 < i_2 < \cdots < i_l$, and $\mathbf{e}_S = \mathbf{e}_{i_1} \mathbf{e}_{i_2} \cdots \mathbf{e}_{i_l}$. It is easily seen that $< x, y > = \operatorname{Re}(x\overline{y})$. The norm associated with this inner product is $|x| = < x, x >^{\frac{1}{2}} = (\sum_S |x_S|^2)^{\frac{1}{2}}$.

So, if a transform in $\mathbb{C}^{(n)}$ preserves the norm, then it also preserves the inner product. If $x, y, \ldots, u$ are vectors, then $|xy \cdots u| = |x||y| \cdots |u|$. The angle between two vectors x and y, denoted by $\arg(x, y)$, is defined to be $\arccos \frac{<x,y>}{|x||y|}$, where the inverse function arccos takes values in $[0, \pi)$. The concept of angle can be extended to any two elements in $\mathbb{R}^{(n)}$ with the same definition, as both the inner product and the norm are available to elements in $\mathbb{R}^{(n)}$. The unit sphere $\{\underline{x} \in \mathbb{R}^n \mid |\underline{x}| = 1\}$ in $\mathbb{R}^n$ is denoted by $\mathbf{S}^{n-1}$. A general vector $\underline{x} \in \mathbb{R}^n$ is sometimes written as $\underline{x} = x' + x_n \mathbf{e}_n$, where $x' = x_1 \mathbf{e}_1 + \cdots + x_{n-1} \mathbf{e}_{n-1} \in \mathbb{R}^{n-1}$.

We shall be working with $\mathbb{R}^n$-variable and $\mathbb{C}^{(n)}$-valued functions. The results of this paper also hold for functions defined in $\mathbb{R}_1^n$. In particular, the theory for the complex plane is identical to that for $\mathbb{R}^2$ through the correspondence $a + bi \to a\mathbf{e}_1 + b\mathbf{e}_2$. It is an example of the general correspondence between $\mathbb{R}_1^{n-1}$ and $\mathbb{R}^n$ as cited, for instance, in [Q6]. The concepts of left- and right-monogeneity are introduced in the usual way via the Dirac operator $\underline{D} = \frac{\partial}{\partial x_1}\mathbf{e}_1 + \cdots + \frac{\partial}{\partial x_n}\mathbf{e}_n = D' + \frac{\partial}{\partial x_n}\mathbf{e}_n$. In this paper, a function is said to be *monogenic* if it is both left- and right-monogenic. The Cauchy kernel stands for $\underline{E}(\underline{x}) = \frac{-x}{|x|^n}$. We assume the reader to be familiar with Cauchy's theorem and Cauchy's formula in the forms as exhibited in, e.g., [BDS], or [LMcQ], or [DSS].

We recall that *Clifford group* Γ_n is defined to be the multiplicative group of all elements in the Clifford algebra which can be written as products of non-zero vectors in $\mathbb{R}^n$.

By *Möbius group* we mean the group of the orientation-preserving transformations acting in the one point compactification, $\mathbb{R}^n \cup \{0\}$, of $\mathbb{R}^n$, generated by rigid motions, reflections, dilations, and inversions ([A1-4]). All Möbius transformations, ϕ, from $\mathbb{R}^n \cup \{\infty\}$ to $\mathbb{R}^n \cup \{\infty\}$ are exactly those of the form

$$\phi(\underline{x}) = (a\underline{x} + b)(c\underline{x} + d)^{-1},$$

where $a, b, c, d \in \Gamma_n \cup \{0\}$ and

$$ad^* - bc^* \in \mathbb{R} \setminus \{0\}, \quad a^*c, cd^*, d^*b, ba^* \in \mathbb{R}^n$$

(see [A2], for instance). Below, we shall call the vector $-c^{-1}d = \phi^{-1}(\infty)$ the *singular point of* ϕ.

The Clifford matrices,

$$\begin{pmatrix} a & b \\ c & d \end{pmatrix},$$

are called *Vahlen matrices*. The identification between ϕ and the associated Vahlen matrix exhibits a homomorphism under 2×2 block matrix multiplication.

Notations C, C_ν, will be used for constants which may vary from one occurrence to the next. Subscripts, such as ν in C_ν and Σ in C_Σ, are used to stress the dependence of the constants.

3 Möbius transformations and Lipschitz surfaces

We shall only speak about $(n-1)$-dimensional surfaces in $\mathbb{R}^n$ for $n > 2$ although all results and arguments are valid for curves in the complex plane as well. In the complex plane, using our method, we can deal with the conformal mappings arising from the Riemann Mapping Theorem. By this, we map a simply connected Lipschitz domain to the interior part of the unit disc while the Lipschitz boundary is mapped onto a closed star-shaped Lipschitz curve. For the case $n > 2$, however, in order to obtain the Hardy spaces results (see [Ry2]) we need to restrict ourselves to only the Möbius transformations.

A function $A : \mathbb{R}^{n-1} \to \mathbb{R}$ is said to be a Lipschitz function if A is continuous, and there exists a positive number M such that $|f(\underline{x})-f(\underline{y})|/|\underline{x}-\underline{y}| \leq M$. The minimal value of such M is called the Lipschitz constant of A, denoted by $Lip(A)$. Since $|\nabla A|$ is the maximum value of all directional derivatives at a point, we have $Lip(A) \leq \|\nabla A\|_\infty$. Let Γ_A be the graph of a **bounded** continuous Lipschitz function A in $\mathbb{R}^{n-1}$; then the Lipschitz constant of A is defined to be the pseudo-Lipschitz constant of Γ_A, denoted by $pLip(\Gamma_A)$, ie., $pLip(\Gamma_A) = Lip(A)$.

An $(n-1)$-dimensional surface is said to be *orientable* if it has a tangent plane almost everywhere on the surface with respect to the surface area measure, and there exists a vector field almost everywhere defined on the surface consisting of unit normals of the tangent planes such that every point on the surface has a neighborhood in which the angles between the defined unit normals are less than $\pi/2$.

Let Ω be a simply connected bounded or unbounded domain. We say that Ω is a *Lipschitz domain* if its boundary surface $\partial\Omega$ is orientable, and it can be covered by a finite number of coordinate cylinders, constituting a coordinate cylindrical covering, and the part of $\partial\Omega$ in each of the coordinate cylinders can be extended to become a bounded Lipschitz graph on $\mathbb{R}^{n-1}$ (see [V]). Each such extension has a pseudo-Lipschitz constant and the greatest lower bound of all pseudo-Lipschitz constants corresponding to all possible extensions is defined to be the *local pseudo-Lipschitz constant* of this piece of $\partial\Omega$. For a coordinate cylindrical covering of $\partial\Omega$, the maximum value of all the local pseudo-Lipschitz constants of the covering is defined to be the *pseudo-Lipschitz constant of Ω of the coordinate cylindrical covering*. We define the *Lipschitz constant of Ω*, or $\partial\Omega$, to be the greatest lower bound of the pseudo-Lipschitz constants corresponding to all possible coverings, denoted by $Lip(\partial\Omega)$.

According to the definition, the Lipschitz constant of a given Lipschitz graph of a bounded Lipschitz function A is, in general, less than the pseudo-Lipschitz constant, ie., $Lip(\Gamma_A) \leq pLip(\Gamma_A) = Lip(A) \leq \|\nabla A\|_\infty$. According to the definition, the Lipschitz constant of the boundary of a C^1 domain is zero.

In the sequel, we shall be concerned with only four types of surfaces: Lipschitz graphs ([McQ1,2], [Mc3], [LMcQ], [LMcS]), starlike Lipschitz curves and surfaces (see [Q1-Q6]), and their Möbius image (Sections 4 and 5 below).

We shall denote by Σ the boundary surface of a general simply connected bounded or unbounded Lipschitz domain. We shall briefly call such surfaces *Lipschitz surfaces*. If a Lipschitz surface is actually a graph of a Lipschitz function in $\mathbb{R}^{n-1}$; and if we wish to emphasis this fact, then we denote it by Γ. If ϕ is a Möbius transformation, then the images of Σ or Γ under the transformation is denoted by $\phi(\Sigma)$ or $\phi(\Gamma)$, respectively. The fact that Möbius transformations form a group implies that all Lipschitz surfaces are divided into equivalent classes; and, if two such surfaces are in the same class, then one is the Möbius image of the other.

Proposition 1. *If ϕ is a Möbius transformation and Σ is a Lipschitz surface, then $Lip(\Sigma) = Lip(\phi(\Sigma))$.*

Proof. According to the definition, to find the Lipschitz constant of a Lipschitz surface is to minimize the maximal slopes of chords on the surface using finer and finer coordinate cylindrical coverings and rigid movement of the underlying spaces $\mathbb{R}^{n-1}$ for the variable $\underline{x}$ of the Lipschitz functions whose graphs have common pieces with the surface. The chords, however, are very close to tangent lines of the surface if the covering is fine enough. Since Möbius transformations preserve angles of tangent lines at each point, and the Lipschitz constant is the measurement of discontinuity of the slopes of the tangent lines, we conclude that Möbius transformations preserve Lipschitz constants of surfaces. $\square$

Let Σ be a Lipschitz surface. The space of square integrable functions on Σ is defined by

$$L^2(\Sigma) = \{f : \Sigma \to \mathbb{C}^{(n)} \mid \|f\|_{L^2(\Sigma)} = (\int_{\Sigma} |f(\underline{x})|^2 d\sigma(\underline{x}))^{\frac{1}{2}} < \infty\},$$

where $d\sigma$ is the surface area measure on Σ. It can be easily verified that the space with the norm $\| \ \|_{L^2(\Sigma)}$ forms a Banach space. In [C] (also see [QR]) it is obtained that

Proposition 2. *If Σ is a Lipschitz surface, then for each pair $f, g \in L^2(\Sigma)$ we have*

$$\int_{\Sigma} f(\underline{y})g(\underline{y})d\sigma(\underline{y}) = \pm \int_{\phi^{-1}(\Sigma)} f(\phi(\underline{x}))g(\phi(\underline{x}))|c\underline{x} + d|^{-2n+2}d\sigma(\underline{x}).$$

The plus or minus sign depends on whether the diffeomorphism ϕ is orientation preserving or reversing, respectively.

As it was pointed out in [QR], we have

$$|c\underline{x} + d|^{-2n+2} = \overline{J(\phi, \underline{x})}J(\phi, \underline{x})$$

where

$$J(\phi, \underline{x}) = \frac{(c\underline{x} + d)^*}{|c\underline{x} + d|^n}.$$

It follows that if $f \in L^2(\Sigma)$, then $J(\phi, x)f(\phi(x))$ is in $L^2(\phi^{-1}(\Sigma))$. Moreover, according to Proposition 2, the mapping that maps f to $J(\phi, \cdot)f(\phi(\cdot))$ is an isometry from $L^2(\Sigma)$ to $L^2(\phi^{-1}(\Sigma))$.

In terms of a different paring involving a chosen normal vector field on the surface, the role of the plus and minus sign in Proposition 2 may be eliminated (see [Ry1]). In that case, we have

Proposition 3. *If Σ is a Lipschitz surface, then for each pair $f, g \in L^2(\Sigma)$ we have*

$$\int_\Sigma f(\underline{y})n(\underline{y})g(\underline{y})d\sigma(\underline{y}) = \int_{\phi^{-1}(\Sigma)} f(\phi(\underline{x}))J^*(\phi, \underline{x})n(\underline{x})J(\phi, \underline{x})g(\phi(\underline{x}))d\sigma(\underline{x}),$$

where $n(\underline{y})$ is an almost everywhere defined unit normal vector field to the surface Σ indicating the orientation, and $n(\underline{x})$ is the corresponding orientation of the surface $\phi^{-1}(\Sigma)$ induced by the Möbius transformation ϕ.

It is an immediate consequence of Möbius covariance of Dirac operators (see [B] or [PQ]) that $f(\underline{y})$ is left-monogenic in U if and only if $J(\phi, \underline{x})f(\phi(\underline{x}))$ is left-monogenic in $\phi^{-1}(U)$. Recalling the definition of Hardy $H^p, 1 < p < \infty$, spaces (see, e.g., [K], [Mi], [Ry2]) and the conformal invariance of the Cauchy kernel

$$\underline{E}(\underline{x} - \underline{y}) = J(\phi, \underline{x})\underline{E}(\phi(\underline{x}) - \phi(\underline{y}))J^*(\phi, \underline{y}),$$

we have

Proposition 4. *If Σ is a Lipschitz surface and ϕ a Möbius transformation, then $H^p(\phi^{-1}(\Sigma)) = \{J(\phi, \cdot)f(\phi(\cdot)) \mid f \in H^p(\Sigma)\}, 1 < p < \infty$.*

4   Singular integrals and Fourier multipliers on Möbius images of Lipschitz graphs

From now on, we assume that $\Gamma = \Gamma_A$ is the graph of a bounded Lipschitz function $A : \mathbb{R}^{n-1} \to \mathbb{R}$ whose pseudo-Lipschitz constant $pLip(\Gamma) = Lip(A)$. A comprehensive singular integral theory on Γ has been developed in [McQ1], [McQ2], [LMcQ], [LMcS], [GLQ], and [T]. The starting point of the theory is Coifman-McIntosh-Meyer's Theorem (CMcM's Theorem); therefore, on any Lipschitz graph Γ the Cauchy integral operator

$$C_\Gamma f(\underline{x}) = p.v. \frac{1}{\omega_{n-1}} \int_\Gamma \underline{E}(\underline{y} - \underline{x})n(\underline{y})f(\underline{y})d\sigma(\underline{y})$$

can be extended to become a bounded operator in $L^2(\Gamma)$ ([CMcM], [Mc1]), where $n(\underline{y})$ is the outward normal to Γ at $\underline{y} \in \Gamma$, ω_{n-1}, which is the surface area of the $(n-1)$-dimensional unit sphere $\mathbf{S}^{n-1}$.

The theory on graphs has the following features.

(i) The singular integral operators with monogenic (holomorphic) kernels of the Calderón-Zygmund type are bounded in $L^p(\Gamma), 1 < p < \infty$. They form an operator algebra.

(ii) The Fourier transforms of the monogenic kernels in the distribution sense are virtually bounded holomorphic functions defined in certain sectors in the $(n-1)$-dimensional complex space. Explicit formulas connecting the kernels and their Fourier transforms are given.

(iii) The Fourier multiplier form and the singular integral form are related by a variation of Parseval's formula. By virtue of this relation the algebraic operations on the singular integral operators are reduced to pointwise operations on the Fourier multipliers.

In [QR], [Ry1] and [Ry2] it is shown that on $\phi^{-1}(\Gamma)$ one may introduce, based on the above mentioned study, a class of singular integrals with monogenic kernels. In [Ry3] the Fourier transform aspect is concerned. The present study is to point out that the introduced class of singular integrals on $\phi^{-1}(\Gamma)$ also has the features (i) to (iii).

To state our results we first need to go through the settings as set out in [LMcQ] and [Mc3].

Let N be a compact set on the unit sphere $\mathbf{S}^{n-1}$ that is starlike about $\mathbf{e}_n$. Denote by μ_N the maximum angle between the vector $\mathbf{e}_n$ and the vectors with initial point 0 and terminal points in N. For any $\mu \in (0, \frac{\pi}{2} - \mu_N)$, denote by N_μ the μ-neighborhood of N, i.e.,

$$N_\mu = \{\underline{y} \in \mathbb{R}^n \mid |\underline{y}| = 1, \angle(\underline{y}, \underline{n}) < \mu \text{ for some } \underline{n} \in N\}.$$

Clearly N_μ is contained in the upper hemisphere. For each unit vector in $\mathbb{R}^n$ let $C_{\underline{n}}^+$ be the open half space $\{\underline{x} \in \mathbb{R}^n \mid \langle \underline{x}, \underline{n} \rangle > 0\}$, and define the open cones associated with N_μ to be

$$C_{N_\mu}^+ = \cup\{C_{\underline{n}}^+ \mid \underline{n} \in N_\mu\}, C_{N_\mu}^- = -C_{N_\mu}^+,$$

and

$$S_{N_\mu} = C_{N_\mu}^+ \cap C_{N_\mu}^-.$$

Let $M(C_{N_\mu}^+)$ be the space of right- and left-monogenic functions Φ from $C_{N_\mu}^+$ to $\mathbb{C}^{(n)}$ for which

$$\|\Phi\|_{(\mu')} = \frac{1}{2}\omega_{n-1}\sup\{|\underline{x}|^{n-1}|\Phi(\underline{x})| \mid \underline{x} \in C_{N_{\mu'}}^+\} < \infty, 0 < \mu' < \mu.$$

The class $M(C_{N_\mu}^-)$ is similarly defined.

Denote by T_{N_μ} the set

$$T_{N_\mu} = \{y \in \mathbb{R}^n \mid y^\perp \subset S_{N_\mu}\},$$

where $y^\perp$ is the $(n-1)$-dimensional plane passing through 0 and orthogonal to the vector with the initial point 0 and the terminal point y.

Let $\Phi_\pm \in M(C_{N_\mu}^\pm)$. A relevant function, $\underline{\Phi} : T_{N_\mu} \to \mathbb{C}^{(n)}$, is defined to be

$$\underline{\Phi}_\pm(y) = \pm \int_{H_{\underline{y}}^\pm} \Phi_\pm(\underline{x}) n(\underline{x}) d\sigma(\underline{x}),$$

where $H_{\underline{y}}^\pm$ is the hemisphere $H_{\underline{y}}^\pm = \{\underline{x} \in \mathbb{R}^n \mid \pm \langle \underline{x}, \underline{y} \rangle > 0, |\underline{x}| = |\underline{y}|\}$.

We introduce $M(S_{N_\mu})$ to be the space of the function pairs

$$(\Phi, \underline{\Phi}) = (\Phi_+ + \Phi_-, \underline{\Phi}_+ + \underline{\Phi}_-)$$

for which

$$\|(\Phi, \underline{\Phi})\|_{\mu'} = \frac{1}{2}\omega_{n-1} \sup\{|\underline{x}|^{n-1}|\Phi(\underline{x})| \mid \underline{x} \in S_{N_{\mu'}}\}$$
$$+ \sup\{|\underline{\Phi}(re_n)| \mid r > 0\} < C_{\mu'}, \quad 0 < \mu' < \mu.$$

In [LMcQ], [LMcS], and [T], kernels of only one-sided monogeneity are studied that induce a wider class of singular integrals than what is concerned in this paper. In this paper we restrict ourself only to two-sided monogenic kernels, i.e., those in $M(C_{N_\mu}^+)$, $M(C_{N_\mu}^-)$ and $M(S_{N_\mu})$. The good thing about those kernels is that the convolution of two kernels in each case is still a kernel of the same type of the space with any smaller index $\mu' \in (0, \mu)$. To be precise, we have (see [LMcQ])

Proposition 5. *If* $(\Phi, \underline{\Phi}) = (\Phi^+ + \Phi^-, \underline{\Phi}^+ + \underline{\Phi}^-)$ *and* $(\Psi, \underline{\Psi}) = (\Psi^+ + \Psi^-, \underline{\Psi}^+ + \underline{\Psi}^-)$ *are two kernels in* $M(S_{N_\mu})$, *then*

$$(\Phi, \underline{\Phi}) * (\Psi, \underline{\Psi}) = (\Phi^+ * \Psi^+ + \Phi^- * \Psi^-, \underline{\Phi}^+ * \underline{\Psi}^+ + \underline{\Phi}^- * \underline{\Psi}^-)$$

is a kernel in $M(S_{N_{\mu'}}), 0 < \mu' < \mu$, *and*

$$\|(\Phi, \underline{\Phi}) * (\Psi, \underline{\Psi})\|_\nu \le C_{\mu,\mu',\nu} \|(\Phi, \underline{\Phi})\|_{\mu'} \|(\Psi, \underline{\Psi})\|_{\mu'}, \quad \nu < \mu' < \mu.$$

Similar propositions hold for kernels in $M(C_{N_\mu}^\pm)$.

The functions Φ and $\underline{\Phi}$ will be used as kernels on Lipschitz surfaces. In order to have them well-defined, we have required them to be monogenically defined in $S_{N_{\mu'}}$, $0 < \mu' < \mu$, where μ is defined in relation to μ_N, where $\tan \mu_N = p\dot{L}ip(\Gamma_A)$. Below, we shall introduce the function spaces for Fourier multipliers. They are related to $pLip(\Gamma_A)$ in a similar way.

Now we turn to the Fourier multipliers side. Consider the $(n-1)$-dimensional complex space $\mathbb{C}^{n-1}$. A general element of $\mathbb{C}^{n-1}$ is denoted by $\zeta = \xi + i\eta$, where

$$\xi = \xi_1 e_1 + \cdots + \xi_{n-1} e_{n-1}, \quad \eta = \eta_1 e_1 + \cdots + \eta_{n-1} e_{n-1} \in \mathbb{R}^{n-1}.$$

For any unit vector $n = n' + n_n e_n, n_n > 0$, in $\mathbb{R}^n$, we define

$$n(\mathbb{C}^{n-1}) = \{\zeta = \xi + i\eta \mid \xi \neq 0 \text{ and } n_n \eta = (n_n^2 |\xi|^2 + \langle \xi, n' \rangle^2)^{\frac{1}{2}} n'\}.$$

For the open set N_μ defined above we associate it with the open cones $N_\mu(\mathbb{C}^{n-1})$ in $\mathbb{C}^{n-1}$, defined by

$$N_\mu(\mathbb{C}^{n-1}) = \cup\{n(\mathbb{C}^{n-1}) \mid n \in N_\mu\}.$$

Denote by $H^\infty(N_\mu(\mathbb{C}^{n-1}))$ the space of all the holomorphic functions from $N_\mu(\mathbb{C}^{n-1})$ to $\mathbb{C}^{(n)}$ for which

$$\|b\|_{\infty,\mu'} = \sup\{|b(\zeta)| \mid \zeta \in N_{\mu'}(\mathbb{C}^{n-1})\} < C_{\mu'}, 0 < \mu' < \mu.$$

The generalization of $\exp(x', \xi) = e^{i\langle x', \xi \rangle}$ to $\underline{x} = x' + x_n e_n \in \mathbb{R}^n$ and $\zeta = \xi + i\eta \in \mathbb{C}^{n-1}$ is

$$e(\underline{x}, \zeta) = e_+(\underline{x}, \zeta) + e_-(\underline{x}, \zeta),$$

where

$$e_\pm(\underline{x}, \zeta) = e^{i\langle x', \zeta \rangle} e^{-(\pm x_n)|\zeta|_\mathbb{C}} \chi_\pm(\zeta),$$

and

$$\chi_\pm(\zeta) = \frac{1}{2}(1 \pm i\zeta e_n |\zeta|_\mathbb{C}^{-1}),$$

and $|\zeta|_\mathbb{C}^2$ is the holomorphic extension of $|\xi|^2$:

$$|\zeta|_\mathbb{C}^2 = \sum_{j=1}^{n-1} \zeta_j^2 = |\xi|^2 - |\eta|^2 + 2i\langle \xi, \eta \rangle,$$

where $\pm|\zeta|_\mathbb{C}$ are the two complex square roots of $|\zeta|_\mathbb{C}^2$. The functions $\chi_\pm$ satisfy the conditions

$$\chi_+(\zeta) + \chi_-(\zeta) = 1, \chi_\pm(\zeta)^2 = \chi_\pm(\zeta), \chi_\pm(\zeta)\chi_\mp(\zeta) = 0,$$

and

$$i\zeta e_n = |\zeta|_\mathbb{C}\chi_+(\zeta) - |\zeta|_\mathbb{C}\chi_-(\zeta).$$

The above properties imply that every function $b \in H^\infty(N_\mu(\mathbb{C}^{n-1}))$ can be uniquely decomposed as $b = b_+ + b_-$, with $b_\pm = b\chi_\pm \in H^\infty_\pm(N_\mu(\mathbb{C}^{n-1}))$.

It is proved in [LMcQ] that a right-monogenic kernel is also left-monogenic if and only if its Fourier transform b satisfies

$$\zeta e_n b(\zeta) = b(\zeta)\zeta e_n. \tag{4.1}$$

Below, we denote by $cH_{\pm}^{\infty}(N_{\mu}(\mathbb{C}^{n-1}))$ and $cH^{\infty}(N_{\mu}(\mathbb{C}^{n-1}))$ the subspaces of the functions in $H_{\pm}^{\infty}(N_{\mu}(\mathbb{C}^{n-1}))$ and $H^{\infty}(N_{\mu}(\mathbb{C}^{n-1}))$, respectively, of the above specified commutative property. We also use the class $cH_N^{\infty} = \cup_{\mu \in (0, \frac{\pi}{2} - \mu_N)} cH^{\infty}(N_{\mu}(\mathbb{C}^{n-1}))$.

In [LMcQ] it is proved that the functions in $cH_{\pm}^{\infty}(N_{\mu}(\mathbb{C}^{n-1}))$ are the Fourier transforms of the functions in $M(C_{N_{\mu}}^{\pm})$, and that the functions in $M(C_{N_{\mu}}^{\pm})$ are the inverse Fourier transforms of the functions in the space $cH_{\pm}^{\infty}(N_{\mu}(\mathbb{C}^{n-1}))$. Here the Fourier transform and its inverse are with respect to the generalised functions $e_{\pm}(\underline{x}, \zeta)$. In this way, there is a one-to-one corresponding relationship between functions in the spaces of kernel functions and the spaces of Fourier multipliers. There exists Parseval's identity between the function pairs $(\Phi, \underline{\Phi})$ in $M(S_{N_{\mu}})$ and their associated Fourier transforms b in $cH^{\infty}(N_{\mu}(\mathbb{C}^{n-1}))$, as follows (see [LMcQ]):

$$(2\pi)^{n-1} \int_{\mathbb{R}^{n-1}} b(\xi)\hat{u}(-\xi)d\xi$$
$$= \lim_{\epsilon \to 0+} \{ \int_{|x'|>\epsilon} \Phi(x')e_n u(x')dx' + \underline{\Phi}(\epsilon e_n)u(0) \},$$

where u is any function in the Schwartz space $\mathcal{S}(\mathbb{R}^n)$. Denote by $\mathcal{A}(\Gamma)$ the space of functions left-monogenic in some open set

$$\{\underline{x} + t e_n \mid \underline{x} \in \Gamma, |t| < h, h > 0\}$$

and square-integrable when restricted on Γ. Using CMcM's theorem we can show that the class $\mathcal{A}(\Gamma)$ forms a dense subspace of $L^2(\Gamma)$ ([CM], or [GQW], or [Q5]).

As in [CM], for functions in $\mathcal{A}(\Gamma)$ one can introduce Fourier transformation: If $f \in \mathcal{A}(\Gamma)$, then its Fourier transform, as a function defined in $\mathbb{R}^{n-1}$, is defined to be

$$\hat{f}(\xi) = \int_{\Gamma} e(-\underline{x}, \xi)n(\underline{x})f(\underline{x})d\sigma(\underline{x}). \tag{4.2}$$

Recalling that A is a bounded function, from the decomposition $e(\underline{x}, \xi) = e_+(\underline{x}, \xi) + e_-(\underline{x}, \xi)$ and the expression

$$e_{\pm}(\underline{x}, \xi) = e^{i<x', \xi>} e^{-(\pm A(x')|\xi|)} \chi_{\pm}(\xi),$$

we see that $e_{\pm}(-\underline{x}, \xi)$ are bounded functions in x'. As in the standard cases it can be proven that the above integral is convergent in the L^2-norm sense and, thus, defines a square-integrable function.

It is noted that the Fourier inverse formula holds for test functions in $\mathcal{A}(\Gamma)$ (see [CM]), while the Plancherel Theorem does not hold in this case (see [McQ2], [Q1]).

So far the compact set N on the unit sphere has not yet been specified. In the context we define N to be the closure of the set

$$\{\underline{n} \in \mathbf{S}^{n-1} \mid \underline{n} = n(\underline{x}), \underline{x} \in \Gamma\}$$

under the natural topology of the sphere.

For a given $b \in cH^\infty(N_\mu(\mathbb{C}^{n-1}))$, one can introduce the Fourier multiplier operator M_b on $\mathcal{A}(\Gamma)$: For $f \in \mathcal{A}(\Gamma)$, define, for $\underline{x} \in \Gamma$,

$$M_b f(\underline{x}) = (2\pi)^{n-1} \lim_{\alpha \to 0+} \{\int_{\mathbb{R}^{n-1}} b_+(\xi) e_+(\underline{x} + \alpha \mathbf{e}_n, \xi) \hat{f}(\xi) d\xi +$$

$$\int_{\mathbb{R}^{n-1}} b_-(\xi) e_-(\underline{x} - \alpha \mathbf{e}_n, \xi) \hat{f}(\xi) d\xi\}.$$

Let $(\Phi, \underline{\Phi})$ be the function pair in $M(S_{N_\mu})$ associated with b in the Fourier transformation sense specified above. Introduce the operator on $\mathcal{A}(\Gamma)$:

$$T_{(\Phi, \underline{\Phi})} f(\underline{x}) = \lim_{\epsilon \to 0+} \{\int_{|\underline{x}-\underline{y}|>\epsilon, \underline{y} \in \Gamma} \Phi(\underline{x} - \underline{y}) n(\underline{y}) f(\underline{y}) d\sigma(\underline{y}) + f(\underline{x}) \underline{\Phi}(\mathbf{e} n(\underline{x}))\},$$

where $\underline{x} \in \Gamma$.

It can be shown that on $\mathcal{A}(\Gamma)$ we have

$$M_b = T_{(\Phi, \underline{\Phi})}$$

which is Parseval's identity on the surface. The operator can be extended to become a bounded operator in $L^p(\Gamma)$, and the above identity is extended to $L^p(\Gamma), 1 < p < \infty$, as well. The proof of the L^2-boundedness part is hard (see [LMcS], [GLQ] and [T]), as the Plancherel Theorem fails to hold in the case. The L^2-boundedness is essential in the sense that, once it is proven, then the L^p-boundedness follows from the standard argument for general Calderón-Zygmund type operators ([St]), based on the fact that the surface measure satisfies the doubling measure condition.

The class $\{M_b \mid b \in cH_N^\infty\}$ forms a bounded holomorphic functional calculus of the Dirac operator $-iD'$ in $L^p(\Gamma), 1 < p < \infty$, in the following sense:

If $b_0 \in cH^\infty(N_{\mu_0}(\mathbb{C}^{n-1})), b_1, b_2 \in cH_N^\infty$ and $\alpha_1, \alpha_2 \in \mathbb{C}$, then

(i) $M_{\alpha_1 b_1 + \alpha_2 b_2} = \alpha_1 M_{b_1} + \alpha_2 M_{b_2}$;

(ii) $M_{b_1 b_2} = M_{b_1} \circ M_{b_2}$;

(iii) $\|M_{b_0}\|_{L^2(\Gamma) \to L^2(\Gamma)} \leq C \|b\|_{\infty, \mu'}, 0 < \mu' < \mu$; and

(iv) $(-iD')f(\underline{x}) = (2\pi)^{n-1} \int_{\mathbb{R}^{n-1}} \xi e(\underline{x},\xi)\hat{f}(\xi)d\xi,$ $\underline{x} \in \Sigma.$

Let ϕ be a given Möbius transformation. Now we introduce the corresponding Fourier multiplier and singular integral operators on the surface $\phi^{-1}(\Gamma)$.

We denote the mapping $f \to J(\phi,\cdot)f(\phi(\cdot))$ introduced in Section 2 by τ and write $g = \tau(f)$ as a typical candidate of test functions on $\phi^{-1}(\Gamma)$. Define

$$\mathcal{A}(\phi^{-1}(\Gamma)) = \{g = \tau(f) \mid f \in \mathcal{A}(\Gamma)\}.$$

Since τ is an isometry form $L^2(\Gamma)$ to $L^2(\phi^{-1}(\Gamma))$, the density of $\mathcal{A}(\Gamma)$ in $L^2(\Gamma)$ implies the density of $\mathcal{A}(\phi^{-1}(\Gamma))$ in $L^2(\phi^{-1}(\Gamma))$.

Proposition 6. *The induced Fourier transformation on the functions g in $\mathcal{A}(\phi^{-1}(\Gamma))$ has the form*

$$\hat{g}(\xi) = \int_{\phi^{-1}(\Gamma)} e(\phi(\underline{x}),\xi)J^*(\phi,\underline{x})n(\underline{x})g(\underline{x})d\sigma(\underline{x}), \quad \xi \in \mathbb{R}^{n-1}.$$

The corresponding Fourier inverse formula is

$$g(\underline{x}) = (2\pi)^{n-1} \int_{\mathbb{R}^{n-1}} J(\phi,\underline{x})e(\phi(\underline{x}),\xi)\hat{g}(\xi)d\xi.$$

Proof. Using Proposition 3 to the right-hand side of (4.2), we deduce the transformation of g by $\hat{g}$. The proof of the inverse formula proceeds as follows:

$$(2\pi)^{n-1} \int_{\mathbb{R}^{n-1}} J(\phi,\underline{x})e(\phi(\underline{x}),\xi)\hat{g}(\xi)d\xi$$

$$= (2\pi)^{n-1} \int_{\mathbb{R}^{n-1}} J(\phi,\underline{x})e(\phi(\underline{x}),\xi) \times$$

$$\int_{\phi^{-1}(\Gamma)} e(\phi(\underline{y}),\xi)J^*(\phi,\underline{y})n(\underline{y})g(\underline{y})d\sigma(\underline{y})d\xi$$

$$= (2\pi)^{n-1} \int_{\mathbb{R}^{n-1}} J(\phi,\underline{x})e(\phi(\underline{x}),\xi) \int_{\Gamma} e(\underline{y},\xi)n(\underline{y})f(\underline{y})d\sigma(\underline{y})d\xi$$

$$= (2\pi)^{n-1} \int_{\mathbb{R}^{n-1}} J(\phi,\underline{x})e(\phi(\underline{x}),\xi)\hat{f}(\xi)d\xi$$

$$= J(\phi,\underline{x})f(\phi(\underline{x}))$$

$$= g(\underline{x}).$$

$\square$

Similar to the definition of M_b, the Fourier multiplier operator on $\phi^{-1}(\Gamma)$ associated with $b \in cH^\infty(N_\mu(\mathbb{C}^{n-1}))$ is accordingly defined to be

$$M_b^\phi g(\underline{x})$$

$$= (2\pi)^{n-1} \lim_{\alpha \to 0+} \{ \int_{\mathbb{R}^{n-1}} J(\phi, \underline{x}) b_+(\xi) e_+(\underline{x} + \alpha e_n, \xi) \hat{g}(\xi) d\xi$$

$$+ \int_{\mathbb{R}^{n-1}} J(\phi, \underline{x}) b_-(\xi) e_-(\underline{x} - \alpha e_n, \xi) \hat{g}(\xi) d\xi \}, \quad \underline{x} \in \Gamma. \quad (4.3)$$

Let $(\Phi, \underline{\Phi})$ be the function pair in $M(S_{N_\mu})$ associated with the function b in $cH^\infty(N_\mu(\mathbb{C}^{n-1}))$. One may introduce the corresponding singular integral operator on $\phi^{-1}(\Gamma)$ in the pattern of the definition of $T_{(\Phi, \underline{\Phi})}$:

$$T_{(\Phi, \underline{\Phi})}^\phi g(\underline{x})$$

$$= \lim_{\epsilon \to 0+} \{ \int_{|\phi(\underline{x}) - \phi(\underline{y})| > \epsilon, \underline{y} \in \phi^{-1}(\Gamma)} \Phi^\phi(\underline{x}, \underline{y}) n(\underline{y}) g(\underline{y}) d\sigma(\underline{y})$$

$$+ \underline{\Phi}^\phi(\phi(\underline{x}), \epsilon n(\phi(\underline{x}))) g(\underline{x}) \}, \quad (4.4)$$

where $\underline{x} \in \phi^{-1}(\Gamma)$ and

$$\Phi^\phi(\underline{x}, \underline{y}) = \Phi_+^\phi(\underline{x}, \underline{y}) + \Phi_-^\phi(\underline{x}, \underline{y}),$$

$$\Phi_\pm^\phi(\underline{x}, \underline{y}) = J(\phi, \underline{x}) \Phi_\pm(\phi(\underline{x}) - \phi(\underline{y})) J^*(\phi, \underline{y}).$$

In the above, $\underline{\Phi}^\phi = \underline{\Phi}_+^\phi + \underline{\Phi}_-^\phi$, where $\underline{\Phi}_\pm^\phi$ are defined in (Γ, T_{N_μ}) :

$$\underline{\Phi}_\pm^\phi(\phi(\underline{x}), t) = \pm \int_{H_t^{\phi(\underline{x}), \pm}} \Phi^\phi(\underline{x}, \underline{y}) n(\underline{y}) d\sigma(\underline{y}),$$

and $H_t^{\phi(\underline{x}), \pm} = \{ \underline{y} \in \mathbb{R}^n \mid \pm \langle \phi(\underline{y}) - \phi(\underline{x}), t \rangle > 0, |\phi(\underline{x}) - \phi(\underline{y})| = |t| \}$.
Our main result is as follows.

Main Theorem. Let $b \in cH^\infty(N_\mu(\mathbb{C}^{n-1}))$ and $(\Phi, \underline{\Phi})$ be the associated function pair in $M(S_{N_\mu}), \mu \in (0, \frac{\pi}{2} - \mu_N)$. Then we have

(i) $M_b^\phi = T_{(\Phi, \underline{\Phi})}^\phi$ on $\mathcal{A}(\phi^{-1}(\Gamma))$;

(ii) M_b^ϕ can be extended to a bounded operator on $L^2(\phi^{-1}(\Gamma))$;

(iii) On $L^2(\phi^{-1}(\Gamma))$ the limits in (4.3) and (4.4) converge in both the L^2-norm sense and the almost everywhere sense on the surface; and

(iv) M_b^ϕ is the bounded holomorphic functional calculus of the induced Dirac operator $-iD_\phi' = \overline{J}^{-1}(\phi, \underline{x})(-i)D'J^{-1}(\phi, \underline{x})$ on $L^2(\phi^{-1}(\Gamma))$, where $D' = \frac{\partial}{\partial x_1} e_1 + \cdots + \frac{\partial}{\partial x_{n-1}} e_{n-1}$.

Proof. Using change of variable, as in Propositions 3 and 2, the assertions (i) and (ii) reduce to the corresponding assertions for Γ. The assertion (iii) is a consequence of (ii).

To prove (iv) we first show that the Fourier multiplier operators are associated with D'_ϕ, the Dirac operator on $\phi^{-1}(\Gamma)$. In fact, applying the operator D'_ϕ to both sides of the inverse Fourier formula

$$g(\underline{x}) = (2\pi)^{n-1} \int_{\mathbb{R}^{n-1}} J(\phi, \underline{x}) e(\phi(\underline{x}), \xi) \hat{g}(\xi) d\xi,$$

we obtain

$$-iD'_\phi g(\underline{x}) = (2\pi)^{n-1} \int_{\mathbb{R}^{n-1}} \overline{J}^{-1}(\phi, \underline{x})(-iD')(e(\phi(\underline{x}), \xi)) \hat{g}(\xi) d\xi$$

$$= (2\pi)^{n-1} \int_{\mathbb{R}^{n-1}} \overline{J}^{-1}(\phi, \underline{x}) \overline{J}(\phi, \underline{x}) J(\phi, \underline{x}) \xi e(\phi(\underline{x}), \xi) \hat{g}(\xi) d\xi$$

$$= (2\pi)^{n-1} \int_{\mathbb{R}^{n-1}} J(\phi, \underline{x}) \xi e(\phi(\underline{x}), \xi) \hat{g}(\xi) d\xi.$$

Next we prove the relation $M^\phi_{b_1 b_2} = M^\phi_{b_1} \circ M^\phi_{b_2}$.

In fact, this holds if and only if it holds simultaneously for $(b_1)_+, (b_2)_+,$ $(b_1)_-$ and $(b_2)_-$. Owing to the commutative property (4.1) of b_1 and b_2, the desired identity follows from the relation

$$(b_1)_\pm (b_2)_\pm = b_1 \chi_\pm b_2 \chi_\pm = b_1 b_2 \chi_\pm^2 = b_1 b_2 \chi_\pm = (b_1 b_2)_\pm.$$

The proof of $b_1 b_2 \in cH^\infty(N_\mu(\mathbb{C}^{n-1}))$ follows from

$$\xi e_n(b_1 b_2) = b_1 \xi e_n b_2 = (b_1 b_2) \xi e_n.$$

The proof is complete. $\square$

The assertion (i) is Parseval's identity on the surface.

The Cayley transformation $\phi(\underline{x}) = (e_n \underline{x} + 1)(\underline{x} + e_n)^{-1}$ takes $\mathbb{R}^{n-1}$ to the sphere $\mathbf{S}^{n-1}$ and the upper half space $\{\underline{x} = x' + x_n e_n \mid x_n > 0\}$ to the interior part of the unit ball $\{\underline{x} \mid |\underline{x}| < 1\}$. The study in the present section, therefore, provides a singular integral and Fourier multiplier theory on the sphere (see [Ry3]). The theory, however, is not the traditional one. On the sphere it would be preferable to have a Fourier series theory, like what we have on the unit circle, rather than a Fourier integral theory. A satisfied Fourier series theory on the sphere and, more generally, on starlike Lipschitz surfaces is developed in [Q7]. The latter theory is associated with spherical Dirac operator as explained in the following section.

5   Theory on Möbius images of starlike Lipschitz surfaces

As the first approach to closed surfaces in Euclidean spaces, we developed a Fourier series theory on starlike Lipschitz surfaces in the quaternionic space (see [Q5]). In [Q7], we extend the theory to Euclidean spaces. In this section, we shall first review the established theory on starlike Lipschitz surfaces in $\mathbb{R}^n$ and then construct an analogous theory on the Möbius images of the surfaces.

Most of the following notation and terminology are referred to in [Q7].

A surface Θ is said to be a *starlike Lipschitz surface* if it is $(n-1)$-dimensional, continuous, and star-shaped about the origin, and there exists a constant $m < \infty$ such that $\underline{x}, \underline{y} \in \Theta$ implies that

$$\frac{|\ln |\underline{x}^{-1}\underline{y}||}{\arg(\underline{x}, \underline{y})} \leq M.$$

The minimum value of M is called the pseudo-Lipschitz constant of Θ, denoted by $pLip(\Theta)$. Since locally

$$\ln |\underline{x}^{-1}\underline{y}| = \ln(1 + (|\underline{x}^{-1}\underline{y}| - 1))|$$
$$\approx (|\underline{x}^{-1}\underline{y}| - 1) \approx |\underline{x}^{-1}|(|\underline{y}| - |\underline{x}|) \approx (|\underline{y}| - |\underline{x}|),$$

the sense of Lipschitz defined here is consistent with the one defined in Section 3.

Denote

$$\rho = \min\{|\underline{x}| : \underline{x} \in \Theta\} \quad \text{and} \quad \iota = \max\{|\underline{x}| : \underline{x} \in \Theta\}.$$

Without loss of generality, we can assume $\rho < 1 < \iota$. We shall be using the class

$$\mathcal{A}(\Theta)$$
$$= \{f : f(\underline{x}) \text{ is left} - \text{monogenic in } \rho - s < |\underline{x}| < \iota + s \text{ for some } s > 0\}.$$

It is a consequence of CMcM's Theorem that $\mathcal{A}(\Theta)$ is dense in $L^2(\Theta)$ (see [CM], or [GQW], or [Q5]).

For $f \in \mathcal{A}(\Theta)$, we have the expansion

$$f(\underline{x}) = \sum_{k=0}^{\infty} P_k(f)(\underline{x}) + \sum_{k=1}^{\infty} Q_k(f)(\underline{x}),$$

where, for $k \geq 0$, $P_k(f)$ belongs to the finite dimensional linear space M_k of k-homogeneous left-monogenic functions in $\mathbb{R}^n$, and, for $k > 0$, $Q_k(f)$ belongs to the finite dimensional linear space M_{-k-n+2} of $-k - n + 2$-homogeneous left-monogenic functions defined in $\mathbb{R}^n \setminus \{0\}$. It is noted that the spaces $M_{-n+2}, M_{-n+3}, \cdots, M_{-1}$ do not exist. The spaces M_k and M_{-k-n+2} are eigenspaces of the *spherical Dirac operator* Γ_ζ, defined through the decomposition

$$\underline{D} = \zeta \partial_r - \frac{1}{r} \partial_{\underline{\zeta}} = \zeta (\partial_r - \frac{1}{r} \Gamma_{\underline{\zeta}}), \qquad (5.1)$$

where $\underline{x} = r\zeta$, $r = |\underline{x}|$ Indeed, we have

$$\Gamma_{\underline{\zeta}} f(\underline{\xi}) = k f(\underline{\xi}), \qquad f \in M_k, \ k = \dots, -n, -n+1, 0, 1, 2, \dots . \qquad (5.2)$$

What plays an important role is the generalization of Fueter's result in the quaternionic space (see [Su]) to Euclidean spaces $\mathbb{R}^n_1$ ([Sc], [Q6], [Q7]). We now briefly recall the generalization. The first generalization was made by Sce in 1967. His result (see [Sc]) asserts that in $\mathbb{R}^n_1$, where $n = odd$ integers, if $f^0(z) = u(x^0, y^0) + iv(x^0, y^0)$ is holomorphically defined in a relatively open set O of the upper half complex plane including the boundary, then the function $\Delta^{\frac{n-1}{2}} \overrightarrow{f^0}(x)$ is Clifford monogenic, where $\overrightarrow{f^0}(x) = u(x_0, |\underline{x}|) + \frac{\underline{x}}{|\underline{x}|} v(x_0, |\underline{x}|)$ and Δ is the Laplacian for $x_0, x_1, \dots, x_n$. Sce's result is extended to $\mathbb{R}^n_1$ for $n =$ any integers in [Q6], by using Fourier multiplier, with crucial applications in [Q7].

This idea suggests to define a sequence of functions $P^{(k)}$, called monomial functions, those give rise, through convolution integrals, to projections of $L^2(\Theta)$-functions into the spaces M_k.

The *monomial functions* in $\mathbb{R}^n_1$ are defined as follows. Define

$$P^{(-k)} = \Delta^{\frac{n-2}{2}} \overline{(\cdot)^{-k}}, \qquad k = 1, 2, \dots ,$$

where $E(x) = \frac{\overline{x}}{|x|^{n+1}}$ is the Cauchy kernel in $\mathbb{R}^n_1$. It is obvious that $P^{(-k)}(x)$ are monogenic with respect to the Dirac operator $D = \frac{\partial}{\partial x_0} + \underline{D}$. Define

$$P^{(k-1)} = I(P^{(-k)}), \qquad k \in \mathbb{Z}^+,$$

where I is the Kelvin inversion : $If(x) = E(x)f(x^{-1})$. The conformal covariance property asserts that $P^{(k-1)}$ are monogenic ([B], [PQ]) with respect to D. We need to deal with $P^{(k)}$ in Euclidean spaces of different dimensions. In Proposition 7 below, the subscript n in the notation $P_n^{(k)}$ is to emphasize that the $P^{(k)}$ is defined in $\mathbb{R}^n_1$.

To summarize, we have (see [Q7]):

Proposition 7. *Let* $k \in \mathbb{Z}^+$. *Then*

(i) $P^{(-1)} = E$;

(ii) $P^{(-k)}(x) = \frac{(-1)^{k-1}}{(k-1)!} (\frac{\partial}{\partial x_0})^{k-1} E(x)$;

(iii) $P^{(-k)}$ *and* $P^{(k-1)}$ *both are monogenic*;

(iv) $P^{(-k)}$ *is homogeneous of degree* $-n+1-k$ *and* $P^{(k-1)}$ *homogeneous of degree* $k-1$;

(v) $c_n P_{n-1}^{(-k)}(x_0 e_0 + \cdots + x_{n-1}e_{n-1}) = \int_{-\infty}^{\infty} P_n^{(-k)}(x)dx_n$, *where* $c_n = \int_{-\infty}^{\infty}(1+t^2)^{-(\frac{n+1}{2})}dt$;

(vi) $P^{(-k)} = I(P^{(k-1)})$;

(vii) *If* n *is odd, then* $P^{(k-1)} = \Delta^{\frac{n-1}{2}}\overrightarrow{(\cdot)^{n+k-2}}$.

As observed in [Q7], in $\mathbb{R}^n$ we may define functions in $y^{-1}x$:

$$P^{(-k)}(y^{-1}x)E(y) = \frac{(-1)^{k-1}}{(k-1)!} < y, \nabla_x >^{k-1} E(x), \qquad k \in Z^+,$$

which can be reduced to the above defined $P_{n-1}^{(-k)}(x)$ when $y = e_n$.

Using these notations, we have convolution forms for the projection operators in the decomposition of $f \in \mathcal{A}(\Theta)$:

$$P_k(f)(x) = \frac{1}{\omega_{n-1}} \int_{\Theta} P^{(k)}(y^{-1}x)E(y)n(y)f(y)d\sigma(y), \quad k = 0, 1, 2, \ldots,$$

and

$$Q_k(f)(x) = \frac{1}{\omega_{n-1}} \int_{\Theta} P^{(-k)}(y^{-1}x)E(y)n(y)f(y)d\sigma(y), \quad k = 1, 2, \ldots.$$

Accordingly, we have

$$f(x) = \sum_{k=-\infty}^{\infty} \frac{1}{\omega_{n-1}} \int_{\Theta} P^{(k)}(y^{-1}x)E(y)n(y)f(y)d\sigma(y).$$

We shall use the heart-shaped regions $\mathbf{H}_{\omega,\pm}$:

$$\mathbf{H}_{\omega,\pm} = \{x \in \mathbb{R}^n \mid \frac{(\pm \ln |e_n x|)}{\arg(e_n, x)} < \tan \omega\},$$

and

$$\mathbf{H}_\omega = \mathbf{H}_{\omega,+} \cap \mathbf{H}_{\omega,-}.$$

That is,

$$\mathbf{H}_\omega = \{x \in \mathbb{R}^n \mid \frac{|\ln |e_n x||}{\arg(e_n, x)} < \tan \omega\}.$$

We shall use the function spaces

$K(\mathbf{H}_{\omega,\pm})$

$= \{\underline{\phi} \mid \mathbf{H}_{\omega,\pm} \to \mathbb{C}^{(n)} \mid \underline{\phi} \text{ is monogenic and } |\underline{\phi}(x)| \leq \frac{C_\mu}{|e_n - x|^{n-1}}, 0 < \mu < \omega\},$

and

$$K(\underline{\mathbf{H}}_\omega) = \{\phi \,|\, \underline{\mathbf{H}}_\omega^c \to \mathbb{C}^{(n)} |\, \phi = \phi^+ + \phi^-,\ \phi^\pm \in K(\underline{\mathbf{H}}_{\omega,\pm})\}.$$

We shall also use the following sets in the complex plane: for $\omega \in (0, \frac{\pi}{2})$,

$$\mathbf{S}_{\omega,\pm}^c = \{z \in \mathbb{C} \,|\, |\arg(\pm z)| < \omega\},$$

where the angle $\arg(z)$ of the complex number z takes values in $(-\pi, \pi]$, and

$$\mathbf{S}_\omega^c = \mathbf{S}_{\omega,+}^c \cup \mathbf{S}_{\omega,-}^c.$$

The following function spaces will be used:

$$H^\infty(\mathbf{S}_{\omega,\pm}^c) = \{b : \mathbf{S}_{\omega,\pm}^c \to \mathbb{C} \,|\, b \text{ is holomorphic and}$$
$$|b(z)| \le C_\mu < \infty \text{ in any } \mathbf{S}_{\mu,\pm}^c, 0 < \mu < \omega\},$$

and

$$H^\infty(\mathbf{S}_\omega^c) = \{b : \mathbf{S}_\omega^c \to \mathbb{C} \,|\, b_\pm = b\chi_{\{z \in \mathbb{C} \,|\, \pm Re z > 0\}} \in H^\infty(\mathbf{S}_{\omega,\pm}^c)\}.$$

Now, for $b \in H^\infty(\mathbf{S}_\omega^c)$, $\omega \in (\arctan(N(\Theta)), \frac{\pi}{2})$, we define the Fourier multiplier operator

$$cM_b f(\underline{x}) = \sum_{k=1}^\infty b(k) P_k(f)(\underline{x}) + \sum_{k=1}^\infty b(-k) Q_k(f)(\underline{x}).$$

Now for $x \in \Theta, r \approx 1$ and $r < 1$, consider the function

$$cM_b^r f(\underline{x}) = \sum_{k=1}^\infty b(k) P_k(f)(r\underline{x}) + \sum_{k=1}^\infty b(-k) Q_k(f)(r^{-1}\underline{x})$$
$$= P^r(\underline{x}) + Q^r(\underline{x}), \qquad \rho - s < |\underline{x}| < \iota + s.$$

Using the convolution expressions of the projections, we have

$$P^r(\underline{x}) = \sum_{k=1}^\infty b(k) \frac{1}{\omega_{n-1}} \int_\Theta P^{(k)}(\underline{y}^{-1} r\underline{x}) \underline{E}(\underline{y}) n(\underline{y}) f(\underline{y}) d\sigma(\underline{y})$$
$$= \frac{1}{\omega_{n-1}} \int_\Theta \left(\sum_{k=1}^\infty b(k) P^{(k)}(\underline{y}^{-1} r\underline{x}) \right) \underline{E}(\underline{y}) n(\underline{y}) f(\underline{y}) d\sigma(\underline{y})$$
$$= \frac{1}{\omega_{n-1}} \int_\Theta c\Phi^+(\underline{y}^{-1} r\underline{x}) \underline{E}(\underline{y}) n(\underline{y}) f(\underline{y}) d\sigma(\underline{y}),$$

where $c\Phi^+ = \sum_{k=1}^\infty b(k) P^{(k)}$. Similarly, we have

$$Q^r(\underline{x}) = \frac{1}{\omega_{n-1}} \int_\Theta c\Phi^-(\underline{y}^{-1} r^{-1}\underline{x}) \underline{E}(\underline{y}) n(\underline{y}) f(\underline{y}) d\sigma(\underline{y}),$$

where $c\Phi^- = \sum_{k=-\infty}^{-1} b(k)P^{(k)}$.

If $f \in \mathcal{A}(\Theta)$, then the projections $P_k(f)$ and $Q_k(f)$ decay rapidly as $k \to \infty$. This justifies the change of order of taking summation and integration in the above argument. It also implies that the series defining $cM_b^r f$ uniformly converges as $r \to 1-$, so that we can exchange the order of taking limit and summation to obtain

$$cM_b f(\underline{x})$$
$$= \lim_{r \to 1-} \frac{1}{\omega_{n-1}} \int_\Theta (c\Phi^+(\underline{y}^{-1}r\underline{x}) + c\Phi^-(\underline{y}^{-1}r^{-1}\underline{x}))\underline{E}(\underline{y})n(\underline{y})f(\underline{y})d\sigma(\underline{y}).$$

Using the last expression, in [Q7] we obtain

$$cM_b f(\underline{x})$$
$$= \lim_{\epsilon \to 0} \frac{1}{\omega_{n-1}}\{\int_{|\underline{y}-\underline{x}|>\epsilon, \underline{y}\in\Theta} c\Phi(\underline{y}^{-1}\underline{x})\underline{E}(\underline{y})n(\underline{y})f(\underline{y})d\sigma(\underline{y}) + \underline{c\Phi}(\epsilon,\underline{x})f(\underline{x})\},$$

where $c\Phi = c\Phi^+ + c\Phi^-$ is the function associated with b as specified above and $\underline{c\Phi}$ is the bounded continuous function: $\underline{c\Phi} = \underline{c\Phi}^+ + \underline{c\Phi}^-$, where

$$\underline{c\Phi}^\pm(\epsilon,\underline{x}) = \int_{S(\epsilon,\underline{x},\pm)} c\Phi^\pm(\underline{y}^{-1}\underline{x})\underline{E}(\underline{y})n(\underline{y})d\sigma(\underline{y}),$$

and $S(\epsilon,\underline{x},\pm)$ is the part of the sphere $|\underline{y}-\underline{x}| = \epsilon$ inside or outside Θ, depending on $\pm$ taking $+$ or $-$, respectively.

Denote by $cT_{(c\Phi,\underline{c\Phi})}$ the above induced singular integral operator from cM_b. In addition to the Parseval's type relation $cM_b = cT_{(c\Phi,\underline{c\Phi})}$, we also obtain, as in the Lipschitz graph case, that for $b \in H^\infty(S_\omega^c), \omega \in (\arctan(pLip(\Theta)), \frac{\pi}{2})$, cM_b can be extended to become a bounded operator from $L^2(\Theta)$ to $L^2(\Theta)$. Moreover, the corresponding class cM_b forms a bounded holomorphic functional calculus of the spherical Dirac operator $\Gamma_{\underline{\varsigma}}$.

The critical technical result to prove the boundedness is the following (see [Q7]).

Proposition 8. *For $b \in H^\infty(S_{\omega,\pm}^c)$ and $c\Phi(\underline{x}) = \sum_{k=\pm 1}^{\pm\infty} b(k)P^{(k)}(\underline{x})$, we have $c\Phi \in K(\mathbf{H}_{\omega,\pm})$.*

In the proof, the assertion is reduced to the analogous result for the complex plane ([Q1]) via the pointwise correspondence of Sce's device and its generalization given in Proposition 7 (see [Q7]).

In the following, we shall show how to induce a corresponding Fourier multiplier and singular integral theory on $\phi^{-1}(\Theta)$.

We shall use a variation of Proposition 3:

Proposition 9. *If Θ is a starlike Lipschitz surface, then for each pair $h, f \in L^2(\Theta)$ we have*

$$\int_\Theta h(\underline{y})\underline{E}(\underline{y})n(\underline{y})f(\underline{y})d\sigma(\underline{y})$$

$$= \int_{\phi^{-1}(\Theta)} h(\phi(\underline{x}))\underline{E}(\phi(\underline{x}))(J^*(\phi,\underline{x})\underline{E}^{-1}(\underline{x}))\underline{E}(\underline{x})n(\underline{x})(J(\phi,\underline{x})f(\phi(\underline{x})))d\sigma(\underline{x}),$$

where $n(\underline{y})$ is a chosen orientation, and $n(\underline{x})$ is the associated orientation on the surface $\phi^{-1}(\Sigma)$ induced by the Möbius transformation ϕ.

Denote again $g(\underline{x}) = J(\phi,\underline{x})f(\phi(\underline{x})) = \tau(f)$. According to Proposition 9, the associated singular integral operator on $L^2_\phi(\phi^{-1}(\Theta))$ is of the form

$$cT^\phi_{(c\Phi,c\underline{\Phi})}g(\underline{x})$$

$$= \lim_{\epsilon \to 0} \frac{1}{\omega_{n-1}} \Big\{ \int_{|\phi(\underline{y})-\phi(\underline{x})|>\epsilon, \phi(\underline{y})\in\Theta} c\Phi^\phi(\underline{x},\underline{y})\underline{E}(\underline{y})n(\underline{y})g(\underline{y})d\sigma(\underline{y})$$

$$+ \underline{c\Phi}^\phi(\phi(\underline{x}), \epsilon n(\phi(\underline{x})))g(\underline{x})\Big\}, \quad (5.3)$$

where

$$c\Phi^\phi(\underline{x},\underline{y}) = J(\phi,\underline{x})c\Phi(\phi(\underline{y})^{-1}\phi(\underline{x}))\underline{E}(\phi(\underline{y}))J^*(\phi,\underline{y})\underline{E}^{-1}(\underline{y}),$$

and $\underline{c\Phi}^\phi = \underline{c\Phi}^\phi_+ + \underline{c\Phi}^\phi_-$,

$$\underline{c\Phi}^\phi_\pm(\underline{x},\underline{t}) = \int_{S^\phi_t(\underline{x},\pm)} c\Phi^\phi_\pm(\underline{x},\underline{y})\underline{E}(\underline{y})n(\underline{y})d\sigma(\underline{y}),$$

where $S^\phi_t(\underline{x},\pm)$ is the part of the surface $\{\underline{y} \mid |\phi(\underline{y}) - \phi(\underline{x})| = |t|\}$ inside or outside $\phi^{-1}(\Theta)$, depending on $\pm$ taking $+$ or $-$, respectively.

Similarly, the projections $P^\phi_k(g)$ and $Q^\phi_k(g)$ of $g(\underline{x}) = J(\phi,\underline{x})f(\phi(\underline{x}))$ onto the spaces

$$M^\phi_k = \{g \mid g = \tau(f), f \in M_k\}$$

and

$$M^\phi_{-k-n+2} = \{g \mid g = \tau(f), f \in M_{-k-n+2}\},$$

respectively, are

$$P^\phi_k(g)(\underline{x}) = \frac{1}{\omega_{n-1}} \int_\Theta P^{(k)}_\phi(\underline{x},\underline{y})\underline{E}(\underline{y})n(\underline{y})g(\underline{y})d\sigma(\underline{y}), \quad k = 0, 1, 2, \ldots,$$

and

$$Q^\phi_k(g)(\underline{x}) = \frac{1}{\omega_{n-1}} \int_\Theta P^{(-k)}_\phi(\underline{x},\underline{y})\underline{E}(\underline{y})n(\underline{y})g(\underline{y})d\sigma(\underline{y}), \quad k = 1, 2, \ldots,$$

where

$$P_\phi^{(k)}(\underline{x},\underline{y}) = J(\phi,\underline{x}))P^{(k)}(\phi(\underline{y})^{-1}\phi(\underline{x}))\underline{E}(\phi(\underline{y}))J^*(\phi,\underline{y})\underline{E}^{-1}(\underline{y}), \quad k \in \mathbb{Z}.$$

There follows

$$g(\underline{x}) = \sum_{k=0}^{\infty} P_k^\phi(g)(\underline{x}) + \sum_{k=1}^{\infty} Q_k^\phi(g)(\underline{x}).$$

The Fourier multiplier operator associated with $cT_{(c\Phi,\underline{c\Phi})}^\phi$ is defined to be

$$cM_b^\phi(g)(\underline{x}) = \lim_{r\to 1-} \{\sum_{k=1}^{\infty} b(k)P_k^\phi(g)(r\underline{x}) + \sum_{k=1}^{\infty} b(-k)Q_k^\phi(g)(r^{-1}\underline{x})\}. \quad (5.4)$$

Similar to the Main Theorem, on the surface $\phi^{-1}(\Theta)$ we have

Theorem 1. *Let* $\omega \in (\arctan(pLip(\Theta)), \frac{\pi}{2}), b \in H^\infty(\mathbf{S}_{\omega,\pm}^c)$ *and* $(c\Phi, \underline{c\Phi})$ *the associated function pair in* $K(\underline{\mathbf{H}}_\omega)$. *Then we have*

(i) $cM_b^\phi = cT_{(c\Phi,\underline{c\Phi})}^\phi$ *on* $\mathcal{A}(\phi^{-1}(\Theta))$;

(ii) cM_b^ϕ *can be extended to become a bounded operator on* $L^2(\phi^{-1}(\Theta))$;

(iii) *On* $L^2(\phi^{-1}(\Theta))$ *the limits in (5.3) and (5.4) converge in both the L^2-norm sense and the almost everywhere sense on the surface; and*

(iv) cM_b^ϕ *is the bounded holomorphic functional calculus of the induced Dirac operator* $\Gamma_{\underline{\zeta}}^\phi = \overline{J}^{-1}(\phi,\underline{x})\Gamma_{\underline{\zeta}}J^{-1}(\phi,\underline{x})$ *on* $L^2(\phi^{-1}(\Theta))$, *where* $\Gamma_{\underline{\zeta}}$ *is the spherical Dirac operator defined in the decomposition (4.3).*

Proof. Using Proposition 9, the Parseval's type identity (i) is reduced to the corresponding one on Θ. The boundedness result (ii) follows from Proposition 2. The convergence results in assertion (iii) are consequences of the boundedness. Assertion (iv) is proved by invoking the counterpart of the relation (5.4) in the star-shaped surface case, and the property (5.2) of the spherical Dirac operator Γ_ζ. $\qquad\square$

REFERENCES

[A1] L. Ahlfors, Old and new in Möbius groups, *Ann. Acad. Sci. Fenn. Ser. A I Math.* **9** (1984), 93–105.

[A2] L. Ahlfors, Möbius transforms and Clifford numbers, in *Differential Geometry and Complex Analysis*, Isaac Chavel and Hershel M. Farkas, eds., Springer-Verlag, Berlin, 1985, 65–73.

[A3] L. Ahlfors, Clifford numbers and Möbius transforms in $\mathbb{R}^n$, in *Clifford Algebras and Their Role in Mathematics and Physics*, J. R. Chisholm and A. K. Common, eds., Canterbury, 1985, Reidel, Dordrecht, 1986, 167–175.

[A4]   L. Ahlfors, Möbius transforms in $\mathbb{R}^n$ expressed through 2×2 Clifford numbers, *Complex Variables Theory Appl.* **5** (1986), 215–224.

[B]   B. Bojarski, Remarks on polyharmonic operators and conformal maps in space, in *Trudyi Vsesoyuznogo Simpoziuma v Tbilisi 21-23 aprelya 1982* (Russian), Tbilisi. Gos. Unic., Tbilisi, 1986, 49–56.

[C]   J. Cnops, Hurwitz pairs and applications of Möbius transformations, Ph.D. Dissertation, University of Gent, Belgium, 1994.

[CM]   R. Coifman and Y. Meyer, Fourier analysis of multilinear convolution, Calderón's theorem, and analysis on Lipschitz curves, *Lecture Notes in Mathematics*, Springer-Verlag **779** (1980), 104–122.

[CMcM] R. Coifman, A. McIntosh, and Y. Meyer, L'intégrale de Cauchy définit un opérateur borné sur L^2 pour les courbes lipschitziennes, *Ann. Math.* **116** (1982), 361–387.

[DSS]   R. Delanghe, F. Sommen, and V. Soucek, *Clifford Algebras and Spinor Valued Functions: A Function Theory for Dirac Operator*, Kluwer, Dordrecht, 1992.

[GLQ]   G. Gaudry, R -L. Long, and T. Qian, A martingale proof of L^2-boundedness of Clifford-valued singular integrals, *Annali di Mathematica Pura Ed Applicata*, Vol. **165**, 1993, 369–394.

[GQW]   G. Gaudry, T. Qian, and S. -L. Wang, Boundedness of singular integral operators with holomorphic kernels on star-shaped Lipschitz curves, *Colloq. Math.* Vol. **LXX**, 1996, 133–150.

[K]   C. E. Kenig, *Harmonic analysis techniques for second order elliptic boundary value problems*, Conference Board of the Mathematics, CBMS, *Regional Conference Series in Mathematics, Number 83*, 1994.

[LMcQ] C. Li, A. McIntosh, and T. Qian, Clifford algebras, Fourier transforms, and singular convolution operators on Lipschitz surfaces, *Revista Matemática Iberoamericana*, Vol. **10**, No. **3** (1994), 665–721.

[LMcS] C. Li, A. McIntosh, and S. Semmes, Convolution singular integrals on Lipschitz surfaces, *J. Amer. Math. Soc.* **5** (1992), 455–481.

[Mc1]   A. McIntosh, Clifford algebras and the high dimensional Cauchy integral, *Approximation and Function Spaces*, Vol. **22**, Banach Center Publications, PWN-Polish Scientific Publishers, Warsaw, 1989.

[Mc2]   A. McIntosh, Operators which have an H^∞-functional calculus, *Miniconference on Operator Theory and Partial Differential Equations, Proc. Centre Math. Analysis*, A.N.U., Canberra, **14** (1986), 210–231.

[Mc3]   A. McIntosh, Clifford algebras, Fourier theory, singular integrals, and harmonic functions on Lipschitz domains, *Clifford Algebras in Analysis and Related Topics*, John Ryan, ed., Studies in Advanced Mathematics Series, CRC Press, Boca Raton, 1996, 33–87.

[McQ1] A. McIntosh and T. Qian, Convolution singular integral operators on Lipschitz curves, *Proc. of the Special Year on Harmonic Analysis at Nankai Inst. of Math.*, Tianjin, China, *Lecture Notes in Math.* **1494** (1991), 142–162.

[McQ2] A. McIntosh and T. Qian, L^p Fourier multipliers on Lipschitz curves, *Trans. Amer. Math. Soc.* **333** (1992), 157–176.

[Mi]   M. Mitrea, Clifford wavelets, singular integrals, and hardy spaces, *Lecture Notes in Mathematics* **1575**, Springer-Verlag, 1994.

[PQ]   J. Peetre and T. Qian, Möbius covariance of iterated Dirac operators, *J. Austral. Math. Soc. Ser. A* **56** (1994), 403–414.

[Q1]   T. Qian, Singular integrals with holomorphic kernels and H^∞-Fourier multipliers on star-shaped Lipschitz curves, *Studia Mathematica* **123** (3) (1997), 195–216.

[Q2]   T. Qian, A holomorphic extension result, *Complex Variables*, Vol. **32**, (1) (1996), 59–77.

[Q3]   T. Qian, Singular integrals on the n-torus and its Lipschitz perturbations, *Clifford Algebras in Analysis and Related Topics, Studies in Advanced Mathematics Series*, John Ryan, ed., CRC Press, Boca Raton, 1996, 94–108.

[Q4]   T. Qian, Transference between infinite Lipschitz graphs and periodic Lipschitz graphs, *Proceedings of the Center for Mathematics and its Applications, ANU*, Vol. **33** (1994), 189–194.

[Q5]   T. Qian, Singular integrals on star-shaped Lipschitz surfaces in the quaternionic space, *Math. Ann.* **310** (4) (April 1998), 601–630.

[Q6]   T. Qian, Generalization of Futer's result in $\mathbb{R}_1^n$, *Rend. Mat. Acc. Lincei*, s.9, Vol. **8**, 1997, 111–117.

[Q7]   T. Qian, Fourier theory on starlike Lipschitz surfaces, preprint.

[QR]   T. Qian and J. Ryan, Conformal transformations and Hardy spaces arising in Clifford analysis, *Journal of Operator Theory* **35** (1996), 349–372.

[Ry1]   J. Ryan, Some applications of conformal covariance in Clifford analysis, *Clifford Algebras in Analysis and Related Topics*, John Ryan, ed., CRC Press, Boca Raton, 1996, 128–155.

[Ry2]   J. Ryan, Dirac operators, conformal transformations, and aspects of classical harmonic analysis, *Journal of Lie Theory*, Vol. **8**, 1998, 67–82.

[Ry3]   J. Ryan, The Fourier transform on the sphere, *Proceedings of the Conference on Quaternionic Structures in Mathematics and Physics*, Trieste, Italy, SISSA, 1996, 277–289.

[Sc]   M. Sce, Osservazioni sulle serie di potenze nei moduli quadratici, *Atti Acc. Lincei Rend. fis.*, s. 8, **23** (1957), 220–225.

[St]   E. Stein, *Singular Integrals and Differentiability Properties of Functions*, Princeton University Press, 1970.

[Su]   A. Sudbery, Quaternionic analysis, *Math. Proc. Camb. Phil. Soc.* **85** (1979), 199–225.

[T]   T. Tao, Convolution operators on Lipschitz graphs with harmonic kernels, *Advances in Applied Clifford Algebras* **6** No. 2 (1996), 207–218.

[V]   G. Verchota, Layer potentials and regularity for the Dirichlet problem for Laplace's equation in Lipschitz domains, *J. of Funct. Anal.* **59**
(1984), 572–611.

Xinhua Ji
Institute of Mathematics, Academia Sinica,
Beijing 100080, P.R. China
E-mail: xhji@math03.math.ac.cn

Tao Qian
School of Mathematical and Computer Sciences
The University of New England, Armidale NSW 2351 Australia
E-mail: tao@turing.une.edu.au

John Ryan
Department of Mathematics, The University of Arkansas,
Fayetteville Arkansas 72701, USA
E-mail: jryan@comp.uark.edu

Received: October 8, 1999; Revised: March 5, 2000

On the Cauchy Type Integral and the Riemann Problem

Juan Bory Reyes and Ricardo Abreu Blaya

ABSTRACT In this paper smoothness properties of the quaternionic Cauchy type integral are considered, especially its continuity up to the boundary of the domain. The Plemelj-Sokhotzkij formulas for the singular integral on Ahlfors regular surfaces are proven, and, using them, we state the solubility of the simplest particular case of the Riemann problem associated to a suitable sub-space of all continuous functions defined over the boundary.

Keywords: Cauchy-type integrals, singular integrals, Ahlfors regular surfaces, quaternionic analysis.

1 Introduction

The Riemann boundary value problem (which we refer to here as the Riemann problem) is one of the principal problems of the theory of boundary value problems of analytic functions. In this note we will concern ourselves with the quaternionic version of this problem. Comprehensive development in the complex case can be found in [BS;Ga;Ka1,2,3;Lu]. In the last decade the Riemann problem in Clifford analysis has been discussed cf. papers of Xu; Shapiro and Vasilievski; Stern and Bernstein; among others, for an excellent up-to-date account of this subject we refer to the papers [Be1,2,3;GS2;Sh;SV1,2;St;Xu1,2] as well as the reference therein. The direct analogue of the Riemann problem in the quaternionic context has been reported by Shapiro and Vasilievski (see [SV1,2]). They consider the Riemann problem in terms of an integral equation. The Fredholm theory for different generalizations of the Riemann problem has been built in these two papers. But in both papers, and also in the other papers mentioned above, the Riemann problem has been only studied in connection with smooth surfaces. In [AB1,2] the explicit solution of the simplest particular case of the Riemann problem over continuous surfaces with little smoothness is established. The method for solving the Riemann problem presented in these papers does

This paper was completed when the first author was visiting Gent State University, Belgium, Department of Mathematical Analysis under the Fellowship Program of the Flemish Minister of Education for Cuban Academics.

AMS Subject Classification: 30E20, 30E25, 30G35, 42B20.

not use boundary integration and can, thus, be used on non-rectifiable and fractal surfaces. Our paper is organized as follows. First we review some basic properties of the real skew field of quaternions, quaternionic function theory, and related function spaces. Then we prove smoothness properties of the quaternionic Cauchy type integral and the related singular integral operator. In the final section our results are then used to solve the Jump Riemann problem involving Ahlfors regular surfaces.

2   Preliminaries

Let us take a look at some basic definitions and properties of function theory corresponding to quaternions. Quaternionic analysis is treated extensively in [GS1] and more recently [GS2;KS]. A basic book for Clifford analysis is [BDS]. Let $\mathbb{H}$ be the skew field of real quaternions, and let $e_0 = 1$, e_1, e_2, e_3 be the quaternion units that fulfill the conditions

$$e_i e_j + e_j e_i = -2\delta_{ij}, \quad i,j = 1,2,3$$
$$e_1 e_2 = e_3, \quad e_2 e_3 = e_1, \quad e_3 e_1 = e_2.$$

For each $a = \sum_{j=0}^{3} a_j e_j$, the norm of a is defined to be $|a| = (\sum_{j=0}^{3} a_j^2)^{\frac{1}{2}}$. The Dirac operator in $\mathbb{R}^3$ is defined by $\mathbf{D} = \sum_{j=1}^{3} e_j \frac{\partial}{\partial x_j}$. Let $\Omega^+ \subset \mathbb{R}^3$ be a bounded and simply connected domain which is bounded by a continuous surface Γ. By Ω^- we denote the complement of $\Omega^+ \cup \Gamma$. A surface Γ is called Ahlfors regular (or sometime called Ahlfors-David regular) if there exists a positive number c such that

$$c^{-1} r^2 \leq \mathcal{H}^2(\Gamma \cap B(z,r)) \leq c r^2$$

for all $z \in \Gamma$, $0 < r < \text{diam } \Gamma$, where $\mathcal{H}^2(\mathcal{F})$, $\mathcal{F} \subset \mathbb{R}^3$ is the 2-dimensional Hausdorff measure of the set $\mathcal{F}$ and $B(z,r)$ stands for the closed ball with center z and radius r. Excellent accounts on Ahlfors regular surfaces can be found in [Da], [Ma], [Se] and in the references quoted there. Throughout the paper Γ denotes an Ahlfors regular surface with diameter d.

For a domain Ω in $\mathbb{R}^3$ we consider functions u defined in Ω with values in $\mathbb{H}$. These functions can be written as

$$u(x) = \sum_{j=0}^{3} u_j(x) e_j, \quad x \in \Omega.$$

We say u has properties such as continuity and differentiability whenever all components u_j have these properties. In this way the usual spaces of these functions are denoted $C^p(\Omega, \mathbb{H})$, $p \in \mathbb{N} \cup \{0\}$.

We introduce the set

$$\mathcal{M}(\Omega) = \text{Ker } \mathbf{D} = \{u \in C^1(\Omega, \mathbb{H}) : \mathbf{D}u = 0\}.$$

The elements of this set will be called (left) monogenic in the domain Ω.

It is well known that the Cauchy kernel $E(x) = \frac{1}{4\pi}\frac{-x}{|x|^3}$ is a fundamental solution for the Dirac operator and then monogenic in $\mathbb{R}^3 \setminus \{0\}$. Let $u \in C(\Omega, \mathbb{H})$; then by using the Cauchy kernel we are going to deal with the operators $\mathcal{C}_\Gamma$, and $\mathcal{S}_\Gamma$ that are given by the formulae

$$(\mathcal{C}_\Gamma u)(x) = \int_\Gamma E(x-y)n(y)u(y)d\mathcal{H}^2(y), \quad x \notin \Gamma.$$

$$(\mathcal{S}_\Gamma u)(z) = 2\int_\Gamma E(z-y)n(y)(u(y)-u(z))d\mathcal{H}^2(y) + u(z), \quad z \in \Gamma,$$

where $n(y) = \sum_{j=1}^3 n_j(y)e_j$ describes the outward pointing normal (unit) vector to Γ at the point y. The integral that defines the operator $\mathcal{S}_\Gamma$ is taken in the sense of the Cauchy's principle value. Since

$$\int_\Gamma E(x-y)n(y)d\mathcal{H}^2(y) = 1, \quad x \in \Gamma,$$

when Γ is a smooth surface, the singular Cauchy operator $\mathcal{S}_\Gamma$ then coincides with the following operator $\widetilde{\mathcal{S}_\Gamma}$:

$$(\widetilde{\mathcal{S}_\Gamma}u)(x) = 2\int_\Gamma E(x-y)n(y)u(y)d\mathcal{H}^2(y), \quad x \in \Gamma.$$

The operator $\mathcal{C}_\Gamma$ is known as the three-dimensional version of the Cauchy type integral, and it was considered by B. W. Bitsadze in 1955 [B]. Later on, Iftimie [I] established properties of $\mathcal{C}_\Gamma$ for Hölder continuous functions defined over compact Liapunov surfaces in Euclidean spaces. These results have been recently considered for many authors, not only on Hölder continuous functions but also on L^2-spaces (see [Mu] and the survey article by A. McIntosh in [Ry]). However, in all the cases some minimal smoothness criterion like Liapunov and Lipschitz are assumed on the surface Γ. In the present work these restrictions are not necessary.

The following section studies sufficient conditions such that $\mathcal{C}_\Gamma$ extends continuously to the boundary Γ, without keeping in mind any control on the modulus of continuity of the function u.

Now we can introduce the class $S(\Omega^\pm, C(\Gamma, \mathbb{H}))$ consisting of $\mathbb{H}$-valued functions $u^\pm$, such that

(i) $u^\pm \in \mathcal{M}(\Omega^\pm)$

(ii) $\lim_{\Omega^\pm \ni x \to z \in \Gamma} u^\pm(x) := u^\pm(z)$ exists everywhere on Γ generating a function from $C(\Gamma, \mathbb{H})$.

(iii) $u^\pm$ is represented by a Cauchy type integral with a density from $C(\Gamma, \mathbb{H})$.

When u is a restriction to Γ of a function from $\mathcal{M}(\Omega^{\pm})$, then of course the Cauchy type integral of u turns into its Cauchy integral.

3   Properties of the Cauchy type integral and the singular integral operator

In this section we want to state some important properties of the integral C_{Γ} and the operator S_{Γ}. In this way we shall assume a priori that Γ is 2-rectifiable, i.e., there exists a Lipschitzian function mapping some bounded subset of $\mathbb{R}^2$ onto Γ (see [Fe, Chapter III p. 251]), but this assumption will be used only in order to guarantee, as a consequence of a theorem of Rademacher, that the normal vector function $n(y)$ is defined almost everywhere on Γ. The estimates we get will not depend on this a priori assumption, in the general case, $n(y)$ should be substituted by the exterior normal vector defined as in [Fe].

The following notation will be used:

$$\Gamma_r(z) = \Gamma \cap B(z, r), \quad z \in \Gamma, \quad 0 < r \le d.$$

We define $\Theta_z(r) = \mathcal{H}^2(\Gamma_r(z))$, $\Theta(r) = \sup_{z \in \Gamma} \Theta_z(r)$, and also use the notation $\rho(x, \Gamma) = \inf_{z \in \Gamma} |z - x|$; additionally for $x \notin \Gamma$, z_x denoted a point of Γ, where $\rho(x, \Gamma) = |z_x - x|$.

For a function $u \in C(\Gamma, \mathbb{H})$ we shall use the notation

$$\omega_u(\tau) = \tau \sup_{\delta \ge \tau} \delta^{-1} \omega(u, \delta),$$

where $\omega(u, \delta)$ is the modulus of continuity of the function u. The domain $\Omega^{\pm}$ will be written in this section, for simplicity in the notation, as $\Omega^{1,2}$ respectively. For the sake of brevity we quote a lemma from [AB2] (Section VIII, Lemma 8.1), which we will use more than once later.

Lemma 1. *Let $\varphi(\tau)$ be a nonnegative function that does not increase in $(0, d]$. Then, for every positive numbers $r', r'' \in (0, d]$, $r'' > r'$, the following formula holds:*

$$\int_{\Gamma_{r''}(z) \setminus \Gamma_{r'}(z)} \varphi(|y - z|) d\mathcal{H}^2(y) = \int_{r'}^{r''} \varphi(\tau) d\Theta_z(\tau).$$

Theorem 1. *For $u \in C(\Gamma, \mathbb{H})$, $\epsilon = \rho(x, \Gamma)$ we have*

$$|\mathcal{L}_{\epsilon}^k(u, z_x, x)| \le \text{cte} \left(\omega_u(\epsilon) + \epsilon \int_{\epsilon}^{d} \frac{\omega_u(\tau)}{\tau^2} d\tau \right),$$

where

$$\mathcal{L}_{\epsilon}^{k}(u, z, x) = (\mathcal{C}_{\Gamma}u)(x) - (2 - k)u(z)$$
$$- \int_{\Gamma \backslash \Gamma_{\epsilon}(z)} E(z - y)n(y)(u(y) - u(z))d\mathcal{H}^{2}(y)$$

for $x \in \Omega^{k}$, $z \in \Gamma$, $k = 1, 2$.

Here, and in the sequel, we shall denote by "cte" a certain generic constant, not necessarily the same in different occurrences.

Proof. Let us consider only the function $\mathcal{L}_{\epsilon}^{1}$; it is clear how one does it for the case $\mathcal{L}_{\epsilon}^{2}$.

$$\mathcal{L}_{\epsilon}^{1}(u, z_{x}, x) = \int_{\Gamma_{\epsilon}(z_{x})} E(x - y)n(y)(u(y) - u(z_{x}))d\mathcal{H}^{2}(y)$$
$$+ \int_{\Gamma \backslash \Gamma_{\epsilon}(z_{x})} (E(x - y) - E(z_{x} - y))n(y)(u(y) - u(z_{x}))d\mathcal{H}^{2}(y).$$

Let us denote the two integrals on the right side of the equality above by I_1 and I_2. Since $|y - z_{x}| \leq \epsilon$, for $y \in \Gamma_{\epsilon}(z_{x})$ and the function ω_{u} increases, then

$$|I_1| \leq \frac{1}{4\pi} \int_{\Gamma_{\epsilon}(z_{x})} \frac{\omega_{u}(|y - z_{x}|)}{|y - x|^{2}} d\mathcal{H}^{2}(y) \leq \frac{\omega_{u}(\epsilon)}{4\pi} \int_{\Gamma_{\epsilon}(z_{x})} \frac{d\mathcal{H}^{2}(y)}{|y - x|^{2}}.$$

But since $|y - x| \geq |z_{x} - x| = \epsilon$, this leads to

$$|I_1| \leq \frac{\omega_{u}(\epsilon)}{4\pi\epsilon^{2}} \Theta_{z_{x}}(\epsilon) \leq \text{cte } \omega_{u}(\epsilon).$$

Since $|y - z_{x}| \leq |y - x| + |x - z_{x}| \leq 2|y - x|$, then in view of [GB] we have

$$|E(x - y) - E(z_{x} - y)| \leq \text{cte } \frac{|x - z_{x}|}{|y - z_{x}|^{3}},$$

hence

$$|I_2| \leq \text{cte}|x - z_{x}| \int_{\Gamma \backslash \Gamma_{\epsilon}(z_{x})} \frac{\omega_{u}(|y - z_{x}|)}{|y - z_{x}|^{3}} d\mathcal{H}^{2}(y).$$

From Lemma 1, we get immediately the desired estimate. $\quad\square$

Theorem 2. *Suppose $u \in C(\Gamma, \mathbb{H})$. If the following integral converges uniformly as $\epsilon \to 0$*

$$\int_{\Gamma \backslash \Gamma_{\epsilon}(z)} E(z - y)n(y)(u(y) - u(z))d\mathcal{H}^{2}(y),$$

for $z \in \Gamma$, *then* $\mathcal{C}_\Gamma u \in C(\Omega^k \cup \Gamma, \mathbb{H})$, $k = 1, 2$.

Let us remark here that the previous condition on uniform convergence of the integral is sufficient for the continuity of $\mathcal{S}_\Gamma$ in Γ as a uniform limit of continuous functions.

Proof of Theorem 2. For the sake of simplicity we consider only $k = 1$. The reasoning is similar in the other case. Let z be a fixed point on Γ and $x \in \Omega^1$. The following inequality holds:

$$|\mathcal{C}_\Gamma u(x) - \frac{1}{2}(\mathcal{S}_\Gamma u(z) + u(z))|$$

$$\leq |\mathcal{C}_\Gamma u(x) - \frac{1}{2}(\mathcal{S}_\Gamma u(z_x) + u(z_x))| + |\mathcal{S}_\Gamma u(z_x) - \mathcal{S}_\Gamma u(z)| + |u(z_x) - u(z)|$$

$$\leq |\mathcal{L}^1_{\rho(x,\Gamma)}(u, z_x, x)| + |\int_{\Gamma_{\rho(x,\Gamma)}(z_x)} E(z_x - y)n(y)(u(y) - u(z_x))d\mathcal{H}^2(y)|$$

$$+ \omega_{\mathcal{S}_\Gamma u}(|z_x - z|) + \omega_u(|z_x - z|).$$

On account of $x \to z$, the first term in the right side of the above inequality tends to zero by Theorem 1, and the second term also tends to zero by using the condition of the theorem. From the continuity in Γ of $\mathcal{S}_\Gamma$ and u the result follows.   □

It should be noted then that the Plemelj-Sokhotzkij formulas have been established for a certain sub-space of all continuous functions defined over Γ: the space $S(\Gamma, \mathbb{H})$, that is, the space of all functions $u \in C(\Gamma, \mathbb{H})$ such that

$$\int_{\Gamma_\epsilon(z)} E(z - y)n(y)(u(y) - u(z))d\mathcal{H}^2(y) \to 0 \text{ when } \epsilon \to 0,$$

uniformly for $z \in \Gamma$.

The operator $\mathcal{P}_\Gamma^\pm := \frac{1}{2}(I \pm \mathcal{S}_\Gamma)$, I being the identical operator, denotes the space of all $\mathbb{H}$-valued functions which allow a left monogenic extension into the domain $\Omega^\pm$ and vanish at infinity. We have the direct decomposition

$$S(\Gamma, \mathbb{H}) = \operatorname{im} \mathcal{P}_\Gamma^+ \cap S(\Gamma, \mathbb{H}) \oplus \operatorname{im} \mathcal{P}_\Gamma^- \cap S(\Gamma, \mathbb{H}),$$

and then each function $u \in S(\Gamma, \mathbb{H})$ admits a unique decomposition of the form $u = u^+ + u^-$, where $u^\pm \in \operatorname{im} \mathcal{P}_\Gamma^\pm \cap S(\Gamma, \mathbb{H})$. Taking into account the decomposition above one can introduce the norm

$$\|u\|_{S(\Gamma, \mathbb{H})} = \|u^+\|_{C(\Gamma, \mathbb{H})} + \|u^-\|_{C(\Gamma, \mathbb{H})}.$$

Moreover, the space $S(\Gamma, \mathbb{H})$ becomes a Banach space. The space $S(\Gamma, \mathbb{H})$ is connected with the Cauchy type integral in the following way. If $u \in$

$S(\Gamma, \mathbb{H})$, then $u^+ = u + \tilde{u}$, where

$$\tilde{u}(z) = \int_{\Gamma} E(z - y)n(y)(u(y) - u(z))d\mathcal{H}^2(y).$$

$S^{\pm}(\Gamma, \mathbb{H}) := \mathcal{P}^{\pm}_{\Gamma}(S(\Gamma, \mathbb{H}))$ is a nearby sub-space of all $C(\Gamma, \mathbb{H})$-traces on Γ of belonging to $S(\Omega^{\pm}, C(\Gamma, \mathbb{H}))$ functions. Particularly, if Γ is a smooth surface these two pairs of spaces coincide. But it is not at all clear what is true for non-smooth surfaces. Unfortunately, we do not know whether the above assertion is true in a general setting.

Theorem 3.

(i) If $U \in S(\Omega^+, C(\Gamma, \mathbb{H}))$, then for $\epsilon > 0$ and $z \in \Gamma$

$$\left| \int_{\Gamma \backslash \Gamma_{\epsilon}(z)} E(z - y)n(y)(U(y) - U(z))d\mathcal{H}^2(y) \right|$$

$$\leq \max_{x \in \partial B(z, \epsilon) \cap \overline{\Omega^+}} |U(x) - U(z)|.$$

(ii) If $U \in S(\Omega^-, C(\Gamma, \mathbb{H}))$ and $U(\infty) = 0$, then for $\epsilon > 0$ and $z \in \Gamma$

$$\left| \int_{\Gamma \backslash \Gamma_{\epsilon}(z)} E(z - y)n(y)(U(y) - U(z))d\mathcal{H}^2(y) + U(z) \right|$$

$$\leq \max_{x \in \partial B(z, \epsilon) \cap \overline{\Omega^-}} |U(x) - U(z)|.$$

Proof. Consider the set $\Omega^+ \backslash B(z, \epsilon)$. The connected components of this set will be denoted by Ω^+_i; $i = 0, 1, \ldots, N < +\infty$. Then $\partial \Omega^+_i \subset \Gamma \cup \partial B(z, \epsilon)$ and

$$\Gamma \backslash \Gamma_{\epsilon}(z) = \bigcup_{i=0}^{N} \partial \Omega^+_i \backslash \partial B(z, \epsilon).$$

Since $z \notin \Omega^+_i$, $i = 0, 1, \ldots, N$, then the function $E(z - y)(U(y) - U(z))$ is monogenic in Ω^+_i and continuous in $\overline{\Omega^+_i}$. In accordance with the Cauchy integral theorem (the Borel-Pompeiu formula is also valid in the case when Γ has a natural geometric condition, i.e., $\mathcal{H}^2(\Gamma) < +\infty$; see [Fe, BA]), we have

$$\int_{\partial \Omega^+_i} E(z - y)n(y)(U(y) - U(z))d\mathcal{H}^2(y) = 0.$$

Consequently,

$$\left| \int_{\Gamma \backslash \Gamma_\epsilon(z)} E(z-y)n(y)(U(y)-U(z))d\mathcal{H}^2(y) \right|$$

$$\leq \frac{1}{4\pi} \int_{\cup_{i=0}^{N} \partial\Omega_i^+ \cap \partial B(z,\epsilon)} \frac{|U(y)-U(z)|}{|y-z|^2} d\mathcal{H}^2(y)$$

$$\leq \frac{1}{4\pi\epsilon^2} \max_{x \in \partial B(z,\epsilon)\cap\overline{\Omega^+}} |U(x)-U(z)|\mathcal{H}^2(\partial B(z,\epsilon))$$

$$= \max_{x \in \partial B(z,\epsilon)\cap\overline{\Omega^+}} |U(x)-U(z)|.$$

The part (ii) may be proved by the same method.   □

From the last two theorems, we obtain immediately the following statements.

Proposition 1. *Let $u \in S(\Gamma, \mathbb{H})$. Then we have*

(i) *The equality $(S_\Gamma u)(x) = \pm u(x)$, $x \in \Gamma$ is valid iff $u \in S^\pm(\Gamma, \mathbb{H})$.*

(ii) *S_Γ is a linear bounded operator acting on the $S(\Gamma, \mathbb{H})$ space, $\|S_\Gamma\| = 1$, and it is an involution, i.e., $S_\Gamma^2 = I$.*

Proposition 2. *Let $U \in S(\Omega^+, C(\Gamma, \mathbb{H}))$ (or $U \in S(\Omega^-, C(\Gamma, \mathbb{H})$ and $U(\infty) = 0$). Then,*

$$\left| \int_{\Gamma \backslash \Gamma_\epsilon(z)} E(z-y)n(y)U(y)d\mathcal{H}^2(y) \right| \leq \text{cte} \, \|U\|_{C(\overline{\Omega^+},\mathbb{H})}.$$

Proof. The case $U \equiv 1$ was already considered in [AB2]. Namely,

$$\left| \int_{\Gamma \backslash \Gamma_\epsilon(z)} E(z-y)n(y)d\mathcal{H}^2(y) \right| \leq 1.$$

Hence, if we consider $U \in S(\Omega^+, C(\Gamma, \mathbb{H}))$, then

$$|\int_{\Gamma \backslash \Gamma_\epsilon(z)} E(z-y)n(y)U(y)d\mathcal{H}^2(y)|$$

$$\leq |\int_{\Gamma \backslash \Gamma_\epsilon(z)} E(z-y)n(y)(U(y)-U(z))d\mathcal{H}^2(y)| +$$

$$|\int_{\Gamma \backslash \Gamma_\epsilon(z)} E(z-y)n(y)d\mathcal{H}^2(y)||U(z)|$$

$$\leq \max_{x \in \partial B(z,\epsilon) \cap \overline{\Omega^+}} |U(x)-U(z)| + |U(z)| \leq 2 \max_{x \in \overline{\Omega^+}} |U(x)| + \max_{z \in \Gamma} |U(z)|$$

$$\leq \text{cte } \|U\|_{C(\overline{\Omega^+}, \mathbb{H})}.$$

A similar proof applies to $U \in S(\Omega^-, C(\Gamma, \mathbb{H}))$ and $U(\infty) = 0$.   □

Further, we will use the Dini space

$$I_0(\Gamma, \mathbb{H}) = \{u \in (\Gamma, \mathbb{H}) : \int_0^d \frac{\omega_u(\tau)}{\tau} d\tau < \infty\}.$$

Theorem 4. *Let* $U \in S(\Omega^+, C(\Gamma, \mathbb{H}))$ *and* $g \in I_0(\Gamma, \mathbb{H})$. *Then*

$$\mathcal{L}^1_{2|z-x|}(Ug, z, x) \longrightarrow \begin{cases} 0, & |x-z| \to 0, x \in \Omega^+ \\ -U(z)g(z), & |x-z| \to 0, x \in \Omega^-. \end{cases}$$

Proof. Our proof is an adaptation of the proof given by Babaev and Salaev in [BS], in order to deduce a similar result for the complex Cauchy type integral on rectifiable curves. To prove the estimate, for $x \in \Omega^+$, $2|z-x| = \epsilon$, we have

$$\mathcal{L}^1_\epsilon(U\,g, z, x)$$

$$= \int_{\Gamma \backslash \Gamma_\epsilon(z)} (E(x-y)-E(z-y))n(y)U(y)(g(y)-g(z_x))d\mathcal{H}^2(y)$$

$$+ \int_{\Gamma \backslash \Gamma_\epsilon(z)} E(z-y)n(y)U(y)d\mathcal{H}^2(y)(g(z)-g(z_x))$$

$$+ \int_{\Gamma_\epsilon(z)} E(x-y)n(y)U(y)(g(y)-g(z_x))d\mathcal{H}^2(y)$$

$$+ (U(x)g(z_x) - U(z)g(z))$$

$$- \int_{\Gamma \backslash \Gamma_\epsilon(z)} E(z-y)n(y)(U(y)-U(z))d\mathcal{H}^2(y)g(z) := \sum_{j=1}^5 I_j.$$

For $y \in \Gamma \setminus \Gamma_\epsilon(z)$ (notice that $|y - z_x| \le 2|y - x|$), $|y - z_x| + \epsilon \le 3|y - z|$, we have

$$|I_1| \le \text{cte} \, \|U\|_{C(\overline{\Omega^+}, \mathbb{H})} \int\limits_{\Gamma \setminus \Gamma_\epsilon(z)} |E(x - y) - E(z - y)| \omega_g(|y - z_x|) d\mathcal{H}^2(y),$$

Following [GB] and by Lemma 1 we get that

$$|I_1| \le \text{cte} \, \epsilon \, \|U\|_{C(\overline{\Omega^+}, \mathbb{H})} \int\limits_0^d \frac{\omega_g(\tau)}{\tau(\tau + \epsilon)} d\tau.$$

Taking into account that $g \in I_0(\Gamma, \mathbb{H})$, it is easy to see that the above integral tends to zero, when $\epsilon \to 0$, then $|I_1| \to 0$ uniformly as $x \to z$.

According to the Proposition 2, we have

$$|I_2| \le \text{cte} \, \|U\|_{C(\overline{\Omega^+}, \mathbb{H})} \omega_g(|z - z_x|)$$

and, therefore, $|I_2|$ converges uniformly to zero as $x \to z$.

Notice that $\Gamma_\epsilon(z) \subset \Gamma_{\frac{5\epsilon}{2}}(z_x)$; then for $\epsilon < \frac{2d}{5}$ we also get

$$|I_3| \le \text{cte} \, \|U\|_{C(\overline{\Omega^+}, \mathbb{H})} \int\limits_{\Gamma_{\frac{5\epsilon}{2}}(z_x)} \frac{\omega_g(|y - z_x|)}{|y - z_x|^2} d\mathcal{H}^2(y)$$

$$\le \text{cte} \, \|U\|_{C(\overline{\Omega^+}, \mathbb{H})} \int\limits_0^{\frac{5\epsilon}{2}} \frac{\omega_g(\tau)}{\tau} d\tau.$$

Hence $|I_3| \to 0$ as $x \to z$. By virtue of the assumption on U and g, we have $|I_4| \to 0$ as $x \to z$. Because g is a bounded function in Γ, by using theorem 3 the latter summand $|I_5|$ converges uniformly to zero as $x \to z$, which completes the proof. $\qquad \square$

Remark. *The aim of this remark is to show that the singular integral $S_\Gamma(U \, g)$ exists in the sense of the Cauchy principal value. Let be $0 < \epsilon \le d$. We have the decomposition*

$$\int\limits_{\Gamma \setminus \Gamma_\epsilon(z)} E(z - y) n(y) (U(y) g(y) - U(z) g(z)) d\mathcal{H}^2(y)$$

$$= \int\limits_{\Gamma \setminus \Gamma_\epsilon(z)} E(z - y) n(y) U(y) (g(y) - g(z)) d\mathcal{H}^2(y)$$

$$+ \left(\int\limits_{\Gamma \setminus \Gamma_\epsilon(z)} E(z - y) n(y) (U(y) - U(z)) d\mathcal{H}^2(y) \right) g(z) := J_1 + J_2.$$

In accordance with Lemma 1, the following estimate holds

$$\left| \int_{\Gamma_{\epsilon_2}(z) \setminus \Gamma_{\epsilon_1}(z)} E(z - y)n(y)U(y)(g(y) - g(z))d\mathcal{H}^2(y) \right|$$

$$\leq \|U\|_{C(\Gamma,\mathbb{H})} \int_{\Gamma_{\epsilon_2}(z) \setminus \Gamma_{\epsilon_1}(z)} \frac{\omega_g(|y - z|)}{|y - z|^2} d\mathcal{H}^2(y)$$

$$\leq \|U\|_{C(\Gamma,\mathbb{H})} \int_{\epsilon_1}^{\epsilon_2} \frac{\omega_g(\tau)}{\tau^2} d\Theta_z(\tau) \leq \text{cte} \, \|U\|_{C(\Gamma,\mathbb{H})} \int_{\epsilon_1}^{\epsilon_2} \frac{\omega_g(\tau)}{\tau} d\tau,$$

where $0 < \epsilon_1 < \epsilon_2 \leq d$. Therefore, the integral J_1 converges uniformly as $\epsilon \to 0$. The integral J_2 can be estimated by making use of the boundedness of the function g and Theorem 3.

4   Jump Riemann boundary value problem

The main goal of this section is to show how certain key results, well known previously in Clifford analysis, can be generalized to a larger class of surfaces with relaxed, to a great extent, smoothness condition of Γ. We base our proof on the smoothness properties of the Cauchy type integral studied previously in this paper.

Let Ω^+ be a bounded and simply connected domain in $\mathbb{R}^3$ which is bounded by an Ahlfors regular surface $\Gamma = \partial\Omega^+$. Furthermore, let $U \in S(\Omega^+, C(\Gamma, \mathbb{H}))$ and $g \in I_0(\Gamma, \mathbb{H})$ be given. The Jump Riemann boundary value problem reads as the problem of finding an $\mathbb{H}$-valued function $u(x)$ that piece wise satisfies the equation

$$\mathbf{D}u = 0, \quad \text{in} \quad \mathbb{R}^3 \setminus \Gamma$$

and the linear conjugation condition

$$u^+ = u^- G + Ug, \quad \text{on} \quad \Gamma$$

and which vanishes at infinity, whereby G is a non-zero quaternion.

Theorem 5. *The unique solution of the jump Riemann problem has the form*

$$u(x) = \left(\int_{\Gamma} E(x - y)n(y)U(y)g(y)G^{-1}d\mathcal{H}^2(y) \right) X(x),$$

where

$$X(x) = \begin{cases} G, & x \in \Omega^+, \\ 1, & x \in \mathbb{R}^3 \setminus \Omega^+. \end{cases}$$

Proof. From Theorem 4 it is evident that the function u is a solution of the problem, and it is easy to prove the uniqueness of this solution. In fact, assuming the existence of the other solution v, we see that the function $u^* = (u-v)X^{-1}$ is monogenic in $\mathbb{R}^3 \setminus \Gamma$; and as u^* can be continuously extended to Γ, we have, by Theorem 2 (Peinleve's theorem) in [BA], that u^* is monogenic in $\mathbb{R}^3$ and vanishes at infinity. Therefore, Liouville's theorem (see [BDS]) implies that it vanishes identically. Consequently, $u = v$. $\square$

Acknowledgment

The authors wishes to thank the referee for suggesting an improved organization of the paper as well as a revision and clarification of some proofs. We are grateful to J. Cnops for useful comments and suggestions for an earlier version of this paper. The first author would like to thank the organizers for arrangement of the financial support during the conference.

REFERENCES

[AB1] R. Abreu and J. Bory, Solvability of a Riemann linear conjugation problem on a fractal surface, *Extracta Matematicae*, Vol. **13**, No. **2** (1998), 239–241.

[AB2] R. Abreu and J. Bory, Boundary value problems for quaternionic monogenic functions on non-smooth surfaces, *Advances in Applied Clifford Algebras* **9** (1), to appear, 1999.

[B] B. W. Bitsadze, On two-dimensional integrals of Cauchy type, *Akademii. Nauk. Grus. SSR* **16**, 1955, 177–184.

[BA] J. Bory and R. Abreu, The quaternionic Riemann boundary value problem with natural geometric condition on the boundary, *Complex Variables, Theory, and Applications*, to appear, 1999.

[BDS] F. Brackx, R. Delanghe, and F. Sommen, *Clifford Analysis, Pitman Research Notes in Math.*, Vol. **76**, 1982.

[BS] A. A. Babaev and V. V. Salaev, Boundary value problem and singular integral equation on a rectifiable contour, *Mat. Zam*, Vol. **31**, (4) (1982), 571–580.

[Be1] S. Bernstein, On the left linear Riemann problem in Clifford analysis, *Bull. Belg. Math. Soc.* **3** (1996), 557–576.

[Be2] S. Bernstein, Left linear and nonlinear Riemann problems in Clifford analysis, *Clifford Algebras and Their Application in Mathematical Physics*, V. Dietrich, et al., eds., Kluwer Academic Publishers, 1998, 17–30.

[Be3] S. Bernstein, On the index of the Clifford algebras valued singular integral operators and left linear Riemann problem, *Complex Variables, Theory And Application*, Vol. **35**, 1998, 33–64.

[Da] G. David, Opérateurs intégraux sur les surfaces reguliéres, *Ann. Sci. École Norm. Sup.* 4(**21**) (1988), 225–258.

[Fe] H. Federer, *Geometric Measure Theory*, Springer-Verlag, Heidelberg/New York, 1969.

[Ga] F. D. Gajov, *Boundary Value Problems, Third Edition*, Nauka, Moscow, 1977; English transl. of 2nd ed., Pergamon Press, Oxford, and Addison-Wesley, Reading, MA, 1966 .

[GB] R. P. Gilbert and J. L. Buchanan, *First Order Elliptic Systems, A Function Theoretic Approach*, N. Y. Academic Press, 1983.

[GS1] K. Gürlebeck and W. Sprössig, *Quaternionic Analysis and Elliptic Boundary Value Problems*, Birkhäuser, Basel, 1990.

[GS2] K. Gürlebeck and W. Sprössig, *Quaternionic and Clifford Calculus for Physicists and Engineers*, Wiley & Sons. Publ., 1997.

[I] V. Iftimie, Fonctions hypercomplexes, *Bull. Math. Soc. Sci. Math. R. S. Rumanie* **9** (1965), 279–332.

[K1] B. A. Kats, The Riemann problem on a closed Jordan curve, *Sov. Math. (Iz VUZ)* **27** (1983), 83–98.

[K2] B. A. Kats, On the Riemann boundary value problem on a fractal curve, *Russ. Acad. Sci. Dokl. Math.* Vol. **48**, (3) (1994), 559–561.

[K3] B. A. Kats, On a version of the Riemann boundary value problem on a fractal curve, *Iz. Vyssh Uchebn. Zaved. Mat.* Vol. **383** (in Russian) (4) (1994), 10–20.

[Ma] P. Mattila, Rectifiability, analytic capacity, and singular integrals, *Doc. Math. J. DMV. Extra Vol. ICM. 1998 II*, 1998, 657–664.

[Mu] M. Murray, The Cauchy integral, Calderon commutation, and conjugation of singular integrals in $\mathbf{R}^n$, *Trans. of the AMS.* **298** (1985), 497–518.

[Ry] J. Ryan, *Clifford Algebras in Analysis and Related Topics*, CRC Press., Boca Raton, New York, London, Tokyo, 1993.

[Se] S. W. Semmes, Finding structure in sets with little smoothness, in *Proceedings of the ICM*, Zurich, Switzerland, Birkhäuser, 1994, 875–885.

[Sh] M. Shapiro, On analogues of the Riemann boundary value problem for a class of hyperholomorphic functions, in *Integral Equation and Boundary Value Problem*, World Scientific, 1991, 184–188.

[SV1] M. Shapiro and N. Vasilievski, Quaternionic ψ-hyperholomorphic functions, singular integral operator, and boundary value problems I, ψ-hypercomplex function theory, *Complex Variables* **27** (1995), 17–46.

[SV2] M. Shapiro and N. Vasilievski, Quaternionic ψ-hyperholomorphic functions, singular integral operator, and boundary value problems II, *Algebras of Singular Integral Operators and Riemann Type Boundary Value Problems, Complex Variables* **27** (1995), 67–96.

[St] I. Stern, Boundary value problems for generalized Cauchy Riemann systems in the space, in *Boundary Value and Initial Value Problem in Complex Analysis*, R. Kühnann and W. Tutschke, eds., Pitman. Res. Notes in Math, 1991, 159–183.

[Xu1] Z. Xu, On linear and nonlinear Riemann-Hilbert problem for regular function with values in a Clifford algebras, *Chin. Am. of Math.* **11B:3** (1990), 349–358.

[Xu2] Z. Xu, Helmholtz equations and boundary value problem, in *Partial Differential Equation with Complex Analysis*, H. Begehr, et al., eds., Harlow, Longman Scientific Technical. Pitman Res. Note Math Ser. **(262)** (1992), 204–214.

Juan Bory Reyes
Department of Mathematics
Faculty of Science
University of Oriente
Santiago of Cuba 90500, Cuba
E-mail: jbory@bioeco.ciges.inf.cu

Ricardo Abreu Blaya
Department of Mathematics
Faculty of Science
University of Oriente
Santiago of Cuba 90500, Cuba
E-mail: rabreu@uho.hlg.edu.cu

Received: September 13, 1999; Revised: March 20, 2000

Convolution and Maximal Operator Inequalities in Clifford Analysis

Mircea Martin

ABSTRACT The main objective of this article is to prove three Hedberg type inequalities for the convolution operator associated with the Euclidean Cauchy kernel in higher dimensions. Each of these inequalities involves a specific maximal operator, and they all provide the best possible constants. A few interesting consequences of the main result and applications in Clifford analysis are highlighted as well. Among them one should single out several higher-dimensional generalizations in the setting of Clifford analysis of a classical inequality in one-variable complex function theory due to Ahlfors and Beurling and some extensions of Alexander's inequality.

Keywords: Convolution, singular integrals, maximal functions, Clifford algebras.

1 Introduction

A major part of harmonic analysis is centered on the study of convolution and maximal operators. The celebrated Hardy-Littlewood-Sobolev fractional integral theorem and its companion, the Hardy-Littlewood-Wiener maximal theorem, are just two far-reaching accomplishments of the early intensive work done in this area. To be specific, we recall that the fractional integral and maximal theorems make apparent some essential properties of two operators associated with the kernel

$$k_\alpha(x) = |x|^{\alpha-n}, \qquad x \in \mathbb{R}^n_0 = \mathbb{R}^n \setminus \{0\},$$

where $0 < \alpha < n$. The former theorem addresses the so-called Riesz potential operator, that is, the convolution operator

$$I_\alpha u(x) = k_\alpha * u(x) = \int_{\mathbb{R}^n} k_\alpha(y)u(x-y)dy, \qquad x \in \mathbb{R}^n,$$

AMS Subject Classification: 42B20, 42B25, 44A35, 15A66.

and the latter deals with the maximal operator

$$Mu(x) = \sup_{t>0} \frac{1}{\text{vol}(t\mathbb{B}^n)} \int_{t\mathbb{B}^n} |u(x-y)| dy, \qquad x \in \mathbb{R}^n.$$

Though Mu does not depend on k_α, it would be proper to observe that the unit ball $\mathbb{B}^n$ in $\mathbb{R}^n$ can be described as $\mathbb{B}^n = \{x \in \mathbb{R}_0^n : k_\alpha(x) \geq 1\} \cup \{0\}$.

Both operators are well defined on $C_0(\mathbb{R}^n, \mathbb{R})$, the space of all continuous real-valued compactly supported functions on $\mathbb{R}^n$. Actually, as it is well-known, I_α and M can be extended to the standard Lebesgue spaces $L^p(\mathbb{R}^n, \mathbb{R})$ for some suitable values of p. According to the Hardy-Littlewood-Wiener theorem, one knows that M yields a continuous operator from $L^p(\mathbb{R}^n, \mathbb{R})$ into $L^p(\mathbb{R}^n, \mathbb{R})$, for every $1 < p \leq \infty$. On the other hand, by the Hardy-Littlewood-Sobolev theorem, it follows that I_α defines a bounded linear operator from $L^p(\mathbb{R}^n, \mathbb{R})$ into $L^q(\mathbb{R}^n, \mathbb{R})$, whenever $1 < p < q < \infty$ and $q^{-1} = p^{-1} - \alpha n^{-1}$. For more details and a full historical account, we refer to the masterfully elaborated monographs by Stein [S1–2] and by Hörmander [Hö], as well as to Lieb [L].

Quite astonishing, it took more than twenty years to pinpoint an inequality that links up the two classical results mentioned above. This inequality, due to Hedberg [He], clearly indicates that the fractional integral theorem is a corollary to the maximal theorem.

Hedberg's Inequality. *If $1 \leq p < q < \infty$ and $q^{-1} = p^{-1} - \alpha n^{-1}$, then*

$$|I_\alpha u(x)| \leq A_{p,q}[Mu(x)]^{p/q} \|u\|_p^{1-p/q}, \qquad x \in \mathbb{R}^n,$$

for every $u \in L^p(\mathbb{R}^n, \mathbb{R})$, where $A_{p,q}$ is a universal constant independent of u.

However, a notable deficiency of Hedberg's inequality, which sets undesired limits on its uses, is that it is not in sharp form and, at the same time, it makes use of a very particular kernel function. Both these imperfections have been completely removed in Martin and Szeptycki [MSz]. The main goal of [MSz] was to generalize Hedberg's inequality for a broader class of kernels and to produce sharp forms of all the inequalities derived for those kernels. More details on the main result of [MSz] are included in Section 2 below where a new, more conceptual proof of that result is outlined.

The primary goal of this article is to formulate and prove a Hedberg type sharp inequality for the convolution and maximal operators associated with the Euclidean Cauchy kernel in higher dimensions. A few essential prerequisites, the precise statement, and a complete proof are brought together in Section 3.

Section 4 discusses some consequences and applications of the main result in the setting of Clifford analysis. Two of them go back to a classical inequality in one-variable complex analysis, due to Ahlfors and Beurling [AB], and a quantitative version of the Hartogs-Rosenthal theorem, due to Alexander [A1–2]. Moreover, our approach enables us to generalize an inequality for the classical Cauchy kernel due to Putinar [P], as well as the higher-dimensional Ahlfors-Beurling type inequalities proved by Martin [M]. To a certain extent, the theme of this article also touches upon some results, due to Gustafsson and Khavinson [GK], Gürlebeck and Sprössig [GS], and Khavinson [K1–2].

2   Convolution and maximal operators associated with homogeneous kernels

Our purpose in this section is to present a new proof of Theorem 1.2 in [MSz]. This theorem establishes links between convolution and maximal operators associated with homogeneous kernels, and, in so doing, it improves and generalizes Hedberg's inequality.

Throughout this section, we assume that $0 < \kappa < 1$, and let $k : \mathbb{R}^n \to \mathbb{R}$ denote a function satisfying the homogeneity condition

$$k(tx) = t^{-\kappa n} k(x), \qquad t \in (0, \infty),\ x \in \mathbb{R}^n_0 = \mathbb{R}^n \setminus \{0\}. \qquad (2.1)$$

We suppose that k is continuous away from the origin. The value of k at $0 \in \mathbb{R}^n$ is irrelevant.

The convolution operator $\mathcal{I} = \mathcal{I}_k$ associated with k is defined as

$$\mathcal{I}u(x) = k * u(x) = \int_{\mathbb{R}^n} k(y)u(x - y)dy, \qquad x \in \mathbb{R}^n. \qquad (2.2)$$

Along with $\mathcal{I}$, we introduce the maximal operator $\mathcal{M} = \mathcal{M}_k$ defined by

$$\mathcal{M}u(x) = \sup_{t>0} \frac{1}{\mathrm{vol}(tX)} \int_{tX} |u(x - y)|dy, \qquad x \in \mathbb{R}^n, \qquad (2.3)$$

where

$$X = \{x \in \mathbb{R}^n_0 : |k(x)| \geq 1\} \cup \{0\}, \qquad (2.4)$$

and

$$tX = \{tx : x \in X\}, \qquad t \in (0, \infty). \qquad (2.5)$$

In the case when $k = k_\alpha$, where k_α is the kernel defined in the introduction, we recover the classical operators I_α and M. All we have already said about these two particular operators remains valid for $\mathcal{I}$ and $\mathcal{M}$. In particular, we have the following generalization of Hedberg's inequality (see Theorem 1.2 in [MSz]).

Theorem 1. *Suppose that* $1 \leq p < (1-\kappa)^{-1}$. *Then*

$$|\mathcal{I}u(x)| \leq A[\mathcal{M}u(x)]^{1-(1-\kappa)p}\|u\|_p^{(1-\kappa)p}, \qquad x \in \mathbb{R}^n, \qquad (2.6)$$

for every $u \in L^p(\mathbb{R}^n, \mathbb{R})$, *with* $A = A(k,p)$ *given by*

$$A(k,p) = \frac{1}{(1-\kappa)p}\left[\frac{\kappa p}{1-(1-\kappa)p} \cdot \mathrm{vol}(X)\right]^\kappa. \qquad (2.7)$$

Moreover, inequality (2.6) is sharp.

Before proceeding with the proof of Theorem 1, we notice that Hedberg's inequality follows from (2.6) by taking $k = k_\alpha$ and $q^{-1} = p^{-1} - (1-\kappa)$. From (2.7) we get that the best value of $A_{p,q}$ is

$$A_{p,q} = \frac{q}{q-p}\left[\frac{pq-q+p}{p} \cdot \mathrm{vol}(\mathbb{B}^n)\right]^{1-p^{-1}+q^{-1}}. \qquad (2.8)$$

Proof of Theorem 1. We start by selecting a positive number τ, and next we define the corresponding inner and outer parts of k as

$$k_{\mathrm{inn}}(x) = \begin{cases} |k(x)| - \tau^{-\kappa n}, & \text{if } x \in \tau X \\ 0, & \text{if } x \in \mathbb{R}^n \setminus \tau X, \end{cases} \qquad (2.9)$$

and

$$k_{\mathrm{out}}(x) = \begin{cases} \tau^{-\kappa n}, & \text{if } x \in \tau X \\ |k(x)|, & \text{if } x \in \mathbb{R}^n \setminus \tau X. \end{cases} \qquad (2.10)$$

The inner part, k_{inn}, is unbounded and compactly supported, and the outer part, k_{out}, is bounded. Since k_{inn} and k_{out} take non-negative values and $|k(x)| = k_{\mathrm{inn}}(x) + k_{\mathrm{out}}(x)$, we clearly have

$$|k * u(x)| \leq k_{\mathrm{inn}} * |u|(x) + k_{\mathrm{out}} * |u|(x) \qquad (2.11)$$

for every $x \in \mathbb{R}^n$. We claim that

$$k_{\mathrm{inn}} * |u|(x) \leq A'(k)\tau^{(1-\kappa)n}\mathcal{M}u(x) \qquad (2.12)$$

where

$$A'(k) = \frac{\kappa}{1-\kappa} \cdot \mathrm{vol}(X), \qquad (2.13)$$

and

$$k_{\mathrm{out}} * |u|(x) \leq A''(k,p)\tau^{-[p^{-1}-(1-\kappa)]n}\|u\|_p \qquad (2.14)$$

where

$$A''(k, p) = \left[\frac{\kappa p}{1 - (1 - \kappa)p} \cdot \text{vol}(X) \right]^{(p-1)/p} \tag{2.15}$$

To prove inequality (2.12) with $A'(k)$ as in (2.13), we assume that $u \in L^p(\mathbb{R}^n, \mathbb{R})$ and $x \in \mathbb{R}^n$ are fixed and let $\mu : (0, \infty) \to [0, \infty)$ be the nondecreasing function given by

$$\mu(t) = \int_{tX} |u(x - y)| dy.$$

We notice that

$$\mu(t) \leq \text{vol}(tX)\mathcal{M}u(x) = t^n \text{vol}(X)\mathcal{M}u(x), \qquad t > 0. \tag{2.16}$$

Since $tX = \{ y \in \mathbb{R}_0^n : k_{\text{inn}}(y) \geq t^{-\kappa n} - \tau^{-\kappa n} \} \cup \{0\}$ for each $0 < t \leq \tau$, we can express $k_{\text{inn}} * |u|(x)$ as a Stieltijes integral, namely,

$$k_{\text{inn}} * |u|(x) = \int_{\tau X} [|k(y)| - \tau^{-\kappa n}] |u(x - y)| dy$$

$$= \int_0^\tau (t^{-\kappa n} - \tau^{-\kappa n}) d\mu(t).$$

An integration by parts, in conjunction with

$$\lim_{t \downarrow 0} t^{-\kappa n} \mu(t) = 0,$$

yields

$$k_{\text{inn}} * |u|(x) = \kappa n \int_0^\tau \mu(t) \cdot t^{-\kappa n - 1} dt.$$

Using (2.16) we easily get

$$k_{\text{inn}} * |u|(x) \leq \kappa n \text{vol}(X)\mathcal{M}u(x) \int_0^\tau t^{(1-\kappa)n - 1} dt,$$

that is,

$$k_{\text{inn}} * |u|(x) \leq \frac{\kappa}{1 - \kappa} \text{vol}(X)\mathcal{M}u(x) \tau^{(1-\kappa)n},$$

an inequality equivalent to (2.12) and (2.13).

We next prove inequality (2.14) with $A''(k, p)$ as in (2.15). By Hölder's inequality we have

$$k_{\text{out}} * |u|(x) \leq \| k_{\text{out}} * |u| \|_\infty \leq \| k_{\text{out}} \|_q \| u \|_p,$$

where $q = p/(p - 1)$ if $p > 1$, and $q = \infty$ if $p = 1$. With regard to our purposes, it clearly suffices to show that

$$\| k_{\text{out}} \|_q = A''(k, p) \tau^{-[p^{-1} - (1-\kappa)]n}. \tag{2.17}$$

To prove (2.17) we set

$$\nu(t) = \mathrm{vol}(tX \setminus \tau X) = (t^n - \tau^n)\mathrm{vol}(X), \qquad t > \tau,$$

and observe that $tX = \{y \in \mathbb{R}^n : k_{\mathrm{out}}(y) \geq t^{-\kappa n}\}$ for each $t > \tau$. We first assume that $p > 1$ and $q = p/(p-1)$. Then

$$\|k_{\mathrm{out}}\|_q^q = \int_{\tau X} (\tau^{-\kappa n})^q dy + \int_{\mathbb{R}^n \setminus \tau X} |k(y)|^q dy$$

$$= \tau^{-\kappa q n}\mathrm{vol}(\tau X) + \int_\tau^\infty t^{-\kappa q n} d\nu(t)$$

$$= \tau^{-(\kappa q-1)n}\mathrm{vol}(X) + n\,\mathrm{vol}(X)\int_\tau^\infty t^{-(\kappa q-1)n-1} dt$$

$$= \tau^{-(\kappa q-1)n}\mathrm{vol}(X) + \frac{1}{\kappa q-1}\tau^{-(\kappa q-1)n}\mathrm{vol}(X)$$

$$= \frac{\kappa q}{\kappa q-1}\mathrm{vol}(X)\tau^{-(\kappa q-1)n}.$$

Equation (2.17) follows by noticing that

$$\frac{\kappa q}{\kappa q-1} = \frac{\kappa p}{\kappa p-(p-1)} = \frac{\kappa p}{1-(1-\kappa)p},$$

and

$$\frac{\kappa q-1}{q} = \frac{\kappa p-(p-1)}{p} = p^{-1}-(1-\kappa).$$

In the case when $p = 1$ and $q = \infty$, we observe that equation (2.17) becomes

$$\|k_{\mathrm{out}}\|_\infty = \tau^{-\kappa n},$$

a property of k_{out} that follows easily from (2.10).

We are now in a position to conclude the proof of Theorem 1. Combining (2.11) and (2.12)–(2.15), we have

$$|k * u(x)| \leq A'(k)\tau^{(1-\kappa)n}\mathcal{M}u(x) + A''(k,p)\tau^{-[p^{-1}-(1-\kappa)]n}\|u\|_p. \qquad (2.18)$$

To minimize the left side of (2.18), we choose

$$\tau = \left[\frac{\kappa p}{1-(1-\kappa)p}\mathrm{vol}(X)\right]^{-1/n}\left[\frac{\mathcal{M}u(x)}{\|u\|_p}\right]^{-p/n}. \qquad (2.19)$$

A straightforward calculation reveals that, for this specific value of τ, inequality (2.18) takes the form (2.6) with a constant $A = A(k,p)$ as in (2.7).

We have yet to show that (2.7) provides the best constant in (2.6). This can be done by checking that the two sides of (2.6), with $A = A(k,p)$ as in (2.7), are equal when $x = 0$ and $u \in L^p(\mathbb{R}^n, \mathbb{R})$ is given by

$$u(x) = \begin{cases} 1 & \text{if } -x \in X \text{ and } k(x) > 0, \\ -1 & \text{if } -x \in X \text{ and } k(x) < 0, \\ 0 & \text{if } -x \in \mathbb{R}^n \setminus X, \end{cases}$$

for $p = 1$, or by

$$u(x) = \begin{cases} 1 & \text{if } -x \in X \text{ and } k(x) > 0, \\ -1 & \text{if } -x \in X \text{ and } k(x) < 0, \\ |k(-x)|^{q/p} & \text{if } -x \in \mathbb{R}^n \setminus X \text{ and } k(x) \geq 0, \\ -|k(-x)|^{q/p} & \text{if } -x \in \mathbb{R}^n \setminus X \text{ and } k(x) < 0, \end{cases}$$

for $p > 1$. The required direct calculations are left to the reader. For more details we refer to Section 2 in [MSz]. □

Inequality (2.6) is no longer sharp if $u \in L^p(\mathbb{R}^n, \mathbb{R})$ is a non-negative function. Theorem 1 can be refined for such functions as follows.

We let $k = k_+ - k_-$ be the so-called Jordan decomposition of k, where $k_\pm(x) = \max\{\pm k(x), 0\}$ for each $x \in \mathbb{R}^n$. Both k_+ and k_- are non-negative functions that satisfy the homogeneity condition (2.1) and are continuous on $\mathbb{R}_0^n$. We next introduce the compact sets

$$X_\pm = \{x \in \mathbb{R}_0^n : \pm k(x) \geq 1\} \cup \{0\} = \{x \in \mathbb{R}_0^n : k_\pm(x) \geq 1\} \cup \{0\},$$

and the corresponding maximal operators $\mathcal{M}_+$ and $\mathcal{M}_-$ associated with k_+ and k_-, respectively. In addition, we define the maximal operator $\mathcal{M}_*$ as

$$\mathcal{M}_* u(x) = \max\{\mathcal{M}_+ u(x), \mathcal{M}_- u(x)\} \tag{2.20}$$

and the constant $A_*(k, p)$ as

$$A_*(k, p) = \max\{A(k_+, p), A(k_-, p)\}. \tag{2.21}$$

Theorem 2. *Suppose that* $1 \leq p < (1 - \kappa)^{-1}$. *If* $u \in L^p(\mathbb{R}^n, \mathbb{R})$ *and* $u \geq 0$, *then*

$$|\mathcal{I}u(x)| \leq A_*(k, p)[\mathcal{M}_* u(x)]^{1-(1-\kappa)p} \|u\|_p^{(1-\kappa)p}, \qquad x \in \mathbb{R}^n. \tag{2.22}$$

Proof. Since $u \geq 0$, we clearly have

$$|k * u(x)| \leq \max\{|k_+ * u(x)|, |k_- * u(x)|\}. \tag{2.23}$$

By applying Theorem 1 to k_+ and k_- we get

$$|k_\pm * u(x)| \leq A(k_\pm, p)[\mathcal{M}_\pm u(x)]^{1-(1-\kappa)p} \|u\|_p. \tag{2.24}$$

It remains to combine (2.23) and (2.24) with (2.20) and (2.21).

The reader can easily check that (2.22) is sharp. Moreover, if we assume that the kernel k is an odd function, then $X_- = (-1)X_+$ and

$$\text{vol}(X_+) = \text{vol}(X_-) = \frac{1}{2}\text{vol}(X);$$

hence,

$$A_*(k,p) = 2^{-\kappa}A(k,p). \tag{2.25}$$

$\square$

It goes without saying that Theorem 2 only applies to kernels with nonzero positive and negative parts.

3   Sharp pointwise estimates for the convolution operator associated with the Euclidean Cauchy kernel

In this section we state and prove two results similar to Theorem 1 and Theorem 2 for the Euclidean Cauchy kernel. We begin with a few prerequisites from Clifford analysis. Excellent accounts on the subject can be found in [BDS], [DSS], [GM], [GS], and [Mi].

Throughout the remainder of this article, we will let $\mathfrak{A}_m$ denote the real Clifford algebra associated with the Euclidean space $\mathbb{R}^m$, $m \geq 1$. The algebra $\mathfrak{A}_m$ can be defined as the unital associative real algebra generated by the standard orthonormal basis $\{e_1, e_2, \ldots, e_m\}$ for $\mathbb{R}^m$, subject to the relations

$$e_j e_j + e_j e_i = -2\delta_{ij}e_0, \qquad 1 \leq i, j \leq m, \tag{3.1}$$

where e_0 stands for the identity of $\mathfrak{A}_m$, and δ_{ij} equals 1 or 0 as $i = j$ or $i \neq j$. The dimension of $\mathfrak{A}_m$ as a real vector space equals 2^m, and the set consisting of e_0 and all reduced products,

$$e_I = e_{i_1} \cdot \cdots \cdot e_{i_p}, \quad I = (i_1, \ldots, i_p), \qquad 1 \leq i_1 < \cdots < i_p \leq m, \tag{3.2}$$

yields a basis for $\mathfrak{A}_m$. The Clifford algebra $\mathfrak{A}_m$ is equipped with an inner product such that the basis just defined is orthonormal. The corresponding norm will be denoted by $|\cdot|$. Further, by regarding $\mathfrak{A}_m$ as an algebra of left multiplication operators on $(\mathfrak{A}_m, |\cdot|)$, we convert $\mathfrak{A}_m$ into a real C^*-algebra. The C^*-algebra norm on $\mathfrak{A}_m$ is denoted by $\|\cdot\|$. The involution on $\mathfrak{A}_m$ is provided by Clifford conjugation, which is uniquely determined by the rule

$$\bar{e}_i = -e_i, \qquad 1 \leq i \leq m. \tag{3.3}$$

We next identify any $x = (x_0, x_1, \ldots, x_m)$ in $\mathbb{R}^{m+1}$ with the element $x = x_0 e_0 + x_1 e_1 + \cdots + x_m e_m$ of $\mathfrak{A}_m$, and thus, we get an embedding of $\mathbb{R}^{m+1}$ into $\mathfrak{A}_m$. The two norms $|\cdot|$ and $\|\cdot\|$ on $\mathfrak{A}_m$ defined above induce the Euclidean norm on $\mathbb{R}^{m+1}$.

For a later use, we let $\mathfrak{H}$ denote a unitary left $\mathfrak{A}_m$-module, that is, a finite-dimensional Hilbert space upon which the algebra $\mathfrak{A}_m$ acts on the left such that every generator e_i of $\mathfrak{A}_m$, $1 \leq i \leq m$, determines a skew-adjoint unitary operator, and e_0 corresponds to the identity operator. For the sake of simplicity, the norm on $\mathfrak{H}$ will be denoted by $|\cdot|$. As a typical example we can take $(\mathfrak{A}_m, |\cdot|)$.

Assume next that $\Omega \subseteq \mathbb{R}^{m+1}$ is an open set. The space $C^\infty(\Omega, \mathfrak{H})$ of smooth $\mathfrak{H}$-valued functions on Ω is a left $\mathfrak{A}_m$-module under pointwise multiplication. Therefore, it makes sense to define a linear first-order differential operator $D : C^\infty(\Omega, \mathfrak{H}) \to C^\infty(\Omega, \mathfrak{H})$ by setting

$$D = e_0 \partial_0 + e_1 \partial_1 + \cdots + e_m \partial_m, \tag{3.4}$$

where $\partial_k = \partial/\partial x_k$, $0 \leq k \leq m$. The so-defined operator D is called the *Euclidean Dirac operator* on $C^\infty(\Omega, \mathfrak{H})$. As another piece of notion and terminology, we set

$$\mathfrak{M}(\Omega, \mathfrak{H}) = \{\varphi \in C^\infty(\Omega, \mathfrak{H}) : D\varphi = 0\}$$

and refer to any function φ in $\mathfrak{M}(\Omega, \mathfrak{H})$ as a monogenic—or Clifford analytic—$\mathfrak{H}$-valued function on Ω.

A basic example of a monogenic $\mathfrak{A}_m$-valued function is provided by the *Cauchy kernel* E on $\mathbb{R}^{m+1}$ defined as

$$E(x) = \frac{1}{\sigma_m} \cdot \frac{\bar{x}}{|x|^{m+1}}, \qquad x \in \mathbb{R}_0^{m+1} = \mathbb{R}^{m+1} \setminus \{0\}, \tag{3.5}$$

where σ_m is the area of the unit sphere $\mathbb{S}^m$ in $\mathbb{R}^{m+1}$, and $\bar{x}$ and $|x|$ stand for the Clifford conjugate and the Euclidean norm of a vector x in $\mathbb{R}^{m+1}$, respectively. Actually, E is the fundamental solution of the Dirac operator D.

We are going to examine the convolution operator associated with the Cauchy kernel (3.5). In the sequel, we will let $L^p(\mathbb{R}^{m+1}, \mathcal{H})$ denote the Lebesgue spaces of p-integrable $(1 \leq p < \infty)$ or essentially bounded $(p = \infty)$ $\mathcal{H}$-valued functions on $\mathbb{R}^{m+1}$, where either $\mathcal{H} = \mathbb{R}$, or $\mathcal{H} = \mathfrak{H}$ with $\mathfrak{H}$ an $\mathfrak{A}_m$-module as discussed earlier. The norms on these spaces are denoted by $\|\cdot\|_p$, $1 \leq p \leq \infty$.

We also employ the space $C_0(\mathbb{R}^{m+1}, \mathcal{H})$ of all continuous compactly $\mathcal{H}$-valued functions on $\mathbb{R}^{m+1}$. We define the convolution operator $\mathfrak{J}$ on the space $C_0(\mathbb{R}^{m+1}, \mathcal{H})$ by setting

$$\mathfrak{J}u(x) = E * u(x) = \frac{1}{\sigma_m} \int_{\mathbb{R}^{m+1}} \frac{\bar{y}}{|y|^{m+1}} u(x - y) dy, \qquad x \in \mathbb{R}^{m+1}. \tag{3.6}$$

If $\mathcal{H} = \mathbb{R}$, then $E * u$ is an $\mathbb{R}^{m+1}$-valued function. In the case when $\mathcal{H} = \mathfrak{H}$ we get an $\mathfrak{H}$-valued function. Actually, $\mathfrak{I}$ can be extended to a bounded operator on the Lebesgue spaces $L^p(\mathbb{R}^{m+1}, \mathcal{H})$ for every $1 \leq p < m + 1$. The existence of such an extension follows from the fractional integral theorem in conjunction with the obvious pointwise estimate

$$|\mathfrak{I}u(x)| \leq \frac{1}{\sigma_m} \int_{\mathbb{R}^{m+1}} |y|^{-m} |u(x-y)| dy, \qquad x \in \mathbb{R}^{m+1}. \qquad (3.7)$$

Moreover, applying Hedberg's inequality to the convolution operator in the right-hand side of (3.7), we get

$$|\mathfrak{I}u(x)| \leq \frac{A_{p,q}}{\sigma_m} [Mu(u)]^{p/q} \|u\|_p^{1-p/q}, \qquad x \in \mathbb{R}^{m+1}, \qquad (3.8)$$

for every $u \in L^p(\mathbb{R}^{m+1}, \mathcal{H})$, where $A_{p,q}$ is the constant given by (2.8), M is the classical maximal operator extended to $L^p(\mathbb{R}^{m+1}, \mathcal{H})$ in a natural way, namely, by setting

$$Mu(x) = \sup_{t>0} \frac{1}{\text{vol}(t\mathbb{B}^{m+1})} \int_{t\mathbb{B}^{m+1}} |u(x-y)| dy, \qquad x \in \mathbb{R}^{m+1}, \qquad (3.9)$$

and $1 \leq p < q < \infty$ with $q^{-1} = p^{-1} - (m+1)^{-1}$.

Eventually, we are going to prove that (3.8) is sharp if $\mathcal{H} = \mathfrak{H}$, that is, for vector-valued functions. The constant in (3.8) is no longer the best for real-valued functions. To carry out the task of finding the sharp form of (3.8) for real-valued functions we need to introduce a specific maximal operator associated to the Cauchy kernel.

More precisely, we are going to define a maximal operator $\mathfrak{M}_{\mathcal{H}}$ for $\mathcal{H}$-valued functions in the two situations we are interested in, $\mathcal{H} = \mathbb{R}$ or $\mathcal{H} = \mathfrak{H}$.

To begin with, suppose that e is a unit vector in $\mathbb{R}^{m+1}$ and let $X[e]$ be the set defined by

$$X[e] = \{x \in \mathbb{R}_0^{m+1} : |\langle E(x), e \rangle| \geq 1\} \cup \{0\}. \qquad (3.10)$$

We introduce the maximal operator $\mathfrak{M}_{\mathbb{R}}$ for real-valued functions as

$$\mathfrak{M}_{\mathbb{R}} u(x) = \sup_{e \in \mathbb{S}^m} \sup_{t>0} \frac{1}{\text{vol}(tX[e])} \int_{tX[e]} |u(x-y)| dy, \qquad x \in \mathbb{R}^{m+1}. \qquad (3.11)$$

Next, we define the set $X[\mathbb{S}^m]$ by

$$X[\mathbb{S}^m] = \bigcup_{e \in \mathbb{S}^m} X[e] \qquad (3.12)$$

and introduce the maximal operator $\mathfrak{M}_{\mathfrak{H}}$ for $\mathfrak{H}$-valued functions as

$$\mathfrak{M}_{\mathfrak{H}} u(x) = \sup_{t>0} \frac{1}{\text{vol}(tX[\mathbb{S}^m])} \int_{tX[\mathbb{S}^m]} |u(x-y)| dy, \qquad x \in \mathbb{R}^{m+1}. \qquad (3.13)$$

We should notice that the sets $X[e]$, where $e \in \mathbb{S}^m$, have equal Lebesgue measure. In other words,

$$\text{vol}\, X[e] = \text{vol}\, X[e_0], \qquad e \in \mathbb{S}^m, \tag{3.14}$$

where e_0 is the first unit vector in the standard orthonormal basis for $\mathbb{R}^{m+1}$. We recall that as an element of $\mathfrak{A}_m$, e_0 equals the identity. To prove (3.14) we observe that

$$X[e] = \{x \in \mathbb{R}^{m+1} : |\langle x, \bar{e} \rangle| \geq \sigma_m |x|^{m+1}\}. \tag{3.15}$$

This new description of $X[e]$ shows that if T is an orthogonal transformation on $\mathbb{R}^{m+1}$ such that $T\bar{e} = e_0$, then $TX[e] = X[e_0]$, so $X[e]$ and $X[e_0]$ have equal volume.

We claim that $X[\mathbb{S}^m]$ is a ball of radius $\rho = (\sigma_m)^{-1/m}$, namely,

$$X[\mathbb{S}^m] = (\sigma_m)^{-1/m} \mathbb{B}^{m+1}. \tag{3.16}$$

The inclusion $X[e] \subseteq (\sigma_m)^{-1/m} \mathbb{B}^{m+1}$, $e \in \mathbb{S}^m$, follows easily from (3.15). On the other hand, if $x \in (\sigma_m)^{-1/m} \mathbb{B}^{m+1}$ and $x \neq 0$, then using (3.15) once more we get $x \in X[\bar{x}/|x|]$.

Based on (3.16) we conclude that the maximal operator $\mathfrak{M}_{\mathfrak{H}}$ equals the classical maximal operator M defined by (3.9). As an alternative description of $X[\mathbb{S}^m]$, we have

$$X[\mathbb{S}^m] = \{x \in \mathbb{R}_0^{m+1} : |E(x)| \geq 1\} \cup \{0\}. \tag{3.17}$$

Finally, we introduce the constants $A_{\mathcal{H}}(E, p)$, where $1 \leq p < m+1$ and $\mathcal{H} = \mathbb{R}$ or $\mathcal{H} = \mathfrak{H}$ by setting

$$A_{\mathbb{R}}(E, p) = \frac{m+1}{p} \left[\frac{mp}{m+1-p} \cdot \text{vol}(X[e_0]) \right]^{m/(m+1)} \tag{3.18}$$

and

$$A_{\mathfrak{H}}(E, p) = \frac{m+1}{p} \left[\frac{mp}{m+1-p} \cdot \text{vol}(X[\mathbb{S}^m]) \right]^{m/(m+1)}. \tag{3.19}$$

We are now in a position to prove the following Hedberg type inequality.

Theorem 3. *Suppose that $1 \leq p < m+1$. Then,*

$$|\mathfrak{I}u(x)| \leq A_{\mathcal{H}} [\mathfrak{M}_{\mathcal{H}} u(x)]^{(m+1-p)/(m+1)} \|u\|_p^{p/(m+1)}, \qquad x \in \mathbb{R}^{m+1}, \tag{3.20}$$

for every $u \in L^p(\mathbb{R}^{m+1}, \mathcal{H})$, with $A_{\mathcal{H}} = A_{\mathcal{H}}(E, p)$. Moreover, inequality (3.20) is sharp.

Proof. Assume first that $u \in L^p(\mathbb{R}^{m+1}, \mathbb{R})$. Since $\mathfrak{J}u(x) \in \mathbb{R}^{m+1}$, we clearly have

$$|\mathfrak{J}u(x)| = \sup_{e \in \mathbb{S}^m} |\langle \mathfrak{J}u(x), e \rangle|. \tag{3.21}$$

Next we notice that

$$\langle \mathfrak{J}u(x), e \rangle = k * u(x), \tag{3.22}$$

where k is the real-valued homogeneous kernel on $\mathbb{R}^{m+1}$ defined by $k(x) = \langle E(x), e \rangle$, $x \in \mathbb{R}^{m+1}$. The compact set X associated with k by (2.4) in Section 2 equals the set $X[e]$. Applying Theorem 1 to this kernel, and observing that $\kappa = m/(m+1)$ and $\mathrm{vol}\, X = \mathrm{vol}\, X[e] = \mathrm{vol}\, X[e_0]$, we end up with

$$|k * u(x)| \leq A(k,p)[\mathcal{M}u(x)]^{(m+1-p)/(m+1)}\|u\|_p^{p/(m+1)}, \tag{3.23}$$

where $A(k,p) = A_{\mathbb{R}}(E,p)$, and

$$\mathcal{M}u(x) = \sup_{t>0} \frac{1}{\mathrm{vol}(tX[e])} \int_{tX[e]} |u(x-y)|dy, \qquad x \in \mathbb{R}^n. \tag{3.24}$$

Inequality (3.20) for $\mathcal{H} = \mathbb{R}$ follows by assembling (3.21)–(3.24) and (3.11).

The proof of (3.20) for $u \in L^p(\mathbb{R}^{m+1}, \mathfrak{H})$ is even shorter. It also relies on Theorem 1. From (3.7) we get

$$|\mathfrak{J}u(x)| \leq k * |u|(x), \tag{3.25}$$

where k is the real-valued homogeneous kernel defined by $k(x) = |E(x)|$, $x \in \mathbb{R}^{m+1}$. By (3.17) we have that the compact set X associated with k equals $X[\mathbb{S}^m]$. Therefore, the maximal operator $\mathcal{M}$ and the constants $A(k,p)$ corresponding to k coincide with $\mathfrak{M}_{\mathfrak{H}}$ and $A_{\mathfrak{H}}(E,p)$. Inequality (3.20) follows now from (2.6).

We can prove that (3.20) is sharp for real-valued functions merely based on the fact that (2.6) is sharp.

To check that (3.20) is sharp for vector-valued functions and $p > 1$, we take the $\mathfrak{A}_m$-valued function u defined as

$$u(x) = \begin{cases} -x/|x|, & \text{if } |x| \leq 1 \text{ and } x \neq 0, \\ e_0, & \text{if } x = 0, \\ -x/|x|^{1+m/(p-1)}, & \text{if } |x| > 1, \end{cases}$$

and observe that $\bar{x} \cdot x = |x|^2$; hence,

$$\mathfrak{J}u(0) = \frac{1}{\sigma_m} \left[\int_{\mathbb{B}^{m+1}} |y|^{-m} dy + \int_{\mathbb{R}^{m+1} \setminus \mathbb{B}^{m+1}} |y|^{-mp/(p-1)} dy \right] e_0.$$

Using polar coordinates we have

$$\Im u(0) = \left[\int_0^1 1 d\rho + \int_1^\infty \rho^{-m/(p-1)} d\rho \right] e_0 = \frac{m}{m+1-p} \cdot e_0,$$

that is,

$$|\Im u(0)| = \frac{m}{m+1-p}. \tag{3.26}$$

Since $|u(x)| = 1$ if $|x| \le 1$ and $|u(x)| = |x|^{-m/(p-1)} < 1$ for $|x| > 1$, a few direct calculations show that

$$\mathfrak{M}_{\mathfrak{H}} u(0) = 1 \tag{3.27}$$

and

$$\|u\|_p^p = \mathrm{vol}(\mathbb{B}^{m+1}) + \sigma_m \cdot \frac{p-1}{m+1-p} = \frac{mp}{(m+1)(m+1-p)} \cdot \sigma_m. \tag{3.28}$$

From (3.16) we get

$$\mathrm{vol}\, X[\mathbb{S}^m] = (\sigma_m)^{-(m+1)/m} \mathrm{vol}(\mathbb{B}^{m+1})$$

$$= (\sigma_m)^{-(m+1)/m} \cdot \frac{\sigma_m}{m+1} = \frac{(\sigma_m)^{-1/m}}{m+1};$$

hence, by using (3.19),

$$A_{\mathfrak{H}}(E,p) = \frac{m+1}{p} \left[\frac{mp}{m+1-p} \cdot \frac{(\sigma_m)^{-1/m}}{m+1} \right]^{m/(m+1)}. \tag{3.29}$$

We still must check that

$$|\Im u(0)| = A_{\mathfrak{H}}(E,p) [\mathfrak{M}_{\mathfrak{H}} u(0)]^{(m+1-p)/(m+1)} \|u\|_p^{p/(m+1)}.$$

To show that (3.20) is sharp in the case $p = 1$, we can use as an extreme function the $\mathfrak{A}_m$-valued function u defined as

$$u(x) = \begin{cases} -x/|x|, & \text{if } |x| \le 1 \text{ and } x \ne 0, \\ e_0, & \text{if } x = 0, \\ 0, & \text{if } |x| > 1. \end{cases}$$

Actually, for this function we get all the equations (3.25)–(3.26) with $p = 1$.
$\square$

We conclude this section by refining Theorem 3 for non-negative functions.

Since the kernel E is odd, we can take advantage of the comments made at the end of Section 2. To be specific, motivated by these comments we start by defining the compact sets

$$X_{\pm}[e] = \{x \in \mathbb{R}_0^{m+1} : \pm\langle E(x), e\rangle \geq 1\} \cup \{0\} \qquad (3.30)$$

for any $e \in \mathbb{S}^{m+1}$, and the maximal operators

$$\mathfrak{M}_{\pm}u(x) = \sup_{e\in\mathbb{S}^m} \sup_{t>0} \frac{1}{\text{vol}(tX_{\pm}[e])} \int_{tX_{\pm}[e]} |u(x - y)|dy, \quad x \in \mathbb{R}^{m+1}. \quad (3.31)$$

Since $X_-[e] = (-1)X_+[e] = X_+[-e]$, we get that $\mathfrak{M}_+ = \mathfrak{M}_-$, and let $\mathfrak{M}_*$ denote whichever operator $\mathfrak{M}_+$ or $\mathfrak{M}_-$.

In addition, we introduce the constant $A_*(E,p)$ as

$$A_*(E,p) = 2^{-m/(m+1)}A_{\mathbb{R}}(E,p). \qquad (3.32)$$

The line of argument development in the first part of the proof of Theorem 3 can be now adjusted to derive the following analog of Theorem 2 for the Euclidean Cauchy kernel. The proof is left to the reader.

Theorem 4. *Suppose that $1 \leq p < m + 1$. If $u \in L^p(\mathbb{R}^n, \mathbb{R})$ and $u \geq 0$; then*

$$|\mathfrak{J}u(x)| \leq A_*(E,p)[\mathfrak{M}_*u(x)]^{(m+1-p)/(m+1)}\|u\|_p^{p/(m+1)}, \quad x \in \mathbb{R}^{m+1}. \quad (3.33)$$

Moreover, inequality (3.33) is sharp.

Before concluding this section, we want to mention that a great deal of inequalities similar to (3.20) and (3.33) are proved in Chapter 3 of the monograph by Gürlebeck and Sprössig [GS], with a particular emphasis on properties of the Teodorescu transform.

4   Consequences and applications

This section points out a few consequences of Theorems 3 and 4 that are of interest in their own. The notation is the same as in Sections 1 and 2.

It is obvious that the maximal operator $\mathfrak{M}_{\mathcal{H}}$ is a contraction on the Lebesgue space $L^\infty(\mathbb{R}^{m+1}, \mathcal{H})$. Using this remark and Theorems 3 and 4, we get the following uniform estimates for the convolution operator $\mathfrak{J}$.

Proposition 1. *Suppose that $1 \leq p < m + 1$.*

(i) *If $u \in L^p(\mathbb{R}^{m+1}, \mathcal{H}) \cap L^\infty(\mathbb{R}^{m+1}, \mathcal{H})$, then*

$$\|\mathfrak{J}u\|_\infty \leq A_{\mathcal{H}}(E,p)\|u\|_\infty^{(m+1-p)/(m+1)}\|u\|_p^{p/(m+1)}. \qquad (4.1)$$

(ii) If $u \in L^p(\mathbb{R}^{m+1}, \mathbb{R}) \cap L^\infty(\mathbb{R}^{m+1}, \mathbb{R})$ and $u \geq 0$, then

$$\|\mathfrak{I}u\|_\infty \leq A_*(E, p)\|u\|_\infty^{(m+1-p)/(m+1)}\|u\|_p^{p/(m+1)}. \qquad (4.2)$$

In the case $p = 1$, from Proposition 1 we recover the higher-dimensional Ahlfors-Beurling type inequalities proved by Martin [M]. If $p = 1$ and $m = 1$, then E is the classical Cauchy kernel used in one-variable complex analysis, and assertion (ii) in Proposition 1 reduces to a result due to Putinar [P]. Under the same assumptions, if u is the characteristic function of a compact subset K of the complex plane $\mathbb{C}$, then assertion (ii) provides the original Ahlfors-Beurling inequality (see [AB]).

As a direct generalization of the Ahlfors-Beurling inequality, we indicate the following corollary to Proposition 1.

Corollary 1. *Suppose that K is a compact subset of $\mathbb{R}^{m+1}$. Then*

$$\left| \int_K E(x-y)dy \right|$$
$$\leq 2^{-m/(m+1)}(m+1)[\mathrm{vol}(X[e_0])]^{m/(m+1)}[\mathrm{vol}(K)]^{1/(m+1)} \qquad (4.3)$$

for any $x \in \mathbb{R}^{m+1}$.

The Ahlfors-Beurling inequality has been used by Alexander [A1–2] to prove a quantitative version of the Hartogs-Rosenthal theorem. Theorems 3 and 4 can be used to derive higher-dimensional analogues of Alexander's result. We proceed with some prerequisites.

We let $K \subset \mathbb{R}^{m+1}$ be a compact set and let $C(K, \mathfrak{H})$ denote the Banach space of all $\mathfrak{H}$-valued continuous functions on K. The norm on $C(K, \mathfrak{H})$ is given by

$$\|\varphi\|_{\infty, K} = \sup_{x \in K} |\varphi(x)|, \qquad \varphi \in C(K, \mathfrak{H}).$$

For later use we also introduce the norms $\|\cdot\|_{p,K}$, $1 \leq p < m+1$, defined in the standard way, namely,

$$\|\varphi\|_{p,K} = \left[\int_K |\varphi(x)|^p dx \right]^{1/p}.$$

In addition, we consider the space $C^\infty(K, \mathfrak{H})$, consisting of restrictions to K of smooth $\mathfrak{H}$-valued functions defined on open neighborhoods of K in $\mathbb{R}^{m+1}$, and the subspace $\mathfrak{M}(K, \mathfrak{H})$ of $C^\infty(K, \mathfrak{H})$, consisting of restrictions to K of functions from $\mathfrak{M}(\Omega, \mathfrak{H})$, where $\Omega \subset \mathbb{R}^{m+1}$ is an arbitrary open neighborhood of K. We recall that $\mathfrak{M}(\Omega, \mathfrak{H})$ stands for the space of $\mathfrak{H}$-valued monogenic functions on Ω.

Suppose now that Ω and Δ are two bounded open sets, with $K \subset \Delta$ and $\bar{\Delta} \subset \Omega$, assume that Δ has a smooth boundary $\partial\Delta$, and let $\eta(x) \in \mathbb{R}^{m+1} \subseteq \mathfrak{A}_m$ be the inward pointing unit normal to $\partial\Delta$ at $x \in \partial\Delta$. To every function $\varphi \in C^\infty(\Omega, \mathfrak{H})$, we associate two $\mathfrak{H}$-valued smooth functions $C_{\partial\Delta}\varphi$ and $C_\Delta\varphi$ on Δ by setting

$$C_{\partial\Delta}\varphi(x) = \int_{\partial\Delta} E(x-y)\eta(y)\varphi(y)d\sigma(y), \qquad x \in \Delta, \qquad (4.4)$$

and

$$C_\Delta\varphi(x) = \int_\Delta E(x-y)D\varphi(y)dy, \qquad x \in \Delta, \qquad (4.5)$$

where D is the Dirac operator and σ stands for the surface area measure on $\partial\Delta$. The two so-defined functions are related to the original function φ by the Borel-Pompeiu formula, namely,

$$\varphi|\Delta = C_{\partial\Delta}\varphi + C_\Delta\varphi. \qquad (4.6)$$

It should be pointed out that $C_{\partial\Delta}\varphi \in \mathfrak{M}(\Delta, \mathfrak{H})$ for any $\varphi \in C^\infty(\Omega, \mathfrak{H})$. Therefore, $\varphi \mid \Delta - C_\Delta\varphi \in \mathfrak{M}(\Delta, \mathfrak{H})$ for any $\varphi \in C^\infty(\Omega, \mathfrak{H})$.

Our objective is to estimate the distance in $C(K, \mathfrak{H})$ from such a function φ to the subspace $\mathfrak{M}(K, \mathfrak{H})$ or, equivalently, to the closure of this subspace in $C(K, \mathfrak{H})$.

Proposition 2. *Suppose that $\varphi \in C^\infty(K, \mathfrak{H})$. Then*

$$\text{dist}_{C(K,\mathfrak{H})}(\varphi, \mathfrak{M}(K, \mathfrak{H}))$$
$$\leq A_\mathfrak{H}(E, p)\|D\varphi\|_{\infty,K}^{(m+1-)/(m+1)}\|D\varphi\|_{p,K}^{p/(m+1)} \qquad (4.7)$$

for each $1 \leq p < m+1$.

Proof. With no loss of generality we may assume that $\varphi \in C^\infty(\Omega, \mathfrak{H})$, where Ω is an open neighborhood of K. We let Δ be an open set as in (4.4)–(4.6) and define $\psi \in \mathfrak{M}(\Delta, \mathfrak{H})$ as $\psi = \varphi \mid \Delta - C_\Delta\varphi$. Since $\varphi \mid \Delta - \psi = C_\Delta\varphi$, we get

$$\text{dist}_{C(K,\mathfrak{H})}(\varphi, \mathfrak{M}(K, \mathfrak{H})) \leq \|C_\Delta\varphi\|_{\infty,K}. \qquad (4.8)$$

Further, let $u \in L^p(\mathbb{R}^{m+1}, \mathfrak{H}) \cap L^\infty(\mathbb{R}^{m+1}, \mathfrak{H})$ be the function defined as

$$u(x) = \begin{cases} D\varphi(x), & \text{if } x \in \bar{\Delta}, \\ 0, & \text{if } x \notin \bar{\Delta}, \end{cases} \qquad (4.9)$$

and observe that $C_\Delta\varphi(x) = E * u(x)$ for any $x \in \Delta$. By Proposition 1 we get

$$\|C_\Delta\varphi(x)\|_{\infty,K} \leq A_\mathfrak{H}(E, p)\|u\|_\infty^{(m+1-p)/(m+1)}\|u\|_p^{p/(m+1)}. \qquad (4.10)$$

Inequality (4.7) follows from (4.8), (4.9), and (4.10) by letting Δ approach K. $\qquad\square$

Corollary 2. *Under the same assumptions as in Proposition 1, we have*

$$\text{dist}_{C(K,\mathfrak{H})}(\varphi, \mathfrak{M}(K,\mathfrak{H})) \leq A_{\mathfrak{H}}(E,1)[\text{vol}(K)]^{1/(m+1)}\|D\varphi\|_{\infty,K}. \quad (4.11)$$

Proof. Set $p = 1$ in (4.7) and use the obvious estimate

$$\|D\varphi\|_{1,K} \leq \|D\varphi\|_{\infty,K} \cdot \text{vol}(K).$$

$\square$

Corollary 3. *If $K \subset \mathbb{R}^{m+1}$ is compact with Lebesgue measure zero, then $\mathfrak{M}(K,\mathfrak{H})$ is dense in $C(K,\mathfrak{H})$.*

Proof. From (4.11) we have

$$\text{dist}_{C(K,\mathfrak{H})}(\varphi, \mathfrak{M}(K,\mathfrak{H})) = 0$$

for any $\varphi \in C^\infty(K,\mathfrak{H})$. On the other hand, by the Stone-Weierstrass theorem we get that $C^\infty(K,\mathfrak{H})$ is dense in $C(K,\mathfrak{H})$, an observation that concludes the proof. $\square$

This corollary is just a substitute for the Hartogs-Rosenthal theorem in Clifford analysis. For a different proof we refer to [BDS, 8].

We next refine Proposition 2 for a specific $\mathfrak{A}_m$-valued function.

Proposition 3. *Considering $\bar{x}$ as a function in $C(K, \mathfrak{A}_m)$, we have*

$$\text{dist}_{C(K,\mathfrak{A}_m)}(\bar{x}, \mathfrak{M}(K, \mathfrak{A}_m))$$
$$\leq 2^{-m/(m+1)}(m+1)^2[\text{vol}(X[e_0])]^{m/(m+1)}[\text{vol}(K)]^{1/(m+1)}.$$

Proof. We let φ denote the function defined as $\varphi(x) = \bar{x}$ and observe that $D\varphi(x) = (m+1)e_0$ for each $x \in \mathbb{R}^{m+1}$. Arguing as in the proof of Proposition 2 above, we get that

$$\text{dist}_{C(K,\mathfrak{A}_m)}(\bar{x}, \mathfrak{M}(K, \mathfrak{A}_m)) \leq |\mathfrak{I}u|_{\infty,K}, \quad (4.12)$$

where u is given by

$$u(x) = \begin{cases} (m+1)e_0 & \text{if } x \in K, \\ 0 & \text{if } x \notin K. \end{cases}$$

We next notice that

$$\mathfrak{I}u(x) = (m+1)\left[\int_K E(x-y)dy\right]e_0, \qquad x \in K;$$

hence,

$$|\mathfrak{I}u(x)|_{\infty,K} = (m+1)\sup_{x \in K}\left|\int_K E(x-y)dy\right|. \quad (4.13)$$

Proposition 3 follows from (4.12) and (4.13) in conjunction with (4.3). $\square$

Alexander's inequality results from Proposition 3 by assuming that $m = 1$. In this particular situation, $\mathfrak{M}(K, \mathfrak{A}_m)$ is the space of all rational functions analytic on open neighborhoods of K. This inequality supplies an important tool in studying rational approximation (see, for instance, [G] or [B]). More general results on approximation by solutions of elliptic equations can be found in Browder [B1–2], Gustafsson and Khavinson [GK], Khavinson [K1–2], Martin and Szeptycki [MSz], Tarkhanov [T], and Weinstock [W].

REFERENCES

[AB] Ahlfors, L. and A. Beurling, A., Conformal invariants and function theoretic null sets, *Acta Math.* **83** (1950), 101–129.

[A1] Alexander, H., Projections of polynomial hulls, *J. Funct. Anal.* **13** (1973), 13–19.

[A2] Alexander, H., On the area of the spectrum of an element of a uniform algebra, *Complex Approximation Proceedings, Quebec, July 3–8, 1978*, Birkhäuser, 1980, 3–12.

[B] Browder, A., *Introduction to Function Algebras*, Benjamin, New York, 1969.

[B1] Browder, F., Approximation by solutions of partial differential equations, *Amer. J. Math.* **84** (1962), 134–160.

[B1] Browder, F., Functional analysis and partial differential equations. II, *Math. Ann.* **145** (1961), 81–226.

[BDS] Brachx, F., Delanghe, R., and F. Sommen, *Clifford Analysis*, Pitman Research Notes in Mathematics Series, 76, 1982.

[DSS] Delanghe, R., Sommen, F., and V. Souček, *Clifford Algebra and Spinor-Valued Functions*, Kluwer Academic Publishers, 1992.

[G] Gamelin, T. W., *Uniform Algebras*, Prentice Hall, 1969.

[GM] Gilbert, J. E. and Murray, M. A. M., *Clifford Algebras and Dirac Operators in Harmonic Analysis*, Cambridge Studies in Advanced Mathematics, 26, Cambridge University Press, 1991.

[GK] Gustafsson, B. and Khavinson, D., On approximation by harmonic vector fields, *Houston J. Math.* **20** (1994), 75–92.

[GS] Gürlebeck, K. and Sprössig, W., *Quaternionic and Clifford Calculus for Physicists and Engineers*, John Wiley & Sons, New York, 1997.

[He] Hedberg, L., On certain convolution inequalities, *Proc. Amer. Math. Soc.* **36** (1972), 505–510.

[Hö] Hörmander, L., *The Analysis of Linear Partial Differential Operators, vol. I: Distribution Theory and Fourier Analysis*, Springer-Verlag, Berlin, 1983.

[K1] Khavinson, D., On uniform approximation by harmonic functions, *Mich. Math. J.* **34** (1987), 465–473.

[K2] Khavinson, D., Duality and uniform approximation by solutions of elliptic equations, *Operator Theory: Advances and Applications* **35** (1988), Birkhäuser Verlag, Basel, 129–141.

[L]   Lieb, E. H., Sharp constants in the Hardy-Littlewood-Sobolev and related inequalities, *Annals of Math.* **118** (1983) 349–379.

[M]   Martin, M., Higher-dimensional Ahlfors-Beurling inequalities *Proc. Amer. Math. Soc.* **126** (1998), 2863–2871.

[MSz] Martin, M. and Szeptycki, P., Sharp inequalities for convolution operators with homogeneous kernels and applications, *Indiana Univ. Math.* **46** (1997), 975–988.

[Mi]   Mitrea, M., *Singular Integrals, Hardy Spaces, and Clifford Wavelets*, Lecture Notes in Mathematics, 1575, Springer-Verlag, Heidelberg, 1994.

[P]   Putinar, M., Extreme hyponormal operators, *Operator Theory: Advances and Applications* **28** (1988), 249–265.

[S1]   Stein, E. M., *Singular Integrals and Differentiability Properties of Functions*, Princeton Univ. Press, Princeton, 1970.

[S2]   Stein, E. M. *Harmonic Analysis: Real-Variable Methods, Orthogonality, and Oscillatory Integrals*, Princeton Univ. Press, Princeton, NJ, 1993.

[T]   Tarkhanov, N. N., *The Cauchy Problem for Solutions of Elliptic Equations*, Akademie Verlag, Berlin, 1995.

[W]   Weinstock, B. M., Uniform approximations by solutions of elliptic equations, *Proc. Amer. Math. Soc.* **41** (1973), 513–517.

Mircea Martin
Department of Mathematics
Baker University
Baldwin City, KS 66006 U.S.A.
E-mail: mmartin@harvey.bakeru.edu

Received: September 30, 1999; Revised: January 3, 2000

3.

APPLICATIONS IN GEOMETRY AND PHYSICS

A Borel-Pompeiu Formula in $\mathbb{C}^n$ and Its Application to Inverse Scattering Theory

Swanhild Bernstein

ABSTRACT We develop a Borel-Pompeiu formula for functions in several complex variables using Clifford analysis. The obtained formula contains the Bochner-Martinelli formula and additional information. The Borel-Pompeiu formula will be used for a new inverse scattering transform in multidimensions.
Keywords: Clifford analysis, higher dimensional complex analysis, inverse scattering theory.

1 Introduction

1. In one-dimensional complex analysis the following Borel-Pompeiu formula (sometimes also called Cauchy-Green formula or generalized Cauchy formula)

$$\frac{1}{2\pi i}\int_G \frac{\bar{\partial}m(z)}{z-z_0}dz\wedge d\bar{z} + \frac{1}{2\pi i}\int_\Gamma \frac{m(z)}{z-z_0}dz = \begin{cases} m(z_0), & z_0\in G, \\ 0, & z_0\in\mathbb{C}\backslash\{0\} \end{cases}$$

is used. Here Γ is a closed curve in $\mathbb{C}$ and boundary of the smooth domain G, m is a smooth complex-valued function. This formula is important for a lot of applications because it allows us to reconstruct a function from its $\bar{\partial}$-derivative and boundary values. Unfortunately, there exists no formula of this type in $\mathbb{C}^n$. But in real Clifford analysis there exists such a formula which is completely analogous:

$$\frac{1}{\omega_n}\int_\Gamma \frac{(\bar{x}-\bar{y})}{|x-y|^n}n(y)u(y)d\Gamma_y$$

$$-\frac{1}{\omega_n}\int_G \frac{(\bar{x}-\bar{y})}{|x-y|^n}(Du)(y)dy = \begin{cases} u(x), & x\in G \\ 0, & x\in\mathbb{R}^n\backslash\{0\}, \end{cases}$$

AMS Subject Classification: 32A26, 30G35, 47A40.

where G is a bounded domain in $\mathbb{R}^n$ bounded by Γ and D denotes the so-called Dirac operator. Both operators, the $\bar{\partial}$-operator and the Dirac operator, are elliptic operators and both formulas can be obtained by partial integration. In complex Clifford analysis there exist several analogues to this type of formula (see [21]). One possible way is to use a complexified Dirac operator $D_x + iD_y$ in $\mathbb{C}^n \sim \mathbb{R}^{2n}$. Unfortunately, this operator is not an elliptic operator. But we will be able to obtain a Borel-Pompeiu formula by partial integration in $\mathbb{C}^n \sim \mathbb{R}^{2n}$.

2. In the theory of several complex variables the role of the Cauchy kernel is played by Koppelman's or the Bochner-Martinelli kernel. The Bochner-Martinelli kernel is not holomorphic itself but universal. Koppelman's kernel is holomorphic but not universal. Our investigations lead to the Bochner-Martinelli formula.

3. One-dimensional complex analysis can be extended to higher dimensions in different ways. One way are several complex variables which preserves the fact that the multiplication is commutative and the product of holomorphic functions is again a holomorphic function. Another way is Clifford analysis in $\mathbb{R}^n$ where the multiplication is non-commutative and the product of monogenic functions need not be a monogenic function, but there exists a Cauchy kernel which is reproducing, universal and monogenic.

4. Considering Clifford analysis helps to understand a lot of difficult problems in several complex variables from a very natural viewpoint. For example, if G is a domain in $\mathbb{C}^n \sim \mathbb{R}^{2n}$ with C^2-boundary and defining function ρ such that the gradient of ρ vanishes at no point of $\partial G = \Gamma$, i.e., the vector

$$N(z) = (\bar{\partial}_1 \rho(z), \dots, \bar{\partial}_n \rho(z)) \sim \sum_{j=1}^{n} \bar{\partial}\rho(z)e_j$$

satisfies $N(z) \neq 0$ for every $z \in \Gamma$. The integral along the boundary

$$\int_{\Gamma} m(\xi, \eta)d\sigma(KD_\xi)((\xi, \eta) - (x, y)) - i(D_\eta K)((\xi, \eta) - (x, y))d\sigma m(\xi, \eta)$$

$$= \frac{1}{\omega_{2n-1}} \int_{\Gamma} m(\xi, \eta) \frac{(\langle \xi - x, n_\xi + in_\eta \rangle - i\langle \eta - y, n_\xi + in_\eta \rangle)}{|(\xi, \eta) - (x, y)|^{2n-1}} d\Gamma_{\xi, \eta}$$

$$+ \int_{\Gamma} K((\xi, \eta) - (x, y)) \frac{1}{|\overline{\mathcal{D}}\rho(\xi, \eta)|} L_{jk} m(\xi, \eta) d\Gamma_{\xi, \eta}$$

where $K(\xi, \eta)$ is a fundamental solution of the Laplacian in $\mathbb{R}^{2n}$ and

$$L_{jk} = \bar{\partial}_k \rho \bar{\partial}_j - \bar{\partial}_j \rho \bar{\partial}_k, \ 1 \leq j < k \leq n,$$

denotes the tangential Cauchy-Riemann operator. Thus, the first integral is the Bochner-Martinelli formula, and the second integral vanishes if the function m can be extended holomorphically from Γ into the domain. First steps toward the theory of complex Clifford analysis were done in [18].

5. It is an interesting question to ask for generalizations of derivatives and their meaning in higher dimensions. This will not be a part of our considerations. We recommend the paper [14] which explains the interaction among several notions of "derivatives" in $\mathbb{C}^2$ and the quaternions. In fact the ideas developed there can be extended also to higher dimensions.

6. Another topic we don't want to discuss in detail is the existence and behavior of singular integrals on Lipschitz surfaces. This was considered in [12], [13]. They found and used methods from Clifford analysis and also several complex variables to get their results.

2 Preliminaries

We want to start with the real Clifford algebra $C\ell_{0,n}$ which is generated by the elements $e_1, e_2, \ldots, e_n$. These generating elements fulfill the *non-commutative* multiplication rule

$$e_i e_j + e_j e_i = -2\delta_{ij}, \ i, j = 1, 2, \ldots, n,$$

the unit element of this algebra is denoted by e_0 and commutes with all generating elements, i.e., $e_0 e_i = e_i e_0$, obviously $e_0^2 = 1$. Thus, we can identify $e_0 \equiv 1$. An arbitrary element of the algebra is now a linear combination of all possible products of generating elements, more exactly,

$$a = \sum_\alpha a_\alpha e_\alpha,$$

where $e_\alpha = e_{\alpha_1 \alpha_2 \ldots \alpha_k} = e_{\alpha_1} \cdot e_{\alpha_2} \cdot \ldots \cdot e_{\alpha_k}$ and $1 \leq \alpha_1 < \alpha_2 < \ldots \alpha_k \leq n$, $\alpha_j \in \{1, 2, \ldots, n\}$. We identify $e_\emptyset = e_0$. The part $Sc(a) = a_0 e_0 = a_0$ is called the *scalar-part* and $Vec(a) = a - Sc(a)$ is called the *vector-part* of a.

In the real Clifford algebra there is a conjugation also called *main involution*. We have

$$\bar{a} = \sum_\alpha a_\alpha \bar{e}_\alpha,$$

and $\bar{e}_\alpha = \bar{e}_{\alpha_1 \alpha_2 \ldots \alpha_k} = \bar{e}_{\alpha_k} \cdot \bar{e}_{\alpha_{k-1}} \cdot \ldots \cdot \bar{e}_{\alpha_1}$. For the generating elements we have $\bar{e}_j = -e_j$ and for the unit element $\bar{e}_0 = e_0$. We will denote by f an arbitrary function defined on $\mathbb{R}^n$ as

$$f(x) = \sum_\alpha f_\alpha(x) e_\alpha,$$

where $x \in \mathbb{R}^n$ and $f(x) \in \mathbb{R}$. Again, we denote by $Sc(f) = f_0(x)e_0 = f_0(x)$ the scalar-part of f. All properties we address to f should be fulfilled by all functions f_α. We introduce a first order differential operator, the Dirac operator, which plays the role of the $\bar{\partial}$ operator in complex analysis:

$$D = \sum_{j=1}^{n} e_j \frac{\partial}{\partial x_j}$$

Due to the non-commutative multiplication we have to distinguish among functions with $Df = 0$, which will be called *left-monogenic*, and functions with $fD = 0$, which will be called *right-monogenic*. Due to

$$D\bar{D} = \bar{D}D = \left(\sum_{j=1}^{n} \bar{e}_j \frac{\partial}{\partial x_j} \right) \left(\sum_{j=1}^{n} e_j \frac{\partial}{\partial x_j} \right) = \sum_{j=1}^{n} \frac{\partial^2}{\partial x_j^2} = \Delta$$

(left- and right-) monogenic functions are harmonic. We want to give an overview of the properties of functions in one complex variable, several complex variables and (real) Clifford analysis in the following table. A fundamental solution of D is $\frac{1}{\omega_n} \frac{-x}{|x|^n}$. Using the fundamental solution we are able to state a Borel-Pompeiu formula, and we can compute the Plemelj-Sokhotzkij formulas. Let G be a smooth domain in $\mathbb{R}^n$ with boundary Γ and $u \in C^1(\bar{G})$ scalar-valued or with values in a Clifford algebra. Then a left-inverse operator to D is given by

$$Tu = \frac{-1}{\omega_n} \int_G \frac{(\bar{x} - \bar{y})}{|x - y|^n} u(y) dy.$$

We also use the Cauchy-type operators

$$Fu = \frac{1}{\omega_n} \int_\Gamma \frac{(\bar{x} - \bar{y})}{|x - y|^n} n(y) u(y) d\Gamma, \; x \notin \Gamma,$$

and

$$Su = \frac{2}{\omega_n} \int_\Gamma \frac{(\bar{x} - \bar{y})}{|x - y|^n} n(y) u(y) d\Gamma, x \in \Gamma.$$

Then the Borel-Pompeiu formula can be written as

$$Fu + TDu = \begin{cases} u, & \text{in } G, \\ 0, & \text{in } \mathbb{R}^n \backslash \bar{G}. \end{cases}$$

We have the Plemelj-Sokhotzkij formulas

$$\lim_{x \in G \to x_0 \in \Gamma} (Fu)(x) = \frac{1}{2}(I + S)u(x_0)$$

and

$$\lim_{x \notin G \to x_0 \in \Gamma} (Fu)(x) = -\frac{1}{2}(I - S)u(x_0).$$

A very important property of the singular Cauchy-operator S is that $S^2 = I$. The Bochner-Martinelli kernel does not in general lead to an integral operator with this property! The existence of appropriate operators in Clifford analysis allows one to study the Riemann-Hilbert problems using singular integral equations. Unfortunately, the results there are not as nice as in the one dimensional complex analysis. For details see [2], [3], [19].

The literature about Clifford analysis is divided into books and papers concerning the quaternions which are a special case of Clifford analysis but has also its own features. An introduction into quaternionic analysis is given in the books [7] and [15]. On the other hand, there are books related to Clifford analysis based on Clifford algebras. An introduction into Clifford analysis can be found in the books [5], [6] and [8] as well as in several other books on this topic. There are also a huge number of books concerned with Clifford algebras and their applications to physics, especially to theoretical physics. See, for example, [9], [10] and [11]. A comparison of basic facts of the theory of one-dimensional complex analysis, analysis of several complex variables and (real) Clifford analysis is given in Table 1.1.

For our considerations we will need complex Clifford analysis in the following way. Now, we assume that an arbitrary functions f is given as

$$f(z) = \sum_{\alpha} f_\alpha(z) e_\alpha,$$

where $z = x + iy$, $x, y \in Cl_{0,n}$ and $f_\alpha \in \mathbb{C}$. We introduce an analogue to the $\bar{\partial}$ operator $\overline{\mathcal{D}} = D_x + iD_y$, where $D_x = \sum_{j=1}^{n} e_j \frac{\partial}{\partial x_j}$ and $D_y = \sum_{j=1}^{n} e_j \frac{\partial}{\partial y_j}$. Then a holomorphic function $f(x, y)$ in $\mathbb{C}^n$ fulfills the relation $\overline{\mathcal{D}} f = 0$ because $\left(\frac{\partial}{\partial x_j} + i \frac{\partial}{\partial y_j} \right) f(x, y) = 0$, $j = 1, 2, \ldots, n$.

The main properties of complex Clifford analysis are contained in Table 1.2.

For our considerations we will identify

$$(\vec{x}, \vec{y}) \in \mathbb{R}^{2n} \sim \vec{x} + i\vec{y} \in \mathbb{C}^n$$
$$\sim (x, y) \in Cl_{0,n}(\mathbb{R}) \times Cl_{0,n}(\mathbb{R})$$
$$\sim x + iy \in Cl_{0,n}(\mathbb{C}).$$

This allows combining methods from real, complex and Clifford analysis.

3   Borel-Pompeiu formula

In this section we want to demonstrate how it is possible to get an integral formula in $\mathbb{C}^n$.

Several complex variables	One complex variable	Real Clifford algebras
$z_j = x_j + iy_j$	$z = x + iy$	$x = \sum\limits_{j=1}^{n} x_j e_j$
$f(z_1,\ldots,z_n) =$ $= u + iv$	$f(z) = u(z) + iv(z),$ $u, v \in \mathbb{R}$	$f(x) = f_0 e_0 + \sum\limits_{\alpha} f_\alpha e_\alpha,$ $f_0, f_\alpha \in \mathbb{R}$
$i^2 = -1$	$i^2 = -1$	$e_j^2 = -1$
$\bar{\partial}_j = \frac{\partial}{\partial x_j} + i\frac{\partial}{\partial y_j}$	$\bar{\partial} = \frac{\partial}{\partial x} + i\frac{\partial}{\partial y}$	$D = \sum\limits_{j=1}^{n} e_j \frac{\partial}{\partial x_j}$
$\bar{\partial}_j f = 0 \ \forall j$ f holomorphic	$\bar{\partial} f = 0$ f holomorphic	$Df = 0$ f (left-) monogenic
	$\partial\bar{\partial} = \bar{\partial}\partial = \Delta$	$D\bar{D} = \bar{D}D = \Delta$
holomorphic functions are harmonic	holomorphic functions are harmonic	monogenic functions are harmonic
product of holomorphic functions is holomorphic	product of holomorphic functions is holomorphic	product of monogenic functions is in general NOT monogenic
Koppelman's formula holomorphic but NOT universal kernel, Bochner-Martinelli formula NOT holomorphic but universal kernel	universal and holomorphic Cauchy kernel	universal and monogenic Cauchy kernel

TABLE 1.1.

Lemma 1. *Let G be a bounded strongly Lipschitz domain in $\mathbb{R}^{2n}$ with boundary $\partial G = \Gamma$ and $f, g \in C^1(\overline{G})$. Then we have*

$$\int_G [(fD_\xi)g + f(D_\xi g)]d\xi\, d\eta = \int_\Gamma f d\sigma_\xi g,$$

$$\int_G [(fD_\eta)g + f(D_\eta g)]d\xi\, d\eta = \int_\Gamma f d\sigma_\eta g,$$

Proof. On a bounded strongly Lipschitz domain, the exterior normal is defined almost everywhere, and Gauss' theorem is valid. $\square$

We will use this lemma to prove the following theorem.

$$z = \sum_{j=1}^{n}(x_j + iy_j)e_j = x + iy$$

$$f(z) = f_0(z)e_0 + \sum_{\alpha} f_\alpha(z)e_\alpha, \ \ f_0(z), f_\alpha(z) \in \mathbb{C}$$

$$\overline{D} = \sum_{j=1}^{n} e_j\left(\frac{\partial}{\partial x_j} + i\frac{\partial}{\partial y_j}\right) = D_x + iD_y$$

$$\overline{D}D \neq \Delta \text{ but } Sc(\overline{D}D) = Sc(\overline{D}D) = \Delta$$

TABLE 1.2.

If we take

$$K((\xi,\eta) - (x,y)) = \frac{1}{2(1-n)\omega_{2n-1}}\frac{1}{((\xi-x)^2 + (\eta-y)^2)^{n-1}},$$

where ω_{2n-1} denotes the surface area of the unit sphere in $\mathbb{R}^{2n}$, and denote by $\overline{D}$ the expression $D_\xi + iD_\eta$, we get

Theorem 1. *Let $G \in \mathbb{R}^{2n}$ be a bounded strongly Lipschitz domain with boundary Γ and exterior unit normal $n = n_x + in_y$ (we identify $\mathbb{R}^{2n}$ with $\mathbb{C}^n$, thus our normal has a real part in x-direction and an imaginary part in y-direction) defined almost everywhere on Γ and $m \in C^1(\overline{G})$. Then we have for $(x,y) \in G$*

$$\int_G (\overline{D}m)(\xi,\eta)(KD_\xi)((\xi,\eta) - (x,y))$$
$$- i(D_\eta K)((xi,\eta) - (x,y))(\overline{D}m)(\xi,\eta)d\xi\, d\eta$$
$$= \int_\Gamma m(\xi,\eta)d\sigma(KD_\xi)((\xi,\eta) - (x,y))$$
$$- i(D_\eta K)((\xi,\eta) - (x,y))d\sigma m(\xi,\eta) + m(x,y).$$

Proof. We apply Lemma 1 several times. First, let be $f = m$ and $g = KD_\xi$. Then we get

$$\int_G [(mD_\xi)(KD_\xi) + m(D_\xi(KD_\xi))]d\xi\, d\eta = \int_\Gamma md\sigma_\xi(KD_\xi),$$
$$i\int_G [(mD_\eta)(KD_\xi) + m(D_\eta(KD_\xi))]d\xi\, d\eta = i\int_\Gamma md\sigma_\eta(KD_\xi).$$

Second, we take $f = D_\eta K$ and $g = m$

$$-i\int_G [((D_\eta K)D_\xi)m + (D_\eta K)(D_\xi m)]d\xi\, d\eta = -i\int_\Gamma (D_\eta K)d\sigma_\xi m,$$
$$\int_G [((D_\eta K)D_\eta)m + (D_\eta K)(D_\eta m)]d\xi\, d\eta = -i\int_\Gamma (D_\eta K)id\sigma_\eta m.$$

Now, summing up all four equations we obtain

$$\int_G ((D_\xi + iD_\eta)m)(KD_\xi) - i(D_\eta K)((D_\xi + iD_\eta)m)d\xi d\eta$$

$$+ \int_G m(D_\xi KD_\xi) + (D_\eta KD_\eta)md\xi d\eta$$

$$= \int_\Gamma m(d\sigma(KD_\xi) - i(D_\eta K)d\sigma).$$

Because of

$$(D_\xi KD_\xi) + D_\eta KD_\eta) = (D_\xi D_\xi + D_\eta D_\eta)K = (-\Delta_\xi - \Delta_\eta)K$$
$$= -\Delta_{(\xi,\eta)}K = -\delta((\xi,\eta) - (x,y)),$$

we get that the second integral is equal to $-m(x,y)$. Thus, we get the desired formula. $\square$

Remark 1. *Due to the density of the Sobolev space $W_2^1(G)$ in $C^1(G)$ and the continuity of weakly singular integral operators, the theorem is also valid for functions from $W_2^1(G)$.*

Corollary 1. *Let $f(x,y) \in \mathbb{C}^n$ be a holomorphic function; then*

$$f(x,y)$$

$$= -\int_\Gamma f(\xi,\eta)(d\sigma(KD_\xi)((\xi,\eta) - (x,y)) - i(D_\eta K)((\xi,\eta) - (x,y))d\sigma)$$

$$= \frac{1}{\omega_{2n-1}} \int_\Gamma f(\xi,\eta) \frac{(\langle \xi - x, n_\xi + in_\eta \rangle - i\langle \eta - y, n_\xi + in_\eta \rangle)}{|(\xi,\eta) - (x,y)|^{2n-1}} dS^{2n-1}.$$

Remark 2. *The formula of Corollary 1 is nothing but the Martinelli-Bochner formula. In [14] the reader can find the relation between this formula and the singular Cauchy integral in the biquaternions which shows a deep connection between complex analysis, complex analysis in $\mathbb{C}^n$ and Clifford analysis.*

As a special case we consider the situation when G is the whole space $\mathbb{C}^n \sim \mathbb{R}^{2n}$.

Theorem 2. *Let $m \in C^1(\overline{\mathbb{R}^{2n}})$. Then we have*

$$\int_{\mathbb{R}^{2n}} (\overline{\mathcal{D}}m)(\xi,\eta)(KD_\xi)((\xi,\eta) - (x,y))$$

$$- i(D_\eta K)((\xi,\eta) - (x,y))(\overline{\mathcal{D}}m)(\xi,\eta)d\xi\, d\eta = m(x,y) - m(\infty)$$

where $m(\infty) = \lim_{|x|^2 + |y|^2 \to \infty} m(x,y)$.

Proof. Fix the point $(x, y) \in \mathbb{R}^{2n}$, let $R > 0$ and consider the domain

$$G_R((x, y)) = \{(\xi, \eta) \in \mathbb{R}^{2n} : |(x, y) - (\xi, \eta)|^2 = |x - \xi|^2 + |y - \eta|^2 < R^2\}.$$

Then for all points in G_R the relation of Theorem 1 holds. Now, we consider the integral over the boundary. We have $d\sigma = \frac{\xi + i\eta - x - iy}{|(x, y) - (\xi, \eta)|}$ and thus

$$m(\xi, \eta)d\sigma(KD_\xi) - i(D_\eta K)d\sigma m(\xi, \eta)$$

$$= \frac{m(\xi, \eta)}{\omega_{2n-1}} \left(\frac{(\xi + i\eta - x - iy)}{|(x, y) - (\xi, \eta)|} \frac{(x - \xi)}{|(x, y) - (\xi, \eta)|^{2n}} - \right.$$

$$\left. \frac{(iy - i\eta)}{|(x, y) - (\xi, \eta)|} \frac{(\xi + i\eta - x - iy)}{|(x, y) - (\xi, \eta)|^{2n}} \right)$$

$$= \frac{-1}{\omega_{2n-1}} m(\xi, \eta) \frac{1}{|(x, y) - (\xi, \eta)|^{2n-1}}$$

The result now follows by placing $m(\xi, \eta) = m(\infty) + (m(\xi.\eta) - m(\infty)$ and letting R tend to infinity. $\qquad\square$

Remark 3. *The theorem is also true for functions m from $W_2^1(G)$ which have a limit $\lim_{|x|^2 + |y|^2 \to \infty} m(x, y) = m(\infty) = const.$*

Corollary 2. *If m is twice differentiable in $W_2^2(\mathbb{R}^{2n})$, we get, under the assumptions of theorem 2, that*

$$\frac{\partial}{\partial \overline{z_l}} \frac{\partial}{\partial \overline{z_k}} m = \frac{\partial}{\partial \overline{z_k}} \frac{\partial}{\partial \overline{z_l}} m, \quad l \neq k,$$

almost everywhere.

Proof. From Theorem 2 we see that the vector part of

$$\int_{\mathbb{R}^{2n}} (\overline{D}m)(\xi, \eta)(KD_\xi)((\xi, \eta) - (x, y))$$

$$- i(D_\eta K)((\xi, \eta) - (x, y))(\overline{D}m)(\xi, \eta)d\xi\, d\eta$$

must be zero because the right-hand-side has a vanishing vector part. Now, if we write down the compounds of this vector-part, we get for $k, l =$

$1, 2, \ldots, n, \; k \neq l,$

$$\int_{\mathbb{R}^{2n}} \left(\frac{\partial}{\partial \xi_l} + \frac{\partial}{\partial \eta_l} \right) m(\xi, \eta) \left(\frac{\partial}{\partial \xi_k} K \right) ((\xi, \eta) - (x, y))$$

$$- i \left(\frac{\partial}{\partial \eta_l} K \right) ((\xi, \eta) - (x, y)) \left(\frac{\partial}{\partial \xi_k} + \frac{\partial}{\partial \eta_k} \right) m(\xi, \eta)$$

$$- \left(\frac{\partial}{\partial \xi_k} + \frac{\partial}{\partial \eta_k} \right) m(\xi, \eta) \left(\frac{\partial}{\partial \xi_l} K \right) ((\xi, \eta) - (x, y))$$

$$+ i \left(\frac{\partial}{\partial \eta_k} K \right) ((\xi, \eta) - (x, y)) \left(\frac{\partial}{\partial \xi_l} + \frac{\partial}{\partial \eta_l} \right) m(\xi, \eta) d\xi d\eta = 0.$$

After a partial integration we get

$$\int_{\mathbb{R}^{2n}} K((\xi, \eta) - (x, y)) \left(-\frac{\partial}{\partial \xi_k} \frac{\partial}{\partial \overline{z}_l} m(\xi, \eta) + i \frac{\partial}{\partial \eta_l} \frac{\partial}{\partial \overline{z}_k} m(\xi, \eta) \right.$$

$$\left. + \frac{\partial}{\partial \xi_l} \frac{\partial}{\partial \overline{z}_k} m(\xi, \eta) - i \frac{\partial}{\partial \eta_k} \frac{\partial}{\partial \overline{z}_l} m(\xi, \eta) \right) d\xi d\eta = 0,$$

and after a rearrangement we have

$$\int_{\mathbb{R}^{2n}} K((\xi, \eta) - (x, y)) \left(\frac{\partial}{\partial \overline{z}_l} \frac{\partial}{\partial \overline{z}_k} - \frac{\partial}{\partial \overline{z}_k} \frac{\partial}{\partial \overline{z}_l} \right) m(\xi, \eta) d\xi d\eta = 0.$$

Now, applying the Laplacian gives the desired relation.   □

Corollary 3. *The only entire holomorphic functions in $\mathbb{C}^n$ with $|f(x, y)| < M$ are constants.*

Remark 4. *This Borel-Pompeiu formula shows that there is a lot of interaction between complex analysis, complex analysis in $\mathbb{C}^n$ and Clifford analysis. An application of this Borel-Pompeiu formula to inverse scattering problems will be given elsewhere.*

4   The integral along the boundary

Our Borel-Pompeiu formula contains the integral along the boundary

$$\int_\Gamma m(d\sigma(K D_\xi) - i(D_\eta K) d\sigma).$$

The scalar part of the latter delivers the Bochner-Martinelli formula. But because $d\sigma$ and D_ξ, D_η are vectors in the Clifford algebra, we have also a vector part which provides also information about the function m. More precisely, let G be a region in $\mathbb{C}^n$, let ρ be a real-valued C^2-function with domain G, and put

$$N(w) = \sum_{j=1} e_j \bar{\partial}_j \rho(w), \ \ w \in G.$$

Let $\Gamma = \partial G$ be the set where $\rho = 0$, and assume that

$$N(\zeta) \neq 0, \ \ \ \forall \zeta \in \Gamma.$$

Now, let

$$L_{jk} = \bar{\partial}_k \rho \bar{\partial}_j - \bar{\partial}_j \rho \bar{\partial}_k, \ \ \ j \neq k, \ \ \ j, k = 1, 2, \dots, n.$$

be the tangential Cauchy-Riemann equations. The system reduces to just one equation when $n = 2$.

Example 1. *Let $G = \mathbb{C}^n$ and $\rho(z) = |z|^2 - 1$, so that Γ is the unit sphere in $\mathbb{C}^n$. Then the operators L_{jk} are given by $L_{jk} = \zeta_k \bar{\partial}_j - \zeta_j \bar{\partial}_k$.*

Theorem 3. *Let $m \in C^2(\bar{G})$, $G = \{\rho < 0\}$ with C^2-boundary $\Gamma = \{\rho = 0\}$. Then the vector part of the integral along the boundary can be represented as single layer potentials:*

$$Vec \left(\int_\Gamma m(d\sigma(KD_\xi) - i(D_\eta K)d\sigma) \right)$$

$$= \sum_{j,k=1, j<k}^n \int_\Gamma K(\zeta - z) \frac{1}{|\overline{\mathcal{D}}\rho(\zeta)|} L_{jk} m(\zeta) \, d\Gamma_\zeta e_j e_k$$

$$= \sum_{j,k=1, j<k}^n \int_\Gamma K(\zeta - z) \frac{1}{|\overline{\mathcal{D}}\rho(\zeta)|} \left(\bar{\partial}_k \rho(\zeta) \bar{\partial}_j m(\zeta) - \bar{\partial}_j \rho(\zeta) \bar{\partial}_k m(\zeta) \right) d\Gamma_\zeta e_j e_k.$$

Proof. We start with one part of the Borel-Pompeiu formula

$$\int_G ((D_\xi + iD_\eta)m)(KD_\xi) - i(D_\eta K)((D_\xi + iD_\eta)m) \, d\xi \, d\eta$$

$$= -\int_G (((D_\xi + iD_\eta)m)D_\xi) K - iK (D_\eta((D_\xi + iD_\eta)m)) \, d\xi \, d\eta$$

$$+ \int_\Gamma (D_\xi + iD_\eta)m d\sigma_\xi K - i \int_\Gamma K d\sigma_\eta (D_\xi + iD_\eta)m.$$

Due to $m \in C^2$ we can interchange the order of differentiation, and the integral along G gives

$$\int_G \Delta m \, K d\xi \, d\eta$$

and has only a scalar part. Now, compare our result with the Borel-Pompeiu formula and we get

$$Vec\left(\int_\Gamma m(d\sigma(KD_\xi) - i(D_\eta K)d\sigma)\right)$$
$$= Vec\left(\int_\Gamma (D_\xi + iD_\eta)md\sigma_\xi K - i\int_\Gamma Kd\sigma_\eta(D_\xi + iD_\eta)m\right).$$

Moreover,

$$d\sigma_\xi + id\sigma_\eta = (n_\xi + in_\eta)d\Gamma = \frac{N(\zeta)}{|N(\zeta)|}d\Gamma$$

and $N_j(\zeta) = \bar{\partial}_j\rho(\zeta)$. To finish the proof we have only to write down the components which belong to e_je_k. $\square$

Theorem 4. *Suppose* $G = \{\rho < 0\}$ *is a bounded region in* $\mathbb{C}^n$, *with boundary* $\Gamma = \{\rho = 0\}$, $\rho \in C^4$, *and* $\mathbb{C}^n\backslash\bar{G}$ *is connected. Then a function* $m \in C^4(\Gamma)$ *can be holomorphically extended from* Γ *into* G *if one of the following equivalent conditions is fulfilled:*

(i) $Vec\left(\int_\Gamma m(d\sigma(KD_\xi) - i(D_\eta K)d\sigma)\right) = 0$, $z \in \mathbb{C}^n\backslash\Gamma$,

(ii) $Vec\left(\int_\Gamma m(d\sigma(KD_\xi) - i(D_\eta K)d\sigma)\right) = 0$, $z \in G$,

(iii) $Vec\left(\int_\Gamma m(d\sigma(KD_\xi) - i(D_\eta K)d\sigma)\right) = 0$, $z \in \Gamma$.

Proof. We have seen that

$$Vec\left(\int_\Gamma m(d\sigma(KD_\xi) - i(D_\eta K)d\sigma)\right)$$
$$= \sum_{j,k=1,\, j<k}^n \int_\Gamma K(\zeta - z)\frac{1}{|\overline{\mathcal{D}}\rho(\zeta)|}\left(\bar{\partial}_k\rho(\zeta)\bar{\partial}_jm(\zeta) - \bar{\partial}_j\rho(\zeta)\bar{\partial}_km(\zeta)\right)d\Gamma_\zeta e_je_k.$$

From [20] we know that m can be extended holomorphically if m satisfies the tangential Cauchy-Riemann equations, i.e.,

$$\left(\bar{\partial}_k\rho(\zeta)\bar{\partial}_jm(\zeta) - \bar{\partial}_j\rho(\zeta)\bar{\partial}_km(\zeta)\right) = 0$$

for all $j, k = 1, \dots, n, j \neq k$, on Γ. Then (i) is fulfilled. The integral

$$Vec\left(\int_\Gamma m(d\sigma(KD_\xi) - i(D_\eta K)d\sigma)\right)$$

is a sum of single layer potentials with density

$$\left(\bar{\partial}_k\rho(\zeta)\bar{\partial}_jm(\zeta) - \bar{\partial}_j\rho(\zeta)\bar{\partial}_km(\zeta)\right).$$

If we take the normal derivative of this single layer potential in $\mathbb{R}^{2n}$ from the inside and the outside, then the difference is equal to the density in $z \in \Gamma$. Hence, if (i) is fulfilled, we get that the tangential Cauchy-Riemann equations are fulfilled. Obviously, if (i) holds, then (ii) also holds. Due to the continuity of the single layer potential we get (ii). The single layer potentials represent harmonic functions which are continuous in $\bar{G}$ and $\mathbb{R}^{2n}\backslash G$ and vanish.   □

5   Application to inverse scattering theory

Inverse scattering in one spatial dimension is related to the Riemann-Hilbert problem. An important development was made in [4] where the so-called $\bar{\partial}$-method was used instead of Riemann-Hilbert problems. It is possible to generalize this method to multidimensions. This was done in [16]. But to use this method an appropriate integral formula is needed for the system of $\bar{\partial}$-equations. We will demonstrate that the Borel-Pompeiu formula in $\mathbb{C}^n$ is a useful tool for this problem.

To give an idea of the background we want to start with the inverse scattering problem for the time-independent Schrödinger equation in $\mathbb{R}^n$, $n \geq 3$:

$$-\Delta\phi(x,k) + V(x)\phi(x,k) = k^2\phi(x,k).$$

Here $x \in \mathbb{R}^n$, $k \in \mathbb{C}^n$ and $k^2 = k \cdot k = \sum_{j=1}^n k_j^2$. Let G_+ be the usual outgoing Green function for the unperturbed equation, given by

$$G_+(x-y,k) = -\frac{1}{(2\pi)^n} \int \frac{e^{i(x-y)\cdot\xi}}{\xi^2 - k^2 - i0} d\xi.$$

Let $\phi_+(x,k)$ be the family of solutions of the time-independent Schrödinger equation obtained by solving the Lippman-Schwinger equation

$$\phi_+(x,k) = e^{ix\cdot k} + \int G_+(x-y,k)V(y)\phi_+(y,k)dy.$$

The scattering amplitude is defined as

$$A(\xi,k) = \int e^{-ix\cdot\xi}V(x)\phi_+(x,k)dx,$$

and can be read off "on shell" from the large $|x|$ asymptotics of ϕ_+ :

$$\phi_+(x,k) = e^{ix\cdot k} + c\frac{|k|^{\frac{n-3}{2}}}{|x|^{\frac{n-1}{2}}}e^{i|k||x|}A(\frac{x}{|x|}|k|,k) + o\left(|x|^{\frac{-n+1}{2}}\right).$$

One would like to characterize among all functions $A(\xi,k)$ (given for all ξ, k in $\mathbb{R}^n$ with $|\xi| = |k|$) those which correspond to local (real valued,

short range) potentials V. This problem is discussed in more detail in
[17]. We want to emphasize the application of the Borel-Pompeiu formula.
Therefore we will move to a more general scattering problem.

It was shown in [1] that a way to understand the nonlinear, over-determined
inverse scattering problem in $\mathbb{R}^n$ is to relate it to a linear over-determined
system of Cauchy-Riemann equations in $\mathbb{C}^n$. Following [1] we will change
denotation. We consider the equation

$$\sigma \frac{\partial v}{\partial y} + \sum_{j=1}^{n} \frac{\partial^2 v}{\partial x_j^2} + u(x, y)v,$$

with $\sigma = \sigma_R + i\sigma_I$, $x = (x_1, x_2, \ldots, x_n) \in \mathbb{R}^n$, $y \in \mathbb{R}$. In order to study
the scattering problem, we seek a solution in the form

$$v(x, y; k) = m(x, y; k)e^{ik \cdot x + k^2 y/\sigma},$$

where $k = k_R + ik_I \in \mathbb{C}^n$. Hence the eigenfunctions m satisfies

$$\sigma \frac{\partial m}{\partial y} + \sum_{j=1}^{n} \frac{\partial^2 m}{\partial x_j^2} + 2ik \cdot \nabla m + u(x, y)m = 0.$$

We shall consider the case $\sigma_R \neq 0$. Now, m is constructed from the following
equation. Given that $u(x, y) \to 0$ sufficiently rapidly as $\sqrt{(x^2 + y^2)} \to \infty$, the direct problem (determination of m, potential u known) is given
by

$$m = 1 + \tilde{G}(um), \qquad (5.1)$$

where

$$\tilde{G}f \equiv G * f \equiv \int \int_{\mathbb{R}^{n+1}} G(x - x', y - y'; k_R, k_I) f(x', y') \, dx' \, dy'.$$

We seek a bounded Green function and take the Fourier transform in both
x and y,

$$G(x, y; k_R, k_I) = C_{n+1} \int \int_{\mathbb{R}^{n+1}} \frac{e^{i(x \cdot \xi + y\eta)}}{\xi^2 + 2k \cdot \xi - i\sigma\eta} \, d\xi\eta =$$

$$= sgn(-y)\sigma^{-1} C_n \int_{\mathbb{R}^n} e^{y(\xi^2 + 2k \cdot \xi)\sigma^{-1} + ix \cdot \xi} \times$$

$$\times \Theta(-y[\sigma_R \xi^2 + 2(\sigma_R k_R + \sigma_I k_I) \cdot \xi]) d\xi,$$

where $C_n = (2\pi)^{-n}$ and $\Theta(x) = \begin{cases} 1, & x > 0, \\ 0, & x < 0. \end{cases}$ Taking the $\bar{\partial}$ derivative

with respect to $\bar{k}_j$, $\left(\text{i.e. } \frac{\partial}{\partial \bar{k}_j} = \frac{1}{2}\left(\frac{\partial}{\partial k_{R_j}} + \frac{\partial}{\partial k_{I_j}}\right)\right)$ we find

$$\frac{\partial m}{\partial \bar{k}_j} = \frac{\partial \tilde{G}}{\partial \bar{k}_j}(um) + \tilde{G}\left(u \frac{\partial m}{\partial \bar{k}_j}\right).$$

The first term is evaluated directly as

$$\frac{\partial \tilde{G}}{\partial \bar{k}_j}(um) = -\frac{1}{(2\pi)^n |\sigma_R|} \int_{\mathbb{R}^n} e^{i\beta(x,y;k_R,k_I,\xi)} T(k_R, k_I, \xi)(\xi_j - k_{R_j})\delta(s(\xi)) \, d\xi,$$

where

$$T(k_R, k_I, \xi) \equiv \int_{\mathbb{R}^{n+1}} e^{-i\beta(x,y;k_R,k_I,\xi)} u(x,y)m(x,y;k_R,k_I) \, dx \, dy,$$

$$\beta(x,y;k_R,k_I,\xi) = \left(x + \frac{2y}{\sigma_R}k_I\right) \cdot (\xi - k_R),$$

$$s(\xi) = s(\xi, k_R, k_I) \equiv \left(\xi + \frac{\sigma_I}{\sigma_R}k_I\right)^2 - \left(k_R + \frac{\sigma_I}{\sigma_R}k_I\right)^2.$$

Assuming that Equation 5.1 has no homogenous solution we can readily calculate

$$\frac{\partial m}{\partial \bar{k}_j} = -\frac{C_n}{|\sigma_R|} \int_{\mathbb{R}^n} e^{i\beta(x,y;k_R,k_I,\xi)} T(k_R, k_I, \xi)(\xi_j - k_{R_j})\delta(s(\xi))m(x,y,\xi,k_I) \, d\xi.$$

This is obtained from the fact that with m, also $\frac{\partial m}{\partial \bar{k}_j}$ is a solution and the symmetry condition on Green function

$$e^{-i\beta(x,y;k_R,k_I,\xi)}G(x,y;k_R,k_I) = G(x,y,\xi,k_I)$$

on $s(\xi) = 0$. The inverse problem is: given $T(k_R, k_I, \xi)$ construct $u(x,y)$. But there is a serious redundancy problem. Namely, $T(k_R, k_I, \xi)$ is a function of $3n$ parameters with one restriction (due to $s(\xi) = 0$): thus, T will be given as a function of $3n - 1$ variables, and we wish to construct a function $u(x,y)$ depending on $n + 1$ variables. (If $n = 1$, i.e., for two spatial dimensions, this difficulty disappears!) There are several possible reconstruction formulae for $u(x,y)$ (see for example [1]). However, crucial restrictions on T are imposed by the requirement that u depend only on x, y and decay at ∞. A standard way of reconstructing the potential is to use the one-dimensional generalized Cauchy formula. But this leads in fact to n possible potentials which of course have to be the same! But we can do better! We use our Borel-Pompeiu formula in $\mathbb{C}^n$. Because function m is scalar valued, m has to fulfill the Bochner-Martinelli formula and the compatibility conditions.

The amazing result of [1] is that the compatibility conditions

$$\frac{\partial^2 m}{\partial \bar{k}_i \partial \bar{k}_j} = \frac{\partial^2 m}{\partial \bar{k}_j \partial \bar{k}_i}, \text{ for } i \neq j.$$

are equivalent to nontrivial, nonlinear restrictions on T:

$$\mathcal{L}_{ij}[T] = \mathcal{N}_{ij}[T],$$

$$\mathcal{L}_{ij}[T] = (\xi_j - k_{R_j})\left(\frac{\partial T}{\partial \overline{k}_i} + \frac{1}{2}\frac{\partial T}{\partial \xi_i}\right) - (\xi_i - k_{R_i})\left(\frac{\partial T}{\partial \overline{k}_j} + \frac{1}{2}\frac{\partial T}{\partial \xi_j}\right),$$

$$\mathcal{N}_{ij}[T](k,\xi) = \int_{\mathbb{R}^n} [(\eta_j - k_{R_j})(\xi_i - \eta_i) - (\eta_i - k_{R_i})(\xi_j - \eta_j)]$$
$$\times \delta(s(\eta))T(k_R,k_I,\eta)T(k_R,k_I,\xi)\,d\xi.$$

Now, there exist new variables $(z,w,w_0) \in \mathbb{C}^{n-1} \times \mathbb{R}^n \times \mathbb{R}$ which parameterize the sphere $s(\xi)$, where $z = (z_2, z_3, \dots, z_n)$. If $w_1 \neq 0$, there is a 1-1 map $(k_R, k_I, \xi) \to (z, w, w_0)$ such that

$$w = \xi - k_R, \quad w_0 = \frac{2}{\sigma_R}k_I \cdot (\xi - k_R), \quad \frac{\partial}{\partial \overline{z}_j} = \mathcal{L}_{1j},$$

which for $i = 1$, $j = 2, \dots, n$ yields

$$\frac{\partial T}{\partial \overline{z}_j} = \mathcal{N}_{1j}[T](z,w,w_0), \quad j = 2, \dots, n.$$

This last equation can be written as a $\overline{\mathcal{D}}$-equation in $\mathbb{R}^{2n-2} \sim \mathbb{C}^{n-1}$:

$$\overline{\mathcal{D}}T = \mathcal{N}[T](z,w,w_0),$$

with $\mathcal{N}[T] = \sum_{j=2}^{n} e_j \mathcal{N}_{1j}[T]$. Now, using Theorem 2 we obtain

$$T(z,w,w_0) = T(\infty,w,w_0) + \int_{\mathbb{R}^{2n-2}} \mathcal{N}[T](\zeta,w,w_0)(KD_{\zeta_R})(\zeta - z)$$
$$- i(D_{\zeta_I}K)(\zeta - z)\mathcal{N}[T](\zeta,w,w_0)\,d\zeta_R\,d\zeta_I.$$

We can explicitly calculate $T(\infty,w,w_0)$. We have $|z| \to \infty$ implies $|k| \to \infty$ and hence $m \to 1$. Incorporating these facts we get

$$\lim_{|k|\to\infty} T(k_R,k_I,\xi) = \lim_{|k|\to\infty}\int_{\mathbb{R}^{n+1}} e^{-i\beta(x,y;k_R,k_I,\xi)}u(x,y)m(x,y;k_R,k_I)dxdy =$$

$$\lim_{|k|\to\infty}\int_{\mathbb{R}^{n+1}} e^{-i(x\cdot w + yw_0)}u(x,y)m(x,y;k_R,k_I)dxdy = T(\infty,w,w_0) = \hat{u}(w,w_0).$$

Thus we can conclude that under suitable smoothness conditions the following theorem is valid.

Theorem 5. *Assuming that there are no homogeneous solutions to equation 5.1, if T fulfills the compatibility conditions and $\mathcal{N}[T]$ has suitable* *decay properties for large w, w_0, then T is admissible. The potential is reconstructed from*

$$u(x,y) = \mathcal{F}^{-1}\Big(T(z,w,w_0) - \int_{\mathbb{R}^{2n-2}} \mathcal{N}[T](\zeta,w,w_0)(KD_{\zeta_R})(\zeta - z)$$
$$- i(D_{\zeta_I}K)(\zeta - z)\mathcal{N}[T](\zeta,w,w_0)\, d\zeta_R\, d\zeta_I\Big)$$

where $\mathcal{F}^{-1}$ denotes the inverse Fourier transform.

REFERENCES

[1] M. J. Ablowitz and P.A. Clarkson, Solitons, nonlinear evolution equations, and inverse scattering, *London Mathematical Society Lecture Note Series 149*, Cambridge University Press, 1991.

[2] S. Bernstein, The quaternionic Riemann problem, *Contemporary Mathematics* **232** (1999), 69–83.

[3] S. Bernstein, Index and wrapping number for paravectorvalued symbols, to appear.

[4] R. Beals and R. R. Coifman, Scattering and inverse scattering for first order systems, *Commun. Pure Appl. Math.* **37** (1984), 39–90.

[5] F. Brackx, R. Delanghe, and F. Sommen, *Clifford Analysis*, Research Notes in Mathematics: Vol. 76, Boston, London, Melbourne, Pitman Adv. Publ. Program, 1982.

[6] R. Delanghe, F. Sommen, and V. Soucek, *Clifford Algebra and Spinor-Valued Functions*, Kluwer Acad. Publ., Dordrecht, 1992.

[7] K. Gürlebeck and W. Sprößig, *Quaternionic Analysis and Elliptic Boundary Value Problems*, Birkhäuser Verlag, Basel, 1990.

[8] K. Gürlebeck and W. Sprößig, *Quaternionic and Clifford Calculus for Engineers and Physicists*, Chichester, Wiley & Sons Publ., 1997.

[9] D. Hestenes, *Space-Time Algebra*, Gordon and Breach, New York, 1966.

[10] D. Hestenes, *New Foundations for Classical Mechanics*, Reidel, Dordrecht, Boston, 1985.

[11] D. Hestenes and G. Sobczyk, *Clifford Algebras for Mathematics and Physics*, Reidel, Dordrecht, 1985.

[12] A. McIntosh, C. Li, and S. Semmes, Convolution singular integrals on Lipschitz surfaces, *J. Amer. Math. Soc.* **5** (1992), 455–481.

[13] A. McIntosh, C. Li, and T. Qian, Clifford algebras, Fourier transforms and singular convolution operators on lipschitz surfaces, *Revista Math. Iberoamer.* **10** (1994), 665–721.

[14] I. M. Mitelman and M. V. Shapiro, Differentiation of the Martinelli-Bochner integrals and the notion of hyperderivability, *Math. Nachrichten Bd.* **172** (1995), 211–238.

[15] V. V. Kravchenko and M. V. Shapiro, Integral representations for spatial models of mathematical physics, *Pitman Research Notes in Mathematics, Series 351*, 1996.

[16] A. I. Nachman and M. J. Ablowitz, A multidimensional inverse scattering method, *Stud. Appl. Math.* **71** (1984), 243–250.

[17] R. B. Lavine and A. I. Nachman, On the inverse scattering transform for the n-dimensional Schrödinger operator, in *Topics in Soliton Theory and Exactly Solvable Nonlinear Equations,* Proceedings, Oberwolfach, Germany, M. J. Ablowitz, B. Fuchssteiner, and M. D. Kruskal, eds., World Scientific, Singapore, 1986.

[18] J. Ryan, Complexified Clifford analysis, *Complex Variables, Theory and Appl.* **1** (1982), 119–149.

[19] M. V. Shapiro and N. L. Vasilevski, Quaternionic ψ-holomorphic functions, singular integral operators, and boundary value problems, *Parts I and II. Complex Variables.: Theory and Appl.* **27** (1995), 14–46 and 67–96.

[20] W. Rudin, Function theory in the unit ball of $\mathbb{C}^n$, *Grundlehren der Mathematischen Wisenschaften 241: A Series of Comprehensive Studies in Mathematics,* Springer-Verlag, New York, Heidelberg, Berlin, 1980.

[21] F. Sommen, Martinelli-Bochner type formulae in complex Clifford analysis, *Zeitschrift für Analysis und ihre Anwendungen, Bd.* **6** (1) (1987), 75–82.

Swanhild Bernstein
Institut für Mathematik und Physik
Fakultät B
Bauhaus-Universität Weimar
Coudray-Str. 13
99421 Weimar, Germany
E-mail: swanhild.bernstein@bauing.uni-weimar.de
        sbernstein@sfb.uni-weimar.de

Received: October 19, 1999; Revised: March 20, 2000

Complex-Distance Potential Theory and Hyperbolic Equations

Gerald Kaiser

ABSTRACT An extension of potential theory in $\mathbb{R}^n$ is obtained by continuing the Euclidean distance function holomorphically to $\mathbb{C}^n$. The resulting Newtonian potential is generated by an extended source distribution $\tilde{\delta}(\boldsymbol{z})$ in $\mathbb{C}^n$ whose restriction to $\mathbb{R}^n$ is the point source $\delta(\boldsymbol{x})$. This provides a possible model for extended particles in physics. In $\mathbb{C}^{n+1}$, interpreted as complex *spacetime*, $\tilde{\delta}$ acts as a *propagator* generating solutions of the wave equation from their initial values. This gives a new connection between elliptic and hyperbolic equations that does not assume analyticity of the Cauchy data. Generalized to Clifford analysis, it induces a similar connection between solutions of elliptic and hyperbolic Dirac equations. There is a natural application to the time-dependent, inhomogeneous Dirac and Maxwell equations, and the "electromagnetic wavelets" introduced previously are an example.
Keywords: Potential theory, analyticity, wave equation, Cauchy problem, Dirac equation, Maxwell's equations.

1 Motivation and preliminaries

Most fundamental theories of physics are based on the concept of potentials and fields generated by *point sources*, which presupposes that objects or "particles" can, in principle, be localized within arbitrarily small regions of space and/or time. This is a vast extrapolation from empirical evidence, and it should perhaps not come as a surprise if such theories experience some fundamental difficulties. In Newtonian mechanics, the problem of N point "bodies" interacting through gravitation has, in general, no solution due to the possibility of collisions. This becomes a serious difficulty for $N \geq 3$, where the set of initial conditions leading to collisions is non-trivial [AM78, Chapter 10]. In classical electrodynamics, difficulties arise where the field produced by a point charge unavoidably acts back on the same charge, leading to infinite self-energies and run-away particle trajectories [J99]. A way out of this dilemma was proposed by Wheeler and

---

AMS Subject Classification: 31-XX, 32-XX, 35-XX, 78-XX.

Feynman [WF45, WF49], but their *action-at-a-distance* theory, apart from being highly counter-intuitive [F64, p. 28-8], has resisted quantization and is not generally regarded as being a fundamental description of Nature. In quantum electrodynamics and other quantum field theories, point particles cause divergences which necessitate infinite "renormalization" procedures, a subject of some controversy [C99]. String theory [P98] does, in fact, not need infinite renormalization because its basic objects (strings) are extended in space rather than mathematical points. This is one of the reasons it is regarded with great hope as a possibility for unifying physical theories. However, a full development of (super)string physics is rather difficult and not expected to near completion for many years.

The ideas developed here began in 1966 and were motivated in part by the hope that an extension of physics to *complex spacetime,* justified at the foundational level, might give a way to circumvent the problems associated with point sources by applying residue methods. After some years of study and research this led to papers [K77, K78, K80, K87] and books [K90, K94] whose main thrust has been to develop a direct physical interpretation of the complex spacetime as an *extended phase space,* with the imaginary spacetime parameters carrying information about direction and velocity while the real spacetime parameters describe (approximate!) localization. A major stumbling block in this program has been the construction of extended sources since it seemed that they tend to spoil the holomorphy of the theory globally rather than just locally. Here we propose a natural solution to this problem. It turns out that although holomorphy enters the theory at the level of the fundamental solution or Green's function, and this indeed gives a canonical extension of general fields to complex spacetime through convolution, these fields *need not be holomorphic anywhere* due to the specific structure of the extended sources.

We begin in Section 2 by extending the Euclidean distance $r(x)$ to a function $\gamma(x + iy)$ holomorphic in a domain $\mathcal{O}_n \subset \mathbb{C}^n$. This leads to a natural coordinate system in $\mathbb{R}^n$ which will play an important role in the sequel: the *oblate spheroidal coordinates* adapted to $y \neq 0$.

In Sections 3 and 4 we use $\gamma(z)$ to extend the Newtonian potential $\phi(x)$ (fundamental solution of the Laplacian in $\mathbb{R}^n$) to $\mathcal{O}_n$. The resulting *holomorphic potential* $\phi(z)$ has a source distribution $\tilde{\delta}(z)$ in $\mathbb{C}^n$ which, although nowhere holomorphic, is nevertheless a canonical extension to $\mathbb{C}^n$ of the point source $\delta(x)$ in $\mathbb{R}^n$. For even $n \geq 4$, $\tilde{\delta}(x+iy)$ with fixed $y \neq 0$ is supported in x on the *sphere* $\mathcal{B}(y)$ of codimension 2 and radius $|y|$, centered at the origin and lying in the hyperplane orthogonal to y. For all other $n \geq 2$, it is supported in x on a *membrane* of codimension 1 whose boundary is $\mathcal{B}(y)$. The membrane is determined by a choice of branch cut in γ, and in the simplest case it is a *disk* $E_0(y)$.

In Section 5 we compute the sources $\tilde{\delta}(x + iy)$ explicitly as distributions in $\mathbb{R}^3$ and $\mathbb{R}^4$ supported, respectively, on $E_0(y)$ and $\mathcal{B}(y)$. This is especially interesting in the physical case of $\mathbb{R}^3$, where $\tilde{\delta}(x + iy)$ consists of a uniform *line charge* on the rim of the disk $E_0(y)$ accompanied by a *single layer* (with zero net charge) and a *double layer* distributed on the interior. Hence, the original unit "charge" carried by $\delta(x)$ is now spread over a circle of radius $|y|$, and a variable polarization density is induced on the disk swept out by "blowing up" the point source from the origin to its rim. That so much natural structure results from a simple analytic continuation is remarkable.

In Section 6 we show how to include *time* in this framework. Since we already have a holomorphic extension $\phi(x, s + it)$ of the fundamental solution $\phi(x, s)$ of the Laplacian in $\mathbb{R}^{n+1}$, it is natural to interpret t as a time parameter and look for a connection with the wave operator $\Delta_x - \partial_t^2$, which is the "analytic continuation" to $\mathbb{R}^{n,1}$ (interpreted as *Minkowskian spacetime*) of the Laplacian $\Delta_x + \partial_s^2$ in $\mathbb{R}^{n+1}$ (interpreted as *Euclidean spacetime.*) We show that $\tilde{\delta}(x, s + it)$ acts as a *propagator*, generating solutions of the initial-value problem for the wave equation in $\mathbb{R}^{n,1}$ from Cauchy data at $t = 0$. This gives a connection between the Laplace equation and the wave equation without assuming analyticity of the initial data because the propagator $\tilde{\delta}(x, s + it)$ is not holomorphic in $s + it$.

In Section 7 we extend our method to Clifford analysis, where the Laplacian and the wave operator are replaced by their "square roots," the *elliptic and hyperbolic Dirac operators*. The time-dependent Maxwell equation is a natural application of the ensuing formalism.

The idea of extending harmonic functions to obtain solutions of the wave equation has been studied by Garabedian [G64] and, in a more general context, by Ryan [R90, R90a, R96a]. However, the methods employed in these references depend critically on the assumption that the boundary data for the harmonic functions is holomorphic. As emphasized above, our method is free of this restriction.

2   Complex distance and spheroidal coordinates

For $n \geq 2$, let

$$z = x + iy \in \mathbb{C}^n, \qquad |x| = r, \ |y| = a. \tag{2.1}$$

Define the *complex length* of z as

$$\gamma(z) \equiv \sqrt{z^2} = \sqrt{x^2 - y^2 + 2ix \cdot y} = \sqrt{r^2 - a^2 + 2ix \cdot y} \equiv p + iq. \tag{2.2}$$

The *complex distance* between any two points $z_1, z_2 \in \mathbb{C}^n$ is then defined as $\gamma(z_1 - z_2)$. To complete this definition, a branch of the complex root must be chosen. The branch points form the *null cone*

$$\mathcal{N} = \{z \mid \gamma(z) = 0\} = \{x + iy \mid x^2 = y^2 \text{ and } x \cdot y = 0\}, \tag{2.3}$$

a manifold of real dimension $2n - 2$ in $\mathbb{C}^n \approx \mathbb{R}^{2n}$. In the context of the wave equation, $\mathcal{N}$ will be seen to be related to the *light cone*. We will be interested in the restriction of $\mathcal{N}$ to a fixed nonzero $y \in \mathbb{R}^n$, i.e.,

$$\mathcal{B} \equiv \mathcal{B}(y) = \{x \in \mathbb{R}^n \mid r = a, \ x \cdot y = 0\}. \tag{2.4}$$

$\mathcal{B}$ is the *sphere* of dimension $n - 2$ in $\mathbb{R}^n$ obtained by intersecting the sphere $r = a$ with the hyperplane orthogonal to y. We call $\mathcal{B}(y)$ the *branch sphere with axis vector* y.

We now concentrate on $\gamma(x + iy)$ as a function of x, regarding y as a fixed nonzero vector. In order to make this function single-valued, it is necessary to introduce a branch cut in $\mathbb{R}^n$. To see how this can be done, fix any $x \in \mathcal{B}$. In the plane determined by the two orthogonal vectors x and y, draw a circle of radius $\varepsilon \leq a$ centered at x. Each time we go around this circle, γ changes sign. In order to obtain a single-valued function for γ, it is therefore necessary to cut every such circle. Furthermore, this must be done subject to the requirement that γ must be an extension to $\mathbb{C}^n$ of the usual length function in $\mathbb{R}^n$, i.e.,

$$y \to 0 \ \Rightarrow \ \gamma(x + iy) \to r(x) \geq 0. \tag{2.5}$$

The simplest cut (but certainly not the only one) is obtained by requiring that

$$p \equiv \operatorname{Re} \gamma \geq 0, \tag{2.6}$$

which means that each of the above circles is cut at the point

$$x_c = (a - \varepsilon)\hat{x}, \qquad \hat{x} \equiv x/r.$$

The totality of such points, as x varies over $\mathcal{B}$ and $0 < \varepsilon \leq a$, together with $\mathcal{B}$ itself ($\varepsilon = 0$), forms the set

$$E_0 \equiv E_0(y) = \{x \in \mathbb{R}^n \mid r \leq a, \ x \cdot y = 0\}, \tag{2.7}$$

which is the *disk* of dimension $n - 1$ in $\mathbb{R}^n$ obtained by intersecting the ball $r \leq a$ with the hyperplane through the origin orthogonal to y. The boundary or *rim* of E_0 is $\mathcal{B}$:

$$\partial E_0 = \mathcal{B}.$$

(For $n = 3$, E_0 is indeed the disk of radius a orthogonal to y, and $\mathcal{B}$ is its rim.) In the context of holomorphic potential theory, E_0 or $\mathcal{B}$, depending on whether n is odd or even, will represent the support of a source distribution. As $y \to 0$, both sets contract to the origin, the support of the usual point source.

We now derive the basic properties of the real and imaginary parts p, q of γ, which will be important in our study of the holomorphic potentials. From (2.2) it follows that

$$p^2 - q^2 = r^2 - a^2, \qquad pq = \boldsymbol{x} \cdot \boldsymbol{y} \equiv a\zeta, \qquad (2.8)$$

where ζ is the projection of $\boldsymbol{x}$ onto $\boldsymbol{y} \neq 0$. Define the cylindrical coordinate ρ by

$$\rho^2 = r^2 - \zeta^2 = a^2 + p^2 - q^2 - \frac{p^2 q^2}{a^2} = \frac{(a^2 + p^2)(a^2 - q^2)}{a^2}. \qquad (2.9)$$

This shows that the imaginary part of γ is bounded by $a = |\boldsymbol{y}|$:

$$-a \leq q \leq a. \qquad (2.10)$$

We will use (ρ, ζ) as part of a cylindrical coordinate system in $\mathbb{R}^n$ with $\boldsymbol{y}$ as its "z-axis." Equations (2.8) and (2.9) show that the level surface $E_p = \{\boldsymbol{x} \mid p = \text{constant}\}$ with a given value of $p > 0$ is an *oblate spheroid* given by

$$E_p : \quad \frac{\rho^2}{a^2 + p^2} + \frac{\zeta^2}{p^2} = 1. \qquad (2.11)$$

Similarly, the level surface $\{\boldsymbol{x} \mid q^2 = \text{constant}\}$ with a given value of $0 < q^2 < a^2$ is a *hyperboloid of one sheet* given by

$$H_q : \quad \frac{\rho^2}{a^2 - q^2} - \frac{\zeta^2}{q^2} = 1. \qquad (2.12)$$

Note that we have so far avoided the surfaces with $p = 0$ and $q = 0, \pm a$. These give degenerate forms of E_p and H_q. In fact, as $p \to +0$, E_p contracts to the disk (2.7):

$$p \to 0 \ \Rightarrow \ E_p \to \{\boldsymbol{x} \mid \zeta = 0, \ \rho \leq a\} = E_0. \qquad (2.13)$$

More precisely, since the interior of E_0 is covered twice, we will distinguish between its *front and back sides*:

$$E_0 = E_0^+ \cup E_0^- \cup \mathcal{B}, \qquad (2.14)$$

where

$$E_0^\pm = \{\boldsymbol{x} \mid \zeta = \pm 0, \ \rho < a\} = \{\boldsymbol{x} \mid p = 0, \ \pm q > 0\} \qquad (2.15)$$

are the *interiors* of the front and back sides of E_0. Although E_0^+ and E_0^- coincide as sets, the distinction between them will be very important for the following reason. It can be easily seen that

$$\boldsymbol{x} \in E_0^\pm \ \Rightarrow \ q = \pm\sqrt{a^2 - r^2} = \pm\sqrt{a^2 - \rho^2}. \qquad (2.16)$$

This shows that, while p is continuous across E_0, q has a spherical jump discontinuity there. In the context of holomorphic potential theory, E_0^+ and E_0^- will be seen to form the two sides of a *double layer*.

The degenerate values of q are $q = 0, \pm a$. As $q \to \pm a$, the semi-hyperboloids $H_q^\pm$ with $\pm q > 0$ collapse to the positive and negative ζ-axis, respectively. As $q \to 0$, H_q collapses to the set $H_0 = \{x \mid \zeta = 0, \ \rho \geq a\}$ which, like E_0, is covered twice, once by the limit of each semi-hyperboloid $H_q^\pm$ as $q \to \pm 0$. But in this case we do not distinguish between the two copies because γ is continuous across H_0.

As mentioned above, the parameters ζ and ρ are part of a cylindrical coordinate system in $\mathbb{R}^n$. Note that the intersection of E_p and H_q is

$$S_\gamma = E_p \cap H_q = \left\{ x \in \mathbb{R}^n \mid a\zeta = pq, \ a\rho = \sqrt{a^2 + p^2}\sqrt{a^2 - q^2} \right\}.$$

Here and henceforth, we denote the pair of real variables (p, q) by the single complex variable $\gamma = p + iq$ for convenience. If $q \neq \pm a$, then S_γ is a sphere of dimension $n - 2$ and radius ρ. The variable point on S_γ may be represented as

$$x = \rho\sigma + \zeta\hat{y} \in S_\gamma, \tag{2.17}$$

where σ runs over the unit sphere of dimension $n - 2$ in the hyperplane orthogonal to y. In particular, $\gamma = 0$ gives $\zeta = 0$ and $\rho = a$, so that

$$S_0 = \{a\sigma \mid \sigma \in S^{n-2}\} = \mathcal{B}(y).$$

(If $n = 2$, S_γ consists of just two points. If $n = 3$, S_γ is a circle.) Thus, for fixed $y \neq 0$, a complete set of *cylindrical coordinates* in $\mathbb{R}^n$ is given by (ρ, ζ, σ). Since ζ and ρ are functions of (p, q), we may equivalently use the coordinates

$$x = (p, q, \sigma) \equiv (\gamma, \sigma). \tag{2.18}$$

This is the *oblate spheroidal coordinate system* in $\mathbb{R}^n$, long known as a useful tool in the theory of special functions [T96]. It is remarkable that this system is so intimately related to the holomorphic extension of the distance function.

Now let ∇ be the gradient with respect to x with y held constant, and let $\Delta = \nabla^2$ be the Laplacian in $\mathbb{R}^n$. Then

$$\gamma^2 = z^2 \implies \gamma\nabla\gamma = z; \quad \text{hence,} \quad (\nabla\gamma)^2 \equiv \nabla\gamma \cdot \nabla\gamma = 1. \tag{2.19}$$

Taking the divergence of (2.19) gives

$$\nabla\gamma \cdot \nabla\gamma + \gamma\Delta\gamma = n \implies \Delta\gamma = \frac{n-1}{\gamma}. \tag{2.20}$$

Now

$$\nabla\gamma = \frac{z}{\gamma} = \frac{\bar{\gamma}z}{\bar{\gamma}\gamma} \quad\Rightarrow\quad \nabla p = \frac{p\,x + q\,y}{\bar{\gamma}\,\gamma}, \quad \nabla q = \frac{p\,y - q\,x}{\bar{\gamma}\,\gamma}; \tag{2.21}$$

therefore,

$$\nabla\gamma \cdot \nabla\gamma = 1 \quad\Rightarrow\quad (\nabla p)^2 - (\nabla q)^2 = 1, \quad \nabla p \cdot \nabla q = 0 \tag{2.22}$$

$$\nabla\bar{\gamma} \cdot \nabla\gamma = \frac{|z|^2}{\bar{\gamma}\gamma} \quad\Rightarrow\quad (\nabla p)^2 + (\nabla q)^2 = \frac{r^2 + a^2}{\bar{\gamma}\gamma}, \tag{2.23}$$

which gives

$$(\nabla p)^2 = \frac{a^2 + p^2}{\bar{\gamma}\gamma}, \quad (\nabla q)^2 = \frac{a^2 - q^2}{\bar{\gamma}\gamma}. \tag{2.24}$$

Finally, the real and imaginary parts of (2.20) give

$$\Delta p = \frac{n-1}{\bar{\gamma}\,\gamma}\,p, \quad \Delta q = -\frac{n-1}{\bar{\gamma}\,\gamma}\,q. \tag{2.25}$$

We will need to compute volume integrals in the oblate spheroidal coordinates. Note that the area of the unit sphere $S^{n-1} \subset \mathbb{R}^n$ is [CH62]

$$\omega_n = \frac{2\pi^{n/2}}{\Gamma(n/2)}, \quad n \geq 2. \tag{2.26}$$

Let $d\sigma$ denote the surface measure on the unit sphere S^{n-2} in the hyperplane orthogonal to $y \neq 0$, normalized, so that

$$\int_{S^{n-2}} d\sigma = 1. \tag{2.27}$$

Then, for fixed y, the volume measure dx in $\mathbb{R}^n$ is given in cylindrical coordinates by

$$dx = \omega_{n-1}\,\rho^{n-2}d\rho\,d\zeta\,d\sigma. \tag{2.28}$$

By (2.8) and (2.9),

$$\rho^{n-2}d\rho\,d\zeta = \frac{\rho^{n-3}}{2}\,d[\rho^2]\,d\zeta = \frac{\rho^{n-3}}{2a^3}\,d\left[(a^2 + p^2)(a^2 - q^2)\right]\,d\,[pq]$$
$$= a^{-1}\,\rho^{n-3}(p\,dp - q\,dq)(p\,dq + q\,dp)$$
$$= a^{-1}\,\rho^{n-3}(p^2 + q^2)\,dp\,dq,$$

where $dp\,dq$ denotes the antisymmetric exterior product of differential forms (see [GS64], for example). Therefore, the volume measure in the oblate spheroidal coordinates is given by

$$dx = \frac{\omega_{n-1}}{a}\,\rho^{n-3}\,(p^2 + q^2)\,dp\,dq\,d\sigma = \frac{\omega_{n-1}}{a}\,\rho^{n-3}\,\bar{\gamma}\gamma\,dp\,dq\,d\sigma. \tag{2.29}$$

Expressions equivalent to our "complex distance" $\gamma = \sqrt{z^2}$ have been studied by Ryan [R90, R90a, R96a] and by McIntosh [M96]. They also form the basis for *complex-source pulsed beams* [HF89, HLK00] and *causal acoustic and electromagnetic wavelets* [K94, K00].

3   Holomorphic potentials and their sources

For simplicity, we now assume that $n \geq 3$. The case $n = 2$ is similar but requires some special attention and will be described elsewhere. Consider the *fundamental solution* $\phi(x)$ of Laplace's equation, defined by

$$\Delta\phi(x) = \delta(x), \quad \lim_{r \to \infty} \phi(x) = 0. \tag{3.1}$$

It is given uniquely by [CH62]

$$\phi(x) = \frac{1}{\omega_n} \frac{r^{2-n}}{2-n}, \qquad x \in \mathbb{R}^n, \ n \geq 3. \tag{3.2}$$

For $n = 3$, $\phi(x) = -1/4\pi r$ is the *Newton-Coulomb potential* with unit mass or charge. Define the *holomorphic potential* in $\mathbb{C}^n$ by

$$\phi(z) = \frac{1}{\omega_n} \frac{\gamma^{2-n}}{2-n}, \qquad z \in \mathbb{C}^n, \ n \geq 3. \tag{3.3}$$

For odd n, $\phi(z)$ inherits the branch cut of $\gamma(z)$. For even n, the only singularities occur when $x \in \mathcal{B}(y)$, where $\gamma = 0$. Thus $\gamma(z)$ and $\phi(z)$ are analytic continuations of $r(x)$ and $\phi(x)$ to the domains

$$\mathcal{O}_n = \{z \in \mathbb{C}^n \mid p > 0\} = \{x + iy \mid x \notin E_o(y)\}, \qquad \text{odd } n \geq 3$$
$$\mathcal{O}_n = \{z \in \mathbb{C}^n \mid \gamma \neq 0\} = \{x + iy \mid x \notin \mathcal{B}(y)\}, \qquad \text{even } n \geq 4.$$

Proposition 1. *For fixed y, $\phi(x + iy)$ is harmonic with respect to x when $x + iy \in \mathcal{O}_n$.*

Proof. By (3.3), we have

$$\omega_n \nabla\phi(z) = \gamma^{1-n} \nabla\gamma.$$

Thus by (2.20),

$$\omega_n \Delta\phi(z) = (1-n)\gamma^{-n} + \gamma^{1-n}\Delta\gamma = 0.$$

$\square$

Our objective is to compute the *source distribution* of $\phi(z)$, which we define formally by analogy with (3.1) as

$$\tilde{\delta}(z) = \Delta\phi(z), \qquad z \in \mathbb{C}^n. \tag{3.4}$$

This will be shown to be a *generalized function* [GS64] of x for any fixed y, meaning that given any sufficiently smooth "test" function $f(x)$ the integral

$$\langle \tilde{\delta}, f \rangle = \int_{\mathbb{R}^n} \tilde{\delta}(x + iy)\, f(x)\, dx \tag{3.5}$$

defines a bounded linear functional of f. (Generalized functions are more commonly known as *distributions* [Z65]. We use the former term here in order to avoid confusion with the term "source distribution.")

By Proposition 1, $\tilde{\delta}(x + iy)$ is supported on $x \in \mathcal{B}$ for even $n \geq 4$, and on $x \in E_0$ otherwise. In any case, it has compact support in the variable x, and, as $y \to 0$, this support contracts to the origin. We will show that

$$y \to 0 \;\Rightarrow\; \langle \tilde{\delta}, f \rangle \to f(0); \quad \text{hence,} \quad \tilde{\delta}(x + iy) \to \delta(x).$$

To compute the generalized function $\tilde{\delta}(z)$, we first define the *regularized potential*

$$\phi_\varepsilon(z) = \theta(p - \varepsilon)\,\phi(z), \quad \varepsilon > 0, \quad \text{where} \quad \theta(\xi) = \begin{cases} 1 & \text{if } \xi > 0 \\ 0 & \text{if } \xi < 0 \end{cases} \tag{3.6}$$

is the (Heaviside) unit step function. The regularization (3.6) eliminates the singularities on E_0 and $\mathcal{B}$, replacing them by a discontinuity in ϕ_ε across the spheroid E_ε. Thus, while the source of the *singular potential* $\phi(z)$ is concentrated on E_0 or $\mathcal{B}$, that of the regularized potential $\phi_\varepsilon(z)$ is concentrated on E_ε. The advantage gained is that, while $\phi(x + iy)$ is infinite for $x \in \mathcal{B} \subset E_0$, $\phi_\varepsilon(x + iy)$ remains bounded in a neighborhood of E_ε.

Now $\phi_\varepsilon(z)$ vanishes in the interior of E_ε but is identical to $\phi(z)$ in the exterior. Therefore, its source distribution, defined as a generalized function by

$$\tilde{\delta}_\varepsilon(z) \equiv \Delta\phi_\varepsilon(z), \tag{3.7}$$

represents an *equivalent source distribution* on $x \in E_\varepsilon$ whose potential field simulates that of $\tilde{\delta}(z)$ in the exterior of E_ε but vanishes in the interior. We call $\tilde{\delta}_\varepsilon(z)$ the *regularized source distribution* and will define the *singular* source distribution $\tilde{\delta}(z)$ as the limit of $\tilde{\delta}_\varepsilon(z)$ in the sense of generalized functions, for any fixed $y \in \mathbb{R}^n$. That is,

$$\langle \tilde{\delta}, f \rangle = \lim_{\varepsilon \to 0+} \langle \tilde{\delta}_\varepsilon, f \rangle \tag{3.8}$$

for every test function $f(x)$ in $\mathbb{R}^n$. Using $\theta'(\xi) = \delta(\xi)$, (3.6) gives

$$\nabla \phi_\varepsilon = \delta(p - \varepsilon) \phi \nabla p + \theta(p - \varepsilon) \nabla \phi;$$

hence,

$$\Delta \phi_\varepsilon = \delta'(p - \varepsilon) \phi (\nabla p)^2 + 2\delta(p - \varepsilon) \nabla \phi \cdot \nabla p$$
$$+ \delta(p - \varepsilon) \phi \Delta p + \theta(p - \varepsilon) \Delta \phi. \quad (3.9)$$

By Proposition 1, $\Delta \phi$ vanishes for $p \geq \varepsilon/2$, so the last term in (3.9) vanishes identically. The remaining terms show that $\tilde{\delta}_\varepsilon(x + iy)$ is indeed supported on E_ε as expected. Inserting

$$\nabla \phi = \phi'(\gamma) \nabla \gamma = \frac{\gamma^{1-n} \nabla \gamma}{\omega_n}$$

into (3.9) and using (2.22)–(2.25), we have

$$\tilde{\delta}_\varepsilon(z) = [\delta'(p - \varepsilon) \phi + 2\delta(p - \varepsilon) \phi'] \frac{a^2 + p^2}{\bar{\gamma} \gamma} + \delta(p - \varepsilon) \phi \frac{n-1}{\bar{\gamma} \gamma} p. \quad (3.10)$$

This expression represents a generalized function of x. To make sense of it we must apply it to a test function $f(x)$ in $\mathbb{R}^n$, assumed to be sufficiently smooth. That is, f is assumed to possess all derivatives which the ensuing computation requires it to possess. (As will be seen, the required degree of smoothness increases with n.) For a given value of $y \in \mathbb{R}^n$, $\tilde{\delta}_\varepsilon$ acts on f as in (3.5),

$$\langle \tilde{\delta}_\varepsilon, f \rangle = \int_{\mathbb{R}^n} \tilde{\delta}_\varepsilon(x + iy) f(x) \, dx. \quad (3.11)$$

Using the oblate spheroidal coordinates (2.18), let us write

$$f(x) = f(\rho \sigma + \zeta \hat{y}) = f^\sharp(p, q, \sigma) \equiv f^\sharp(\gamma, \sigma),$$

where the two expressions on the right are obtained by substituting (2.8) and (2.9) for ζ and ρ in terms of (p, q). Let

$$\bar{f}^\sharp(\gamma) = \int_{S^{n-2}} f^\sharp(\gamma, \sigma) \, d\sigma = \int_{S^{n-2}} f(\rho \sigma + \zeta \hat{y}) \, d\sigma \equiv \bar{f}(\rho, \zeta). \quad (3.12)$$

The notations $f^\sharp(\gamma, \sigma)$ and $\bar{f}^\sharp(\gamma) \equiv \bar{f}^\sharp(p, q)$ are used for convenience and are *not* meant to imply analyticity in γ. Because of the normalization (2.27), $\bar{f}^\sharp(\gamma)$ and $\bar{f}(\rho, \zeta)$ are the means of $f^\sharp(\gamma, \sigma)$ and $f(\rho \sigma + \zeta \hat{y})$ over the sphere $S_\gamma = E_p \cap H_q = \{x : |x - \zeta \sigma| = \rho\}$. Using the expression (2.29) for dx, (3.11) becomes

$$\langle \tilde{\delta}_\varepsilon, f \rangle = \frac{\omega_{n-1}}{a} \int_0^\infty dp \int_{-a}^a dq \, \rho^{n-3} \bar{\gamma} \gamma \tilde{\delta}_\varepsilon(\gamma) \bar{f}^\sharp(\gamma),$$

where we have used the fact that $\tilde{\delta}_\varepsilon(z)$ in (3.10) is independent of σ to write it as $\tilde{\delta}_\varepsilon(p,q) \equiv \tilde{\delta}_\varepsilon(\gamma)$. By (2.9),

$$(a^2 + p^2)\partial_p \rho^{n-3} = (n-3)p\,\rho^{n-3}.$$

Hence, the first term in (3.10) gives, upon integrating by parts over p,

$$\int_0^\infty dp \int_{-a}^a dq\,(a^2+p^2)\rho^{n-3}\delta'(p-\varepsilon)\,\phi(\gamma)\,\bar{f}^\sharp(\gamma)$$

$$= -\int_{-a}^a dq\,\rho^{n-3}\left[(n-3)\,\varepsilon\,\phi\,\bar{f}^\sharp + 2\varepsilon\phi\,\bar{f}^\sharp + (a^2+p^2)(\phi'\,\bar{f}^\sharp + \phi\,\bar{f}_p^\sharp)\right]$$

$$= -\int_{-a}^a dq\,\rho^{n-3}\left[(n-1)\,\varepsilon\,\phi\,\bar{f}^\sharp + (a^2+\varepsilon^2)(\phi'\,\bar{f}^\sharp + \phi\,\bar{f}_p^\sharp)\right],$$

where $\bar{f}_p^\sharp = \partial_p \bar{f}^\sharp(p,q)$ and the integrand is to be evaluated at $p=\varepsilon$. Inserting the other terms in (3.10) and simplifying, we obtain

$$\langle\,\tilde{\delta}_\varepsilon\,,f\,\rangle = \frac{a^2+\varepsilon^2}{a}\,\omega_{n-1}\int_{-a}^a dq\,\rho^{n-3}\left[\phi'\,\bar{f}^\sharp - \phi\,\bar{f}_p^\sharp\right]. \tag{3.13}$$

Proposition 2. *For all dimensions $n \geq 3$, the regularized source distribution $\tilde{\delta}_\varepsilon(x+iy)$ is supported on the oblate spheroid $x \in E_\varepsilon(y)$, and its action on a test function $f(x) = f^\sharp(\gamma,\sigma)$ is given by*

$$\langle\,\tilde{\delta}_\varepsilon\,,f\,\rangle = I_\varepsilon(\bar{f}^\sharp), \tag{3.14}$$

where I_ε is the linear functional defined by

$$I_\varepsilon(\bar{f}^\sharp) = \frac{(a^2+\varepsilon^2)^{\nu+1}}{a^{n-2}A_n}\int_{-a}^a \frac{F^\sharp(\varepsilon+iq)}{(\varepsilon+iq)^{n-1}}\,dq, \quad \nu = \frac{n-3}{2}, \tag{3.15}$$

with

$$A_n = \frac{\omega_n}{\omega_{n-1}}, \quad F^\sharp(\gamma) = (a^2-q^2)^\nu\left[\bar{f}^\sharp(\gamma) + \frac{\gamma\bar{f}_p^\sharp(\gamma)}{n-2}\right]. \tag{3.16}$$

Proof. This follows immediately from (3.13), using

$$\rho^2 = \frac{(a^2+\varepsilon^2)(a^2-q^2)}{a^2}. \qquad \square$$

Let us verify that as $y \to 0$ and the source disk shrinks to a point, the source of the singular potential ϕ contracts to the usual point source. Letting $q = a\xi$ in (3.16), we have

$$F^\sharp(\gamma) = a^{n-3}(1-\xi^2)^\nu\left[\bar{f}^\sharp(\gamma) + \frac{\gamma\bar{f}_p^\sharp(\gamma)}{n-2}\right], \quad \text{where} \quad \gamma = \varepsilon + ia\xi.$$

Therefore,

$$I_\varepsilon(\bar{f}^\sharp) = \frac{(a^2 + \varepsilon^2)^{\frac{n-1}{2}}}{A_n} \int_{-1}^{1} \frac{(1 - \xi^2)^\nu}{(\varepsilon + ia\xi)^{n-1}} \left[\bar{f}^\sharp(\gamma) + \frac{\gamma \bar{f}_p^\sharp(\gamma)}{n - 2} \right] d\xi$$

and

$$\lim_{a \to 0} I_\varepsilon(\bar{f}^\sharp) = \left[\bar{f}^\sharp(\varepsilon) + \frac{\varepsilon \bar{f}_p^\sharp(\varepsilon)}{n - 2} \right] \frac{K_n}{A_n},$$

where

$$K_n = \int_{-1}^{1} (1 - \xi^2)^\nu \, d\xi = B\left(\tfrac{1}{2}, \tfrac{n-1}{2}\right) = \frac{\sqrt{\pi}\,\Gamma\left(\frac{n-1}{2}\right)}{\Gamma\left(\frac{n}{2}\right)} = \frac{\omega_n}{\omega_{n-1}} = A_n \,.$$

Thus, by (3.14),

$$\lim_{a \to 0} \langle \tilde{\delta}_\varepsilon, f \rangle = \bar{f}^\sharp(\varepsilon) + \frac{\varepsilon \bar{f}_p^\sharp(\varepsilon)}{n - 2} \,.$$

Now let $\varepsilon \to 0$, and note that, since E_ε contracts to the origin,

$$\lim_{\varepsilon \to 0} \bar{f}^\sharp(\varepsilon) = \lim_{\varepsilon \to 0} \int_{S^{n-2}} f^\sharp(\varepsilon, \sigma) d\sigma = f(0); \quad \text{hence,} \quad \lim_{\varepsilon \to 0} \lim_{a \to 0} \langle \tilde{\delta}_\varepsilon, f \rangle = f(0).$$

If we assume that the two limits can be exchanged (as will be verified later) and use the definition (3.8) of $\tilde{\delta}$, this states that $\lim_{a \to 0} \langle \tilde{\delta}, f \rangle = f(0)$, giving the following important result.

Theorem 1. *The singular source distribution $\tilde{\delta}(z)$ is an extension of the usual point source in $\mathbb{R}^n$ in the sense that*

$$y \to 0 \;\Rightarrow\; \tilde{\delta}(x + iy) \to \delta(x). \tag{3.17}$$

4   Singular source distributions

We are ready at last to compute the singular source distributions. By (3.8) and (3.14), they are given by the limit

$$\langle \tilde{\delta}, f \rangle = \lim_{\varepsilon \to 0} I_\varepsilon(\bar{f}^\sharp). \tag{4.1}$$

However, we cannot simply let $\varepsilon = 0$ in the expression (3.15) for $I_\varepsilon(\bar{f}^\sharp)$ since the resulting integral diverges. Recall the decomposition of E_0 given in Equation (2.14):

$$E_0 = E_0^+ \cup E_0^- \cup \mathcal{B}, \tag{4.2}$$

where $E_0^\pm$ are the interiors of the front and back sides of E_0 and $\mathcal{B}$ is its rim, the branch sphere. This decomposition recognizes the nature of E_0 as a limit of ellipsoids. Although the two open disks $E_0^\pm$ look identical to a *continuous* function, they look distinct to a generalized function like $\tilde{\delta}$ which is singular across E_0. Furthermore, the oblate spheroidal coordinates are an ideal tool for resolving this decomposition since $q > 0$ on E_0^+, $q < 0$ on E_0^-, and $q = 0$ on $\mathcal{B}$. We will compute $\langle \tilde{\delta}, f \rangle$ by decomposing the integral $I_\varepsilon(\bar{f}^\sharp)$ in a way similar to (4.2) and then taking the limit $\varepsilon \to 0$. The integral over $E_0^+ \cup E_0^-$ gives a sum of *single and double layer distributions* of dimension $n - 1$ over the interior of the disk E_0, while the integral over $\mathcal{B}$ gives a *boundary distribution* of dimension $n - 2$ over $\mathcal{B}$. All these distributions are well-defined, giving a finite expression for $\langle \tilde{\delta}, f \rangle$.

The divergence of $I_0(\bar{f}^\sharp)$ is therefore caused by the attempt to represent the boundary distribution on $\mathcal{B}$ as part of an integral of dimension $n - 1$ over E_0, and the above regularization simply amounts to recognizing this fact.

To regularize the integral

$$I_\varepsilon(\bar{f}^\sharp) = \frac{(a^2 + \varepsilon^2)^{\nu+1}}{a^{n-2} A_n} \int_{-a}^{a} \frac{F^\sharp(\varepsilon + iq)}{(\varepsilon + iq)^{n-1}} \, dq, \tag{4.3}$$

define the Taylor coefficients

$$T_m(\varepsilon) = \frac{1}{m!} \partial_q^m F^\sharp(\varepsilon + iq) \big|_{q=0}, \qquad T_m \equiv T_m(0) \tag{4.4}$$

and the Taylor polynomials approximating $F^\sharp(\varepsilon + iq)$ and $F^\sharp(iq)$ to order q^{n-2},

$$F_{n-2}^\sharp(\varepsilon, q) = \sum_{m=0}^{n-2} q^m T_m(\varepsilon), \quad F_{n-2}^\sharp(q) = \sum_{m=0}^{n-2} q^m T_m. \tag{4.5}$$

We rewrite the integral in (4.3) as

$$\int_{-a}^{a} \frac{F^\sharp(\varepsilon + iq)}{(\varepsilon + iq)^{n-1}} \, dq = \int_{-a}^{a} \frac{F^\sharp(\varepsilon + iq) - F_{n-2}^\sharp(\varepsilon, q)}{(\varepsilon + iq)^{n-1}} \, dq$$

$$+ \sum_{m=0}^{n-2} i^{-m} T_m(\varepsilon) \lambda_{n-1}^m(\varepsilon), \tag{4.6}$$

where

$$\lambda_k^m(\varepsilon) = \int_{-a}^{a} \frac{i^m q^m \, dq}{(\varepsilon + iq)^k}, \quad 0 \le m < k. \tag{4.7}$$

If $f(x)$ is continuously differentiable to order $n-2$, so is $F^\sharp(\varepsilon + iq) \equiv F^\sharp(\varepsilon, q)$. Then,

$$F^\sharp(\varepsilon + iq) - F^\sharp_{n-2}(\varepsilon, q) = O(q^{n-1}),$$

and the integral on the right-hand side of (4.6) has a finite limit as $\varepsilon \to 0$. It therefore remains to compute the limit of the sum. We begin by finding

$$\lambda^0_1(\varepsilon) = \int_{-a}^a \frac{dq}{\varepsilon + iq} = 2\varepsilon \int_0^a \frac{dq}{\varepsilon^2 + q^2} = 2\tan^{-1}\frac{a}{\varepsilon}$$

$$= \pi - 2\tan^{-1}\frac{\varepsilon}{a} \equiv \alpha(\varepsilon). \tag{4.8}$$

All the other integrals (4.7) can be computed from the recursion relations

$$\lambda^0_{k+1}(\varepsilon) = -\frac{1}{k}\partial_\varepsilon \lambda^0_k(\varepsilon) = \frac{(-1)^k}{k!}\partial^k_\varepsilon \alpha(\varepsilon)$$

$$\lambda^{m+1}_k(\varepsilon) = i^m \int_{-a}^a \frac{(\varepsilon + iq - \varepsilon)q^m dq}{(\varepsilon + iq)^k} = \lambda^m_{k-1}(\varepsilon) - \varepsilon\lambda^m_k(\varepsilon).$$

We are interested in the limit $\varepsilon \to 0$, where these relations imply

$$\lambda_{k+1} \equiv \lambda^0_{k+1}(0) = \frac{(-1)^k}{k!}\partial^k_\varepsilon \alpha(0) \equiv (-1)^k L_k \tag{4.9}$$

$$\lambda^m_k(0) = \lambda^0_{k-m}(0) = \lambda_{k-m}, \quad 0 \le m < k. \tag{4.10}$$

The Taylor coefficients L_k of $\alpha(\varepsilon)$ are obtained from the expansion

$$\alpha(\varepsilon) = \pi - 2\tan^{-1}\frac{\varepsilon}{a} = \pi + 2\sum_{l=1}^\infty \frac{(-1)^l}{2l-1}\frac{\varepsilon^{2l-1}}{a^{2l-1}}. \tag{4.11}$$

Thus,

$$\lambda_1 = \pi, \quad \lambda_{2l} = \frac{2(-1)^{l+1}}{(2l-1)a^{2l-1}}, \quad \lambda_{2l+1} = 0, \quad l \ge 1. \tag{4.12}$$

By (4.10), this gives finite values for all the coefficients $\lambda^m_k(0)$. Note that the original expression (4.7) diverges if we set $\varepsilon = 0$. The finiteness of the limits depends on delicate cancellations of contributions from positive and negative values of q when $\varepsilon > 0$, just as happens in Cauchy's principal value integral. In fact, (4.8) shows that *the present regularization reduces to the principal value integral when* $k = 1$.

Using (4.6), we can now compute the limit $\varepsilon \to 0$ in (4.1):

$$\langle \tilde{\delta}, f \rangle = V_n(f) + \frac{a}{A_n}\sum_{m=0}^{n-2} i^{-m} T_m \lambda_{n-m-1}, \tag{4.13}$$

$$V_n(f) \equiv \frac{i^{1-n}a}{A_n}\int_{-a}^a \frac{F^\sharp(iq) - F^\sharp_{n-2}(q)}{q^{n-1}}\,dq. \tag{4.14}$$

$V_n(f)$ will be shown to be a *bounded linear functional* of f, and this establishes $\tilde{\delta}$ as a well-defined generalized function.

Since $\bar{f}^\sharp(0)$ is the mean of f over $\mathcal{B}$, the terms in the sum in (4.13) represent means of f and its derivatives over $\mathcal{B}$. On the other hand, (4.14) represents an integral of f and its normal derivative over the *interior* $E_0^+ \cup E_0^-$ since the boundary terms have already been subtracted in the form of $F_{n-2}^\sharp(q)$. Thus, (4.13) is the promised decomposition of the source distribution corresponding to (4.2).

Equation (4.13) can be greatly simplified because of certain symmetries satisfied by $\bar{f}^\sharp(iq)$. We claim that

$$\bar{f}^\sharp(-iq) = \bar{f}^\sharp(iq), \quad \bar{f}_p^\sharp(-iq) = -\bar{f}_p^\sharp(iq), \quad \bar{f}_q^\sharp(-iq) = -\bar{f}_q^\sharp(iq). \quad (4.15)$$

To see this, note that the coordinates $\gamma = \pm iq$ denote *the same point* of E_0, regarded as belonging to $E_0^\pm$. Since the test function $f(x)$ is continuous across E_0, so is its integral $\bar{f}^\sharp(iq)$, and this proves the first relation. On the other hand, p increases in the $\pm\zeta$ direction on $E_0^\pm$, which proves the second relation. Finally, the third relation follows from the first by differentiation.

By (4.15), the function

$$F^\sharp(iq) = (a^2 - q^2)^\nu \left[\bar{f}^\sharp(iq) + i\,\frac{q\bar{f}_p^\sharp(iq)}{n-2} \right] \quad (4.16)$$

is *even*, so its odd Taylor coefficients vanish:

$$m \text{ is odd} \quad \Rightarrow \quad T_m = 0. \quad (4.17)$$

Furthermore, (4.14) shows that

$$n \text{ is even} \quad \Rightarrow \quad V_n(f) = 0. \quad (4.18)$$

Considering the cases of odd and even n separately and inserting the values of λ_k from (4.12) give the following result.

Theorem 2. *The singular source distribution $\tilde{\delta}(z)$ is a bounded linear functional which acts on a test function $f \in C^{n-2}(\mathbb{R}^n)$ as follows. For even $n = 2k + 2 \geq 4$,*

$$\langle \tilde{\delta}, f \rangle = \frac{\pi a}{A_n} (-1)^k T_{2k}, \quad (4.19)$$

where A_n and T_{2k} are given by (3.16) and (4.4). For odd $n = 2k+3 \geq 3$,

$$\langle \tilde{\delta}, f \rangle = V_n(f) + \frac{2(-1)^k}{A_n} \sum_{l=0}^{k} \frac{a^{2l-2k} T_{2l}}{2k - 2l + 1}, \quad (4.20)$$

where $V_n(f)$ is given by (4.14).

Proof. To prove (4.19), note first that $V_n(f)$ vanishes by (4.18). Furthermore, the sum in (4.13) reduces to a single term because $T_m \lambda_{n-m-1} = 0$ unless m is even, which implies that $n - m - 1$ is odd; the only nonvanishing coefficient λ_l with odd l is $\lambda_1 = \pi$.

Equation (4.20) follows directly from (4.13), (4.12) and (4.17). To establish $\tilde{\delta}$ as a distribution, we must still prove that $V_n(f)$ is a bounded linear functional for odd n. Since the integrand in (4.14) is even when n is odd, we have

$$V_n(f) = \frac{2i^{1-n}a}{A_n} \int_0^a \frac{F^\sharp(iq) - F^\sharp_{n-2}(q)}{q^{n-1}} \, dq.$$

By Taylor's theorem,

$$F^\sharp(iq) - F^\sharp_{n-2}(q) = \frac{1}{(n-2)!} \int_0^q (q-w)^{n-2} (F^\sharp)^{(n-1)}(iw) \, dw.$$

Therefore,

$$|F^\sharp(iq) - F^\sharp_{n-2}(q)| \le q^{n-1} M, \qquad M = \frac{\max |(F^\sharp)^{(n-1)}(iw)|}{(n-2)!},$$

and

$$|V_n(f)| \le 2a^2 M.$$

Since M depends boundedly on f, this shows that $V_n(f)$ is a bounded linear functional as claimed. $\qquad \square$

Thus, for even $n \ge 4$, the integral over the interior of E_0 vanishes and the singular source is concentrated on $\mathcal{B}$. This is to be expected since γ^{2-n} has no branch cut; therefore, the only singularity occurs on the boundary.

We can now derive a useful expression for the singular source distribution $\tilde{\delta}(x + iy)$ in the cylindrical coordinates (ρ, ζ, σ) adapted to $y \ne 0$. As a byproduct, it will be seen that for all $n \ge 3$, the test function $f(x)$ need only be in $C^k(\mathbb{R}^n)$ with $k = [\frac{n-1}{2}]$, instead of $C^{n-2}(\mathbb{R}^n)$ as assumed in Theorem 2. Our first task is to rewrite the function $F^\sharp(iq)$ in (4.16) in terms of cylindrical coordinates. Recall that

$$\zeta = \frac{pq}{a}, \qquad \rho = \frac{\sqrt{(a^2 + p^2)(a^2 - q^2)}}{a} \qquad (4.21)$$

$$x = \rho\sigma + \zeta\hat{y}, \quad \sigma \in S^{n-2} \perp y, \quad f(\rho\sigma + \zeta\hat{y}) = f^\sharp(p, q, \sigma). \qquad (4.22)$$

Then the mean of f on $S_\gamma = E_p \cap H_q$ becomes

$$\bar{f}^\sharp(p, q) = \int_{S^{n-2}} f(\rho\sigma + \zeta\hat{y}) \, d\sigma \equiv \bar{f}(\rho, \zeta), \qquad (4.23)$$

which is the mean of f on the sphere $S_{\rho,\zeta} = \{x : |x - \zeta\,\hat{y}| = \rho\}$.

It will be shown elsewhere [K00] that the source distributions for odd n can be derived from those for even n by integrating over one of the coordinates. Specifically, we have:

Theorem 3. *Let $n \geq 3$, and denote points in $\mathbb{C}^{n+1}$ by*

$$z = x + iy = (z, s + it), \qquad z \in \mathbb{C}^n.$$

For fixed $y \neq 0$, the singular source distributions $\tilde{\delta}_n(x + iy)$ in $\mathbb{R}^n$ and $\tilde{\delta}_{n+1}(x + iy)$ in $\mathbb{R}^{n+1}$ are related by

$$\tilde{\delta}_n(z) = \int_{-\infty}^{\infty} ds\, \tilde{\delta}_{n+1}(z, s). \tag{4.24}$$

That is, for a test function $f(x)$ in $\mathbb{R}^n$ we have

$$\langle \tilde{\delta}_n, f \rangle = \int_{-\infty}^{\infty} ds \int_{\mathbb{R}^n} dx\, \tilde{\delta}_{n+1}(x + iy, s) f(x). \tag{4.25}$$

For $y = 0$, (4.24) follows from $\delta_{n+1}(x, s) = \delta_n(x)\,\delta(s)$. For $y \neq 0$, this tensor product decomposition fails, but Theorem 3 still holds. We, therefore, state the following theorem for the simpler case of even $n \geq 4$ though the transformation to cylindrical coordinates derived below will be used later in the explicit computation of the extended source in $\mathbb{R}^3$.

Theorem 4. *For $n = 2k + 2 \geq 4$, the source distribution $\tilde{\delta}$ is a bounded linear functional on test functions $f(x)$ in $C^k(\mathbb{R}^n)$ whose action is given in the cylindrical coordinates (ρ, ζ, σ) by*

$$\langle \tilde{\delta}, f \rangle = \frac{a\sqrt{\pi}}{\Gamma(k + \frac{1}{2})}\, D_\rho^k\, F(\rho)\,\Big|_{\rho=a}, \tag{4.26}$$

where

$$D_\rho = \frac{\partial}{\partial(\rho^2)} = \frac{1}{2\rho}\frac{\partial}{\partial\rho}, \quad F(\rho) = \rho^{2k-1}\left[\bar{f}(\rho,0) + i\frac{a^2 - \rho^2}{2ka}\,\bar{f}_\zeta(\rho,0)\right]. \tag{4.27}$$

Proof. By (4.21), we have

$$\bar{f}_p^\sharp(p, q) = \frac{\rho p}{a^2 + p^2}\,\bar{f}_\rho(\rho,\zeta) + \frac{q}{a}\,\bar{f}_\zeta(\rho,\zeta). \tag{4.28}$$

Since

$$p = 0 \;\Rightarrow\; q^2 = a^2 - \rho^2,$$

it follows from (4.16) that

$$F^{\sharp}(iq) = (a^2 - q^2)^{\nu} \left[\bar{f}^{\sharp}(0, q) + i \frac{q\bar{f}^{\sharp}_p(0, q)}{n-2} \right]$$

$$= \rho^{n-3} \left[\bar{f}(\rho, 0) + i \frac{q^2}{(n-2)a} \bar{f}_{\zeta}(\rho, 0) \right] = F(\rho). \qquad (4.29)$$

Note that (4.29) makes sense only because $F^{\sharp}(iq)$ is even and so does not depend on the sign of q, which cannot be recovered from ρ when $\zeta = 0$ since $\operatorname{sgn} q = \operatorname{sgn} \zeta$. With

$$D_q = \frac{\partial}{\partial(q^2)} = \frac{1}{2q} \frac{\partial}{\partial q} = -D_\rho \quad \text{when} \quad p = 0,$$

we have

$$F^{\sharp}(iq) = \sum_{m=0}^{\infty} T_{2m} q^{2m} \implies$$

$$T_{2m} = \frac{1}{m!} D_q^m F^{\sharp}(iq) \Big|_{q=0} = \frac{(-1)^m}{m!} D_\rho^m F(\rho) \Big|_{\rho=a}.$$

For $n = 2k + 2 \geq 4$,

$$\omega_n = \frac{2\pi^{k+1}}{\Gamma(k+1)} = \frac{2\pi^{k+1}}{k!}, \quad \omega_{n-1} = \frac{2\pi^{k+\frac{1}{2}}}{\Gamma(k+\frac{1}{2})},$$

and (4.19) gives

$$\langle \tilde{\delta}, f \rangle = \frac{\pi a \omega_{n-1}}{\omega_n} (-1)^k T_{2k} = \frac{a\sqrt{\pi}}{\Gamma(k+\frac{1}{2})} D_\rho^k F(\rho) \Big|_{\rho=a}.$$

Now $F(\rho)$ already contains one derivative ($\bar{f}_{\zeta}$,) and D_ρ^k computes k more; hence, it suffices to have $f \in C^{k+1}(\mathbb{R}^n)$. But the highest derivative of f occurring in (4.26) is

$$\frac{\rho^{2k-1}}{(2\rho)^k} \cdot i \frac{a^2 - \rho^2}{2ka} \partial_\rho^k \bar{f}_{\zeta}(\rho, 0),$$

which vanishes at $\rho = a$. Hence, the highest non-vanishing derivative is of order k, so it suffices for $\bar{f}(\rho, \zeta)$ to be k times continuously differentiable as claimed. ☐

An application of (4.25) now shows that for $n = 2k + 1$, we need $f \in C^k(\mathbb{R}^n)$ in order for $\langle \tilde{\delta}_n, f \rangle$ to make sense. Thus, for any $n \geq 3$, f needs to be in $C^k(\mathbb{R}^n)$ with

$$k = \left[\frac{n-1}{2} \right] = \begin{cases} \frac{n-2}{2}, & n \text{ even} \\ \frac{n-1}{2}, & n \text{ odd.} \end{cases}$$

5   Computation of sources in $\mathbb{R}^3$ and $\mathbb{R}^4$

We now compute the singular source distribution in $\mathbb{R}^3$ and $\mathbb{R}^4$ explicitly and interpret the results.

For $n = 3$, Equation (4.16) becomes

$$F^\natural(iq) = \bar{f}^\natural(iq) + iq\bar{f}^\natural_p(iq); \tag{5.1}$$

hence, $T_0 = \bar{f}^\natural(0)$, $T_1 = 0$, and (4.5) becomes

$$F^\natural_1(q) = \bar{f}^\natural(0).$$

Equation (4.20), therefore, gives

$$\langle \tilde{\delta}, f \rangle = -a \int_{-a}^{a} \frac{\bar{f}^\natural(iq) + iq\bar{f}^\natural_p(iq) - \bar{f}^\natural(0)}{q^2} \, dq + \bar{f}^\natural(0).$$

Using the symmetry of the integrand, we obtain the following result.

Proposition 3. *The singular source distribution*

$$\tilde{\delta}(x + iy) = -\Delta \frac{1}{4\pi\gamma}, \quad x + iy \in \mathbb{C}^3$$

is a bounded linear functional whose action on a test function $f(x) = f^\natural(\gamma, \sigma)$ in $C^1(\mathbb{R}^3)$ is given in oblate spheroidal and cylindrical coordinates by

$$\langle \tilde{\delta}, f \rangle = L_0 + L_1 + iL_2, \tag{5.2}$$

where

$$L_0 = \bar{f}^\natural(0) = \bar{f}(a, 0)$$

$$L_1 = -a \int_0^a \frac{\bar{f}^\natural(iq) - \bar{f}^\natural(0)}{q^2} \, dq = -a \int_0^a \frac{\bar{f}(\rho, 0) - \bar{f}(a, 0)}{(a^2 - \rho^2)^{3/2}} \, \rho \, d\rho$$

$$L_2 = -a \int_0^a \frac{\bar{f}^\natural_p(iq)}{q} \, dq = -\int_0^a \frac{\bar{f}_\zeta(\rho, 0)}{\sqrt{a^2 - \rho^2}} \, \rho \, d\rho.$$

Proof. The action in oblate spheroidal coordinates follows immediately from (4.20). To obtain the action in cylindrical coordinates, use (4.28). $\quad\square$

Note that since

$$\lim_{a \to 0} \bar{f}(a, 0) = f(0) \quad \text{and} \quad \lim_{a \to 0} L_1 = \lim_{a \to 0} L_2 = 0,$$

(5.2) shows that

$$y \to 0 \implies \tilde{\delta}(x + iy) \to \delta(x).$$

This was already seen in Theorem 1 for all $n \geq 3$, but that proof was less rigorous because it depended on the assumption that the order of the limits $a \to 0$ and $\varepsilon \to 0$ can be exchanged.

We now state some other interesting properties of the expression (5.2) which will help interpret its three terms.

Proposition 4. *The monopole and dipole moments of* $\tilde{\delta}(x + iy)$ *in* $\mathbb{R}^3$ *are*

$$Q \equiv \int_{\mathbb{R}^3} \tilde{\delta}(x + iy)\, dx = 1, \qquad P \equiv \int_{\mathbb{R}^3} x\, \tilde{\delta}(x + iy)\, dx = -iy.$$

Given a point source with general complex coordinates $z_s = x_s + iy_s \in \mathbb{C}^3$, *the centroid of its charge distribution is*

$$C(z_s) \equiv \int_{\mathbb{R}^3} x\, \tilde{\delta}(x - z_s)\, dx = z_s. \tag{5.3}$$

Proof. To compute Q, apply Proposition 3 with $f(x) \equiv 1$. To find P, apply it to the *vector-valued* test function

$$f(x) = x = \rho\sigma + \zeta\hat{y}.$$

Finally,

$$C(z_s) = x_s \int_{\mathbb{R}^3} \tilde{\delta}(x - x_s - iy_s)\, dx + \int_{\mathbb{R}^3} (x - x_s)\, \tilde{\delta}(x - x_s - iy_s)\, dx$$

$$= x_s + iy_s = z_s.$$ $\square$

Equation (5.3) is a natural extension to $\mathbb{C}^3$ of the formula

$$\int_{\mathbb{R}^3} x\, \delta(x - x_s)\, dx = x_s, \qquad x_s \in \mathbb{R}^3.$$

Proposition 4 sheds some light on the nature of the source distribution $\tilde{\delta}$.

- L_0 is the mean of f over the rim $\mathcal{B}$. Since $L_1 = L_2 = 0$ when $f(x) \equiv 1$, we see that the "charge" Q resides entirely on this rim.

- L_1 is an integral of f over E_0 which does not involve its derivatives, thus, representing a *single layer distribution* on E_0. Actually, since the contributions from the rim are subtracted, the single layer resides on the *interior* of E_0.

- L_2 is an integral of the normal derivative of f over E_0, so it represents a *double layer distribution* which may be regarded as equal and opposite charge distributions on E_0^+ and E_0^-. This is confirmed by the fact that the dipole moment is $P = -iy$.

Proposition 5. *The source distribution in* $\mathbb{C}^4$,

$$\tilde{\delta}(\boldsymbol{z}) = -\Delta \frac{1}{4\pi^2\gamma^2},$$

acts on a test function $f \in C^1(\mathbb{R}^4)$ *in cylindrical coordinates as follows:*

$$\langle \tilde{\delta}, f \rangle = \bar{f}(a, 0) + a\bar{f}_\rho(a, 0) - ia\bar{f}_\zeta(a, 0). \tag{5.4}$$

Proof. From (4.27) we find

$$F(\rho) = \rho \left[\bar{f}(\rho, 0) + i \frac{a^2 - \rho^2}{2a} \bar{f}_\zeta(\rho, 0) \right],$$

and (4.26) becomes $\langle \tilde{\delta}, f \rangle = \partial_\rho F(a)$, which gives (5.4).   □

As $a \to 0$, this gives $\langle \tilde{\delta}, f \rangle \to f(0)$, which again confirms again that $\tilde{\delta}(\boldsymbol{x} + i\boldsymbol{y}) \to \delta(\boldsymbol{x})$. This property, as well as $Q = 1$, $\boldsymbol{P} = -i\boldsymbol{y}$, $C(\boldsymbol{z}_s) = \boldsymbol{z}_s$ as in Proposition 4, can be shown to hold for *all* n [K00].

The holomorphic potential in $\mathbb{C}^4$ was derived from a different point-of-view in [K94, Section 11.2] where it was shown to decompose into *causal* (retarded) and *anticausal* (advanced) *physical wavelets*. These are closely related to the *complex-source pulsed beams* studied in the engineering literature [HF89].

6   Connection to spacetime and wave equations

It was proposed in Section 1 that point sources in physics be replaced by extended sources based on a continuation of physics to complex spacetime. The distribution $\tilde{\delta}(\boldsymbol{z})$ in complex space ($\mathbb{C}^n$) seems like a promising model since it is supported on the disk E_0 (when n is odd) or the sphere $\mathcal{B}$ (when n is even). But to do physics we need to add *time* to this complex-space formalism. We will first give an argument suggesting that time is *already* included in $\mathbb{C}^{n+1}$ and then justify this by proving that, with this interpretation, the extended source distribution $\tilde{\delta}(\boldsymbol{z})$ in $\mathbb{C}^{n+1}$ acts as a *propagator* in spacetime, generating solutions of the Cauchy problem for the wave equation from their initial values.

Let us write

$$\boldsymbol{z} = \boldsymbol{x} + i\boldsymbol{y} = (z, s + it) \in \mathbb{C}^{n+1}, \quad z \in \mathbb{C}^n$$
$$\boldsymbol{x} = (\boldsymbol{x}, s), \quad \boldsymbol{y} = (\boldsymbol{y}, t) \in \mathbb{R}^{n+1}, \quad |\boldsymbol{x}| = r.$$

Analytic potential theory in $\mathbb{C}^{n+1}$ is based on the complex distance function $\gamma(\boldsymbol{z})$ defined by

$$\gamma^2 \equiv \boldsymbol{z}^2 = z^2 + (s + it)^2, \quad \text{Re } \gamma \geq 0,$$

which reduces to the *Euclidean* metric in $\mathbb{R}^{n+1}$ when $\mathbf{y} \to \mathbf{0}$,

$$\gamma^2 \to \mathbf{x}^2 = r^2 + s^2,$$

and to the *Minkowski* metric in $\mathbb{R}^{n,1}$ when $s \to 0$ and $\mathbf{y} \to \mathbf{0}$,

$$\gamma^2 \to r^2 - t^2.$$

This suggests that *the imaginary part t of the complex space coordinate $z_{n+1} = s + it$ should be interpreted as time.* For this reason, we refer to its real part s as the *Euclidean time*.

The idea of time as an imaginary space coordinate dates back to Minkowski [M23], who realized in 1908 that Einstein's new special relativity theory can be based on a unified four-dimensional spacetime. More recently, complex spacetime has become an important concept in quantum field theory [SW64, GJ87], twistor theory [PR86] and string theory [P98]. Even so, it is generally regarded as a useful mathematical tool rather than a fundamental aspect of physical reality. For example, see the discussion in [MTW73, p. 51].

In previous works [K77, K78, K87, K90, K94], I have attempted to make complex spacetime "concrete" by giving detailed physical interpretations of the imaginary as well as the real coordinates. This is also the present motivation for developing holomorphic potential theory, where the geometrical and physical significance of $\mathbf{y}$ has been emphasized. (The *physical* significance of $\mathbf{y}$ is related to the *directivity* of the physical wavelets associated with $\phi(\mathbf{x} + i\mathbf{y})$ [HF89, K94, K00].)

Assume for simplicity that the Euclidean world is completely democratic with respect to the real space variables $\mathbf{x} \in \mathbb{R}^n$ and the Euclidean time s, so that there is no prefered direction in $\mathbb{R}^{n+1}$. Then if $\mathbf{y} \neq \mathbf{0}$, we may choose a coordinate system in which

$$\mathbf{y} = (\mathbf{0}, t), \quad t = |\mathbf{y}| > 0, \ \Rightarrow \ \mathbf{z} = (\mathbf{x}, s + it) \in \mathbb{C}^{n+1}.$$

In that case, the cylindrical coordinates in $\mathbb{R}^{n+1}$ are

$$\zeta = \mathbf{x} \cdot \hat{\mathbf{y}} = s, \qquad \rho = \sqrt{\mathbf{x}^2 - \zeta^2} = \sqrt{x^2} = r,$$

and a vector $\mathbf{x} \in \mathbb{R}^{n+1}$ is represented by

$$\mathbf{x} = r\boldsymbol{\sigma} + s\hat{\mathbf{y}}, \qquad \boldsymbol{\sigma} \in S^{n-1} \subset \mathbb{R}^n,$$

where we do not distinguish between $\boldsymbol{\sigma} \in \mathbb{R}^n$ and $(\boldsymbol{\sigma}, 0) \in \mathbb{R}^{n+1}$. As before, we denote by $d\boldsymbol{\sigma}$ the normalized surface measure on S^{n-1}, so that

$$\bar{f}(r, s) \equiv \int_{S^{n-1}} f(r\boldsymbol{\sigma} + s\hat{\mathbf{y}}) \, d\boldsymbol{\sigma}$$

is the mean of $f(\mathbf{x}, s)$ over the sphere $|\mathbf{x}| = r$.

For simplicity, we assume to begin with that the dimension n of space is *odd*, so that $n+1$ is even and (4.26) can be applied to $\tilde{\delta}(\mathbf{z})$. We will later extend our results to *even* values of n by applying (4.25). Using (4.26) with the substitutions

$$n \to n+1 \equiv 2k+2, \quad \rho \to r, \quad \zeta \to s, \quad a \to t \tag{6.1}$$

gives the action of $\tilde{\delta}(\mathbf{x}+i\mathbf{y})$ on a test function $f(\mathbf{x})$:

$$\langle \tilde{\delta}, f \rangle = \frac{t\sqrt{\pi}}{\Gamma(k+\frac{1}{2})} D_r^k F(r) \Big|_{r=t}, \tag{6.2}$$

where

$$D_r = \frac{1}{2r}\frac{\partial}{\partial r}, \quad F(r) = r^{2k-1}\left[\tilde{f}(r,0) + i\frac{t^2-r^2}{2kt}\tilde{f}_s(r,0)\right]. \tag{6.3}$$

To establish a connection with the wave equation, define the function $\tilde{f}(\mathbf{z})$ on $\mathbb{C}^{n+1}$ by the convolution

$$\tilde{f}(\mathbf{z}) \equiv \int_{\mathbb{R}^{n+1}} \tilde{\delta}(\mathbf{z}-\mathbf{x}') f(\mathbf{x}') \, d\mathbf{x}'. \tag{6.4}$$

This is an *extension* of f from $\mathbb{R}^{n+1}$ to $\mathbb{C}^{n+1}$ since

$$\mathbf{y} \to 0 \;\Rightarrow\; \tilde{\delta}(\mathbf{x}+i\mathbf{y}-\mathbf{x}') \to \delta(\mathbf{x}-\mathbf{x}') \;\Rightarrow\; \tilde{f}(\mathbf{x}+i\mathbf{y}) \to f(\mathbf{x}). \tag{6.5}$$

Now

$$\gamma(-\mathbf{z}) = \gamma(\mathbf{z}) \;\Rightarrow\; \phi(-\mathbf{z}) = \phi(\mathbf{z}) \;\Rightarrow\; \tilde{\delta}(-\mathbf{z}) = \tilde{\delta}(\mathbf{z});$$

hence,

$$\tilde{f}(\mathbf{x}+i\mathbf{y}) = \int_{\mathbb{R}^{n+1}} \tilde{\delta}(\mathbf{x}'-\mathbf{x}-i\mathbf{y}) f(\mathbf{x}') \, d\mathbf{x}'$$

$$= \int_{\mathbb{R}^{n+1}} \tilde{\delta}(\mathbf{x}'-i\mathbf{y}) f^{\times}(\mathbf{x}') \, d\mathbf{x}' = \langle \tilde{\delta}, f^{\times} \rangle, \tag{6.6}$$

where $f^{\times}$ is the test function defined by

$$f^{\times}(\mathbf{x}') = f(\mathbf{x}+\mathbf{x}') = f(\mathbf{x}+\mathbf{x}', s+s').$$

Note that in (6.6), it is $\tilde{\delta}(\mathbf{x}'-i\mathbf{y})$ rather than $\tilde{\delta}(\mathbf{x}'+i\mathbf{y})$ that acts on $f^{\times}$. We will account for this in Equation (6.7) below by letting $i \to -i$ in (6.3). Thus, since $\mathbf{z} = \mathbf{x}+i\mathbf{y} = (\mathbf{x}, s+it)$, (6.2) gives

$$\tilde{f}(\mathbf{x}, s+it) = \frac{t\sqrt{\pi}}{\Gamma(k+\frac{1}{2})} D_r^k F^{\times}(r) \Big|_{r=t},$$

$$F^{\times}(r) = r^{2k-1}\left[\tilde{f}^{\times}(r,0) + i\frac{r^2-t^2}{2kt}\tilde{f}_s^{\times}(r,0)\right] \tag{6.7}$$

with

$$\bar{f}^{\times}(r, s') \equiv \int_{S^{n-1}} f^{\times}(r\boldsymbol{\sigma} + s'\hat{\mathbf{y}}) \, d\boldsymbol{\sigma}$$

$$= \int_{S^{n-1}} f(\mathbf{x} + r\boldsymbol{\sigma} + s'\hat{\mathbf{y}}) \, d\boldsymbol{\sigma} = \int_{S^{n-1}} f(\mathbf{x} + r\boldsymbol{\sigma}, s + s') \, d\boldsymbol{\sigma}.$$

Therefore,

$$\bar{f}^{\times}(r, 0) = \int_{S^{n-1}} f(\mathbf{x} + r\boldsymbol{\sigma}, s) \, d\boldsymbol{\sigma} \qquad (6.8)$$

is the mean of f over the sphere of radius r centered at $\mathbf{x}$ and orthogonal to $\mathbf{y}$, and

$$\bar{f}^{\times}_{s'}(r, 0) \equiv \partial_{s'} \bar{f}^{\times}(r, s') \Big|_{s'=0} = \int_{S^{n-1}} f_s(\mathbf{x} + r\boldsymbol{\sigma}, s) \, d\boldsymbol{\sigma} \equiv \bar{f}^{\times}_s(r, 0) \qquad (6.9)$$

is the mean of f_s over the same sphere.

Theorem 5. *Let $n = 2k+1 \geq 3$ and $f \in C^{k+2}(\mathbb{R}^{n+1})$. Then $\tilde{f}(\mathbf{x}, s+it)$ belongs to $C^2(\mathbb{R}^{n,1})$ as a function of $(\mathbf{x}, t)$, and it is the unique "classical" solution to the following Cauchy problem for the wave equation:*

$$\partial_t^2 \, \tilde{f}(\mathbf{x}, s+it) = \Delta_{\mathbf{x}} \, \tilde{f}(\mathbf{x}, s+it) \qquad (6.10a)$$

$$\lim_{t \to 0} \tilde{f}(\mathbf{x}, s+it) = f(\mathbf{x}, s), \quad \lim_{t \to 0} \partial_t \tilde{f}(\mathbf{x}, s+it) = i f_s(\mathbf{x}, s). \qquad (6.10b)$$

Proof. The first of the initial conditions is just the extension property (6.5), which has already been established. In view of this, the second initial condition can be written as

$$\lim_{t \to 0} (\partial_s + i\partial_t) \, \tilde{f}(\mathbf{x}, s+it) = 0, \qquad (6.11)$$

which is the *Cauchy-Riemann equation* at $t = 0$. If (6.11) were to hold in a complex neighborhood U of s_0, it would imply that $\tilde{f}(\mathbf{x}, s+it)$ is holomorphic in $s + it \in U$. This shows that our extension, while not necessarily analytic, is nevertheless "analyticity-friendly."

Without assuming analyticity, we now prove that $\tilde{f}(\mathbf{x}, s+it)$ solves the above Cauchy problem. Let

$$u(\mathbf{x}, t) = \tilde{f}(\mathbf{x}, s+it), \quad v(\mathbf{x}, r) = \bar{f}^{\times}(r, 0), \quad w(\mathbf{x}, r) = i\bar{f}^{\times}_s(r, 0), \qquad (6.12)$$

where the dependence on the parameter s is suppressed. Then (6.7) becomes

$$u(\mathbf{x}, t) = \frac{t\sqrt{\pi}}{\Gamma(k+\frac{1}{2})} D_r^k \left[r^{2k-1} v(\mathbf{x}, r) + \frac{r^2 - t^2}{2kt} r^{2k-1} w(\mathbf{x}, r) \right]_{r=t}. \qquad (6.13)$$

According to (6.8), $v(x, r)$ is the mean of f over the sphere of radius r centered at x; hence, $v(x, 0) = f(x, s)$. Similarly, (6.9) states that $w(x, r)$ is the mean of if_s over the same sphere, so $w(x, 0) = if_s(x, s)$. We, therefore, need to show that u solves the following Cauchy problem

$$u_{tt}(x, t) = \Delta u(x, t),$$
$$u(x, 0) = v(x, 0), \quad u_t(x, 0) = w(x, 0). \tag{6.14}$$

By the definition of D_r, the left equation in (6.7) is

$$\frac{\sqrt{\pi}}{2^k \Gamma(k + \frac{1}{2})} \, \partial_t \left(\frac{\partial}{t \partial t} \right)^{k-1} t^{2k-1} v(x, t) = c_n \, \partial_t \left(\frac{\partial}{t \partial t} \right)^{k-1} t^{2k-1} v(x, t), \tag{6.15}$$

where

$$c_n = \frac{1}{1 \cdot 3 \cdots (2k - 1)} = \frac{1}{1 \cdot 3 \cdots (n - 2)}.$$

The right equation in (6.7) is

$$c_n \left(\frac{\partial}{r \partial r} \right)^k \left[\frac{r^2 - t^2}{2k} r^{2k-1} w(x, r) \right]_{r=t}. \tag{6.16}$$

Letting $\xi = r^2/2$, a straightforward computation shows that for any function $G(\xi)$,

$$\left(\frac{\partial}{r \partial r} \right)^k [(r^2 - t^2) G(\xi)] = \partial_\xi^k [(2\xi - t^2) G(\xi)]$$

$$= 2k \partial_\xi^{k-1} G(\xi) + (2\xi - t^2) \partial_\xi^k G(\xi)$$

$$= 2k \left(\frac{\partial}{r \partial r} \right)^{k-1} G + (r^2 - t^2) \left(\frac{\partial}{r \partial r} \right)^k G.$$

Thus, (6.16) becomes

$$c_n \left(\frac{\partial}{r \partial r} \right)^{k-1} [r^{2k-1} w(x, r)]_{r=t} = c_n \left(\frac{\partial}{t \partial t} \right)^{k-1} t^{2k-1} w(x, t). \tag{6.17}$$

The sum of (6.15) and (6.17) is precisely the solution $u(x, t)$ of the initial-value problem (6) with $n = 2k+1$, as expressed in terms of *spherical means*. See John [J55], Courant and Hilbert [CH62, pp. 699–703] or Folland [F95, p. 170].

That $u \in C^2(\mathbb{R}^{n,1})$ follows because $v(x, r) \in C^{k+2}(\mathbb{R}^n)$ in x and (6.15) contains k derivatives, while $w(x, r) \in C^{k+1}(\mathbb{R}^n)$ in x and (6.17) contains $k-1$ derivatives. Finally, uniqueness of the solution u is a general property of the Cauchy problem. $\qquad\square$

As mentioned earlier, the above result can be extended to an *even* number n of space dimensions by applying the recursion relation (4.24) between the singular source distributions in $\mathbb{C}^n$ and $\mathbb{C}^{n+1}$. In terms of the solutions $\tilde{f}(x, s + it)$ to the Cauchy problem, this amounts to using Hadamard's *method of descent* [H52, CH62]. Consequently, *the same formula* (6.6) *gives the solution of the Cauchy problem for the wave equation in* $\mathbb{R}^{n,1}$ *for all values of* $n \geq 2$.

The support properties of $\tilde{\delta}$ now imply some important attributes of waves in $\mathbb{R}^{n,1}$. For $\mathbf{z} = \mathbf{x} + i\mathbf{y} \in \mathbb{C}^{n+1}$ with $\mathbf{y} \neq \mathbf{0}$, recall that

$$\operatorname{supp} \tilde{\delta}(\mathbf{x} + i\mathbf{y}) = \begin{cases} \mathcal{B}(\mathbf{y}) & \text{for odd } n \geq 3 \\ E_0(\mathbf{y}) & \text{for even } n \geq 2. \end{cases}$$

From (6.6) we can therefore immediately draw the following conclusions about *waves* (solutions of the wave equation) $u(x, t)$ in n space dimensions:

- For odd $n \geq 3$, $u(x, t)$ depends on the values of $u(x + v, 0)$ and $u_t(x + v, 0)$ only in an arbitrarily thin shell containing the sphere $|v| = t$. (We need a shell, rather than the sphere itself, because of the derivatives appearing in (6.15) and (6.17).) This is the strong form of *Huygens' principle* [BC87], which states that u depends on the initial data only on the *light cone*.

- For even $n \geq 2$, $u(x, t)$ depends on the values of $u(x + v, 0)$ and $u_t(x + v, 0)$ in the *past cone* $|v| \leq t$. This is the *causality principle*, which states that no signal (information, energy) can travel with speed greater than $c = 1$. (If we rescale time by $t \to ct$ with arbitrary $c > 0$, then the maximum propagation speed is c.)

Note: Although $\phi(\mathbf{z})$ is holomorphic, $\tilde{\delta}(\mathbf{z})$ is *not*. Thus, $\tilde{f}(x, s + it)$ need not be holomorphic in $s + it$.

> *The Cauchy data $f(x, s)$ need not be real-analytic in s.*

This distinguishes our results from all similar results in the literature of which I am aware, where analyticity of the Cauchy data is essential. See Garabedian [G64, pp. 191–202] and Ryan [R90, R90a, R96a].

Theorem 5 and its counterpart for even n can be extended in various ways.

- Clearly it is not necessary to assume that $t > 0$ since the support of $\tilde{\delta}(\mathbf{x} + i\mathbf{y})$ is symmetric with respect to $\mathbf{y} \to -\mathbf{y}$. When $t < 0$, $\tilde{f}(x, s + it)$ is a solution of the "final-value problem" in terms of the Cauchy data at $t = 0$.

- The Cauchy data $f(\mathbf{x})$ need not belong to $C^{k+2}(\mathbb{R}^{n+1})$. When f is a distribution belonging, say, to some Sobolev space [F95, Chapter 6], then $\tilde{f}(\boldsymbol{x}, s+it)$ is a *distributional* solution and the derivatives in (6.15) and (6.17) must be interpreted as distributional derivatives.

- To solve the *inhomogeneous* wave equation

$$u_{tt}(\boldsymbol{x}, t) - \Delta u(\boldsymbol{x}, t) = j(\boldsymbol{x}, t)$$

$$u(\boldsymbol{x}, 0) = v(\boldsymbol{x}), \quad u_t(\boldsymbol{x}, 0) = w(\boldsymbol{x}),$$

one can apply Duhamel's principle to solutions of the homogeneous equation [F95]. This involves integration of $\tilde{\delta}(\boldsymbol{z}')$ on the truncated solid light cone with $0 \le t' \le t$. However, we will see that in the Clifford setting, the present formalism leads directly to solutions of inhomogeneous hyperbolic equations, where the time-dependent source is determined by the given function $f(\mathbf{x})$ in Euclidean spacetime.

7   Extension to Clifford analysis

Clifford and quaternionic analyses [BDS82, R96, GS97, RS98] are generalizations to $\mathbb{R}^n$ of one-dimensional complex analysis that are proving to be a unifying and very powerful tool in physics [H66, KS96, MM98, O98, B99]. We now show that all the above constructions generalize naturally to this setting. Let $\mathcal{Cl}_n$ be the *complex* Clifford algebra generated by elements $e_1, \ldots, e_n$ satisfying the anticommutation relations

$$e_k e_l + e_l e_k = 2\delta_{kl}, \qquad 1 \le k, l \le n. \tag{7.1}$$

As a complex vector space, $\mathcal{Cl}_n$ has dimension 2^n with basis vectors

$$e_K \equiv e_{k_1} \cdots e_{k_p}, \quad K = \{k_1, \ldots, k_p\}, \ 0 \le p \le n, \ 1 \le k_1 < \cdots < k_p \le n,$$

where the element labeled by the empty set $K = \varnothing$ ($p = 0$) is by definition $e_\varnothing = 1$. Thus, a general element has the form $v = \sum_K c_K e_K$, where $c_K \in \mathbb{C}$ and the sum runs over the 2^n sets K as above. Interpreting $\{e_k\}$ as a *vector* a basis for $\mathbb{C}^n$, we obtain an injection

$$\mathbb{C}^n \hookrightarrow \mathcal{Cl}_n : \quad \boldsymbol{z} \mapsto \sum_{k=1}^n z_k e_k.$$

Then, by (7.1), products and squares of such vectors satisfy

$$\boldsymbol{zw} + \boldsymbol{wz} = 2 \sum_{k=1}^n z_k w_k \equiv 2\,\boldsymbol{z} \cdot \boldsymbol{w}, \qquad \boldsymbol{z}^2 = \boldsymbol{z} \cdot \boldsymbol{z} \equiv \gamma(\boldsymbol{z})^2, \tag{7.2}$$

where $\gamma(z)$ is the complex Euclidean distance function. This connection will be our basis for generalizing holomorphic potential theory to the Clifford setting.

Clifford analysis is a noncommutative calculus dealing with Clifford-valued functions

$$f : \mathbb{R}^n \to \mathcal{C}l_n, \quad f(x) = \sum_K e_K f_K(x) \quad \text{where} \quad f_K : \mathbb{R}^n \to \mathbb{C}.$$

The primary tool is the *Dirac operator* $D = \sum_k e_k \partial_k$, which is closely related to the exterior derivative [AM78] but in addition incorporates the underlying metric. It acts on Clifford-valued functions from either the left or right by

$$Df \equiv \sum_{k=1}^n e_k \frac{\partial f}{\partial x_k} \neq f\overleftarrow{D} \equiv \sum_{k=1}^n \frac{\partial f}{\partial x_k} e_k \tag{7.3}$$

$$\Rightarrow D^2 f = f\overleftarrow{D}^2 = \sum_K e_K \Delta f_K = \Delta f. \tag{7.4}$$

Thus D is a "square root" of the Laplacian Δ in $\mathbb{R}^n$. It is an *elliptic* operator because the relations (7.1) are based on the Euclidean metric in $\mathbb{R}^n$. In 1928, Dirac formulated a similar operator in the Minkowskian spacetime $\mathbb{R}^{3,1}$, where it is *hyperbolic* and its square is the wave operator. This formed the basis for his relativistic wave equation of the electron [D58], which had a revolutionary impact on physics, including especially the dramatic prediction of antimatter. Mathematicians usually prefer Euclidean Dirac operators because, among other things, they yield powerful methods for solving boundary-value problems generalizing those in one-dimensional complex analysis [GS97]. For this and similar reasons, most mathematical work on Dirac operators is restricted to the elliptic case; see the discussion in [O98, p. 1]. Consequently, rigorous analyses of Maxwell's equations by Clifford methods usually deal with static or time-harmonic fields [KS96, MM98]. By assuming that the boundary/Cauchy data is holomorphic, it is possible to arrive at solutions of hyperbolic Dirac equations through analytic continuation [R90, R90a, R96a], generalizing the method employed by Garabedian for establishing a connection between solutions of the Laplace and wave equations [G64]. As in Section 6, the present method is *not* restricted by this assumption. We will "Cliffordize" the holomorphic potential theories in $\mathbb{C}^n$ and $\mathbb{C}^{n+1}$, interpreted respectively as *complex space* and *complex spacetime*. The relation established in Section 6 between Laplacians and wave operators then yields the desired connection between elliptic and hyperbolic Dirac operators.

The Clifford counterpart of the Newtonian potential $\phi(x)$ is the *Cauchy kernel* $C : \mathbb{R}^n \to \mathcal{C}l_n$, defined as the fundamental solution of D :

$$DC(x) = C(x)\overleftarrow{D} = \delta(x), \qquad \lim_{r \to \infty} C(x) = 0.$$

Because $D^2 = \overleftarrow{D}^2 = \Delta$, the solution is easily expressed in terms of the Newtonian potential:

$$\phi = \frac{r^{2-n}}{\omega_n(2-n)}, \qquad C(x) = D\phi(x) = \frac{x}{\omega_n \, r^n}. \qquad (7.5)$$

The expression on the right remains valid for $n = 2$ if we take $\phi = (2\pi)^{-1}\ln r$. Replacing $r(x)$ with $\gamma(z)$ immediately gives an extension to $\mathbb{C}^n$, where the point source δ becomes the extended source $\tilde{\delta}$:

$$C(z) \equiv D\phi(z) = \phi(z)\overleftarrow{D} = \frac{z}{\omega_n \, \gamma^n}, \qquad z \in \mathbb{C}^n, \; n \geq 2 \qquad (7.6)$$

$$\Rightarrow \quad DC(z) = C(z)\overleftarrow{D} = \Delta\phi(z) = \tilde{\delta}(z). \qquad (7.7)$$

For all even $n \geq 2$, $C(z)$ is holomorphic on the complement of the null cone $\mathcal{N}$. For odd $n \geq 3$, it inherits the branch cut from γ. Thus, for all $n \geq 2$, $C(z)$ is holomorphic on the complement of the set

$$\mathcal{S}_n = \begin{cases} \{x + iy \in \mathbb{C}^n \mid (x+iy)^2 = 0\} & \text{for even } n \geq 2 \\ \{x + iy \in \mathbb{C}^n \mid x \in E_0(y)\} & \text{for odd } n \geq 3. \end{cases}$$

Given a Clifford-valued test function $f : \mathbb{R}^n \to \mathcal{C\!l}_n$, we define its extension to $\tilde{f} : \mathbb{C}^n \to \mathcal{C\!l}_n$ exactly as before:

$$\tilde{f}(z) \equiv \int_{\mathbb{R}^n} \tilde{\delta}(x' - z)\, f(x')\, dx', \quad \Rightarrow \quad \lim_{y \to 0} \tilde{f}(x + iy) = f(x). \qquad (7.8)$$

Recall that for $\tilde{f}$ to be defined, f must be C^k with $k = \left[\frac{n-1}{2}\right]$ if $n \geq 3$. It can also be shown [K00] that f must be C^1 if $n = 2$. Substituting $\tilde{\delta} = C\overleftarrow{D}$ and integrating by parts gives

$$\tilde{f}(z) = \int_{\mathbb{R}^n} C(x' - z)\overleftarrow{D}f(x')\, dx' = -\int_{\mathbb{R}^n} C(x' - z)Df(x')\, dx', \qquad (7.9)$$

where $\overleftarrow{D}$ denotes the left-acting Dirac operator with respect to x'. We will use this expression to derive an extended version of the *Borel-Pompeiu formula*.

Let M be a bounded domain in $\mathbb{R}^n$ with piecewise smooth (C^1) boundary ∂M; let $\bar{M} = M \cup \partial M$ be its closure; let $\bar{M}' \equiv \mathbb{R}^n\backslash\bar{M}$ be exterior, and define

$$f_M(x) = \chi_M(x)\, f(x), \quad \text{where} \quad \chi_M(x) = \begin{cases} 1 & \text{if } x \in M \\ 0 & \text{if } x \in \bar{M}'. \end{cases}$$

We do not need to define f_M on ∂M. Taking the *distributional* derivatives of f_M gives

$$Df_M(x) = (D\chi_M(x))f(x) + \chi_M(x)Df(x). \qquad (7.10)$$

We want to substitute this into (7.9) to obtain an expression for the extension $\tilde{f}_M(z)$ of $f_M(x)$. This will make sense if the singularities of $Df_M(x)$ do not meet the singularities of $C(x-z)$. That will be the case if we assume that

$$z \notin \partial M + S_n \equiv \{x_b + z \mid x_b \in \partial M, \ z \in S_n\}, \qquad (7.11)$$

and we refer to such points $z \in \mathbb{C}^n$ as regular with respect to ∂M. Note that when $z \in \mathbb{R}^n$, this means simply that $z \notin \partial M$. For z regular, we may substitute (7.10) into (7.9) to obtain

$$\tilde{f}_M(z) = - \int_{\mathbb{R}^n} C(x'-z)(D\chi_M(x'))f(x')\,dx'$$

$$- \int_{\mathbb{R}^n} C(x'-z)\chi_M(x')Df(x')\,dx'. \quad (7.12)$$

We claim that the first term can be written in classical (non-distributional) form as

$$\int_{\partial M} C(x'-z)\,n(x')f(x')\,d\sigma(x'), \qquad (7.13)$$

where $n(x)$ is the outgoing unit normal at $x \in \partial M$ and $d\sigma(x)$ is the area measure on ∂M induced from the volume measure dx. To see this, note that there exists a differentiable function $\mu(x)$ such that

$$M = \{x \mid \mu(x) > 0\}, \quad \partial M = \{x \mid \mu(x) = 0\}$$
$$x \in \partial M \ \Rightarrow \ \nabla\mu(x) = -|\nabla\mu(x)|\,n(x).$$

Then, χ_M can be expressed in terms of the Heaviside step function θ by $\chi_M(x) = \theta(\mu(x))$; hence,

$$D\chi_M(x) = \delta(\mu(x))D\mu(x) = \delta(\mu(x))\nabla\mu(x) = -\delta(\mu(x))\,|\nabla\mu(x)|\,n(x).$$

The connection to the expression (7.13) can now be made by using the implicit function theorem [AM78] with $\mu(x)$ as one of the local coordinates. The second term in (7.12) reduces to the integral over M, giving the following result.

Theorem 6 (Extended Borel-Pompeiu Formula). *Let $z \in \mathbb{C}^n$ be regular with respect to ∂M. Then*

$$\tilde{f}_M(z) = \int_{\partial M} C(x'-z)\,n(x')f(x')\,d\sigma(x')$$

$$- \int_M C(x'-z)\,Df(x')\,dx'. \quad (7.14)$$

For $z \to x \in \mathbb{R}^n \backslash \partial M$, Equation (7.8) applied to f_M reproduces $f_M(x)$ since $\tilde{\delta}(x'-x) = \delta(x'-x)$. Thus (7.14) reduces to the usual Borel-Pompeiu formula for Clifford-valued functions [GS97],

$$\int_{\partial M} C(x' - x)\, n(x')\, f(x')\, d\sigma(x')$$

$$- \int_M C(x' - x)\, Df(x')\, dx' = \begin{cases} f(x), & x \in M \\ 0, & x \in \bar{M}'. \end{cases} \quad (7.15)$$

We may interpret $f(x)$ as a field generated by the source function

$$j(x) \equiv Df(x). \quad (7.16)$$

Then (7.15) solves the boundary value problem for (7.16), expressing the field f inside M in terms of its sources in M and its values on ∂M. (To investigate the limit of (7.15) as $x \to \partial M$, one also needs the *Plemelj-Sokhotzkij formula* [GS97].) In particular, f is said to be *left-monogenic* in M if $j(x) = 0$ in M. In that case, the second term in (7.15) vanishes and the Borel-Pompeiu formula reduces to a multidimensional generalization of Cauchy's integral formula, with monogenicity replacing holomorphy.

To describe waves, such as a time-dependent electromagnetic field, we ascend to $\mathbb{C}^{n+1}$ as explained earlier. In the notation of Section 6, let

$$z = x + iy \in \mathbb{C}^n, \quad \mathbf{z} = (z, z_0) = \mathbf{x} + i\mathbf{y} \in \mathbb{C}^{n+1}$$
$$\mathbf{y} = (0, t) \quad \mathbf{x} = (x, s) \;\Rightarrow\; \mathbf{z} = (x, s + it).$$

The generators of $C\ell_{n+1}$ are $\{e_0, \dots, e_n\}$, satisfying relations identical to (7.1) but with $0 \le k, l \le n$. The elliptic Dirac operators in $\mathbb{R}^n$ and $\mathbb{R}^{n+1}$ are

$$D = \sum_{k=1}^n e_k \partial_k, \quad \mathbb{D} \equiv \sum_{k=0}^n e_k \partial_k = D + e_0 \partial_s, \quad \partial_0 \equiv \partial_s.$$

We also define the *hyperbolic* (space-time) Dirac operator by formally substituting $s \to it$ in $\mathbb{D}$:

$$\tilde{\mathbb{D}} \equiv D - ie_0 \partial_t.$$

Then

$$\mathbb{D}^2 = \Delta + \partial_s^2 \quad \text{and} \quad \tilde{\mathbb{D}}^2 = \Delta - \partial_t^2 = \Box,$$

where Δ is the spatial Laplacian in $\mathbb{R}^n$ and $\Box$ is the wave operator in $\mathbb{R}^{n,1}$.

Theorem 7 (Inhomogeneous Dirac/Maxwell Equation).
For $n = 2k+1 \geq 3$, let $f : \mathbb{R}^{n+1} \to C\ell_{n+1}$ be a C^{k+2} function and define the functions $j(\boldsymbol{x}, s)$ and $\tilde{j}(\boldsymbol{x}, s + it)$ by

$$\mathbb{D}f(\boldsymbol{x}, s) = j(\boldsymbol{x}, s), \qquad \widetilde{\mathbb{D}}\tilde{f}(\boldsymbol{x}, s + it) = \tilde{j}(\boldsymbol{x}, s + it). \qquad (7.17)$$

Then $\tilde{j}(\boldsymbol{x}, s + it)$ is a C^1 solution of

$$\widetilde{\mathbb{D}}\tilde{j}(\boldsymbol{x}, s + it) = 0, \qquad (7.18)$$

and the system $\{\tilde{f}, \tilde{j}\}$ satisfies the initial conditions

$$\tilde{f}(\boldsymbol{x}, s) = f(\boldsymbol{x}, s), \qquad \tilde{j}(\boldsymbol{x}, s) = j(\boldsymbol{x}, s). \qquad (7.19)$$

Proof. This follows directly from Theorem 5. Since $\tilde{f}(\boldsymbol{x}, s + it)$ is a C^2 solution of the homogeneous wave equation, $\tilde{j}(\boldsymbol{x}, s + it)$ is C^1 and

$$\widetilde{\mathbb{D}}\tilde{j} = \widetilde{\mathbb{D}}^2 \tilde{f} = \Box \tilde{f} = 0.$$

From the initial conditions on $\tilde{f}$, we have $\tilde{f}(\boldsymbol{x}, s) = f(\boldsymbol{x}, s)$ and

$$\lim_{t \to 0} \tilde{j}(\boldsymbol{x}, s + it) = D\tilde{f}(\boldsymbol{x}, s) - ie_0 \lim_{t \to 0} \partial_t \tilde{f}(\boldsymbol{x}, s + it)$$
$$= Df(\boldsymbol{x}, s) + e_0 \partial_s f(\boldsymbol{x}, s) = \mathbb{D}f(\boldsymbol{x}, s) = j(\boldsymbol{x}, s).$$

$\square$

For $n = 3$, the right side of (7.17) is precisely the Clifford form of the inhomogeneous, time-dependent Maxwell equations [H66, O98, B99], where

$$\tilde{f}(\boldsymbol{x}, s + it) = \sum_{0=\mu<\nu}^{3} e_{\mu\nu} F^{\mu\nu}(\boldsymbol{x}, t) \equiv F(\boldsymbol{x}, t) \qquad (7.20)$$

is the electromagnetic field (a bivector),

$$\tilde{j}(\boldsymbol{x}, s + it) = \sum_{\mu=0}^{3} e_\mu J^\mu(\boldsymbol{x}, t) \equiv J(\boldsymbol{x}, t) \qquad (7.21)$$

is the charge-current density (a four-vector), and the Euclidean time s is a free parameter. As with our distance function, the Minkowskian metric appears in (7.20) and (7.21) when the time components J^0 and F^{0k} ($k = 1, 2, 3$) are imaginary and the spatial components are real.

> As in the scalar case, we do not need to assume that the Cauchy data $\{f(\boldsymbol{x}, s), j(\boldsymbol{x}, s)\}$ are analytic in s.

Note that although $\tilde{j}(x, s + it)$ is a time-evolved extension of $j(x, s)$, it can be shown [K00] that it is *not* the one obtained by convolving with $\tilde{\delta}$. Instead, the time dependence of the charge-current density is governed by (7.18), the scalar part of which is the *continuity equation* implying conservation of the total charge.

In (7.17) and (7.18), we have taken $M = \mathbb{R}^{n+1}$ for simplicity, giving a pure initial-value problem. These equations may also be formulated in a bounded region $M \subset \mathbb{R}^{n+1}$ by using the extended Borel-Pompeiu formula for $\tilde{f}_M(x, s+it)$. This suggests that Theorem 6 may be used to solve *mixed initial/boundary value problems*, provided one is careful about dealing with the singular points $(x, s + it) \in \partial M + \mathcal{S}_{n+1}$. These are in the *domain of influence* of the boundary; that is, they can be reached at time t by signals originating from the boundary ∂M at time $t = 0$. However, this still does not give a mechanism for boundary effects initiated at $t > 0$ (such as reflections) to influence the solution. This and related questions will be treated elsewhere.

Acknowledgements

This work was supported by AFOSR Contract # F49620-98-C-0013. It is a pleasure to thank Arje Nachman for encouragement and Paul Garabedian, Sigurdur Helgason, David Jerison, John Ryan and Frank Sommen for stimulating and enjoyable conversations.

REFERENCES

[AM78]   R. Abraham and J. E. Marsden, *Foundations of Classical Mechanics*, second edition, Benjamin-Cummings, Reading, 1978.

[B99]   W. E. Baylis, *Electrodynamics: A Modern Geometric Approach*, Birkhäuser, Boston, 1999.

[BC87]   B. B. Baker and E.T. Copson, *The Mathematical Theory of Huygens' Principle*, Third Edition, Chelsea, New York, 1987.

[BDS82]   F. Brackx, R. Delanghe and F. Sommen, *Clifford Analysis*, Pitman, Boston, 1982.

[C99]   T. Y. Cao (Editor), *Conceptual Foundations of Quantum Field Theory*, Cambridge University Press, 1999.

[CH62]   R. Courant and D. Hilbert, *Methods of Mathematical Physics, Vol. II*, Interscience, New York, 1962.

[D58]   P. A. M. Dirac, *The Principles of Quantum Mechanics, Fourth Edition*, Oxford, 1958.

[F64]   R. P. Feynman, *Lectures on Physics, Vol II*, Addison-Wesley, Reading, 1964.

[F95]   G. B. Folland, *Introduction to Partial Differential Equations*, Second Edition, Princeton University Press, 1995.

[G64]   P. R. Garabedian, *Partial Differential Equations*, Chelsea, New York, 1964; AMS Chelsea, Providence, 1998.

[GS64]   I. M. Gelfand and G. E. Shilov, *Generalized Functions, Volume 1: Properties and Operations*, Academic Press, New York, 1964.

[GJ87]   J. Glimm and A. Jaffe, *Quantum Physics: A Functional Integral Point of View*, Second Edition, Springer-Verlag, New York, 1987.

[GS97]   K. Gürlebeck and W. Sprössig, *Quaternionic and Clifford Calculus for Physicists and Engineers*, John Wiley & Sons, Chichester, 1997.

[H52]   J. Hadamard, *Lectures on Cauchy's Problem*, Dover, New York, 1952.

[H66]   D. Hestenes, *Space-Time Algebra*, Gordon and Breach, New York, 1966.

[HF89]   E. Heyman and L. B. Felsen, Complex source pulsed beam fields, *J. Optical Soc. America* **6**, 806–817, 1989.

[HLK00]   E. Heyman, V. Lomakin and G. Kaiser, Physical source realization of complex-source pulsed beams, *J. Acoustical Soc. of America* (to appear in April, 2000).

[J55]   F. John, *Plane Waves and Spherical Means*, Interscience, New York, 1955.

[J99]   J. D. Jackson, *Classical Electrodynamics*, third edition, John Wiley & Sons, New York, 1999.

[K77]   G. Kaiser, Phase-space approach to relativistic quantum mechanics, Part I: Coherent-state representation for massive scalar particles, *J. Math. Phys.* **18** (1977), 952–959.

[K78]   G. Kaiser, Phase-space approach to relativistic quantum mechanics, Part II: Geometrical aspects, *J. Math. Phys.* **19** (1978), 502–507.

[K80]   G. Kaiser, Phase-space approach to relativistic quantum mechanics, Part III: Quantization, relativity, localization and gauge freedom, *J. Math. Phys.* **22** (1980), 705–714.

[K87]   G. Kaiser, Quantized fields in complex spacetime, *Annals of Physics* **173** (1987), 338–354.

[K90]   G. Kaiser, *Quantum Physics, Relativity, and Complex Spacetime*, North-Holland, Amsterdam, 1990.

[K94]   G. Kaiser, *A Friendly Guide to Wavelets*, Birkhäuser, Boston, 1994.

[K00]   G. Kaiser, *Physical Wavelets and Wave Equations*, Birkhäuser, Boston, 2000, in preparation.

[KS96]   V. V. Kravchenko and M. V. Shapiro, *Integral Representations for Spatial Models of Mathematical Physics*, Addison Wesley Longman, Ltd., 1996, available from CRC Press.

[M96]   A. McIntosh, Clifford algebras, Fourier theory, singular integrals, and harmonic functions on Lipschitz domains, in J. Ryan (Editor), *Clifford Algebras in Analysis and Related Topics*, CRC, 1996, pp. 33–87.

[M23]   H. Minkowski, Space and Time, translated from the German and reprinted in *The Principle of Relativity*, H. A. Lorentz, A. Einstein, H. Minkowski and H. Weyl (Editors), Dover, 1923.

[MTW73]   C. W. Misner, K. S. Thorne and J. A. Wheeler, *Gravitation*, W.H. Freeman, San Francisco, 1973.

[MM98]   A. McIntosh and M. Mitrea, Clifford algebras and Maxwell's equations in Lipschitz domains, *Mathematical Methods in Applied Sciences* **22** (1999), 1599-1620.

[O98]   E. Obolashvili, *Partial Differential Equations in Clifford Analysis*, Addison Wesley Longman, Ltd., 1998, available from CRC Press.

[PR86]   R. Penrose and W. Rindler, *Spinors and Space-Time*, Volume 2, Cambridge University Press, 1986.

[P98]   J. Polchinski, *String Theory*, Volumes 1 and 2, Cambridge University Press, 1998.

[R90]   J. Ryan, Complex Clifford analysis and domains of holomorphy, *J. Austral. Math. Soc.* (Series A) **48** (1990), 413-433.

[R90a]   J. Ryan, Cells of harmonicity and generalized Cauchy integral formulae, *Proc. London Math. Soc.* **60** (1990), 295-318.

[R96]   J. Ryan (Editor), *Clifford Algebras in Analysis and Related Topics*, CRC, 1996.

[R96a]   J. Ryan, Intrinsic Dirac operators, *Advances in Mathematics* **118** (1996), 99-133.

[RS98]   J. Ryan and D. Struppa (Editors), *Dirac Operators in Analysis*, Addison Wesley Longman, Ltd., 1998, available from CRC Press.

[SW64]   R. F. Streater and A. S. Wightman, *PCT, Spin and Statistics, and All That*, Addison-Wesley, 1964.

[T96]   N. Temme, *Special Functions*, John Wiley & Sons, New York, 1996.

[T75]   F. Treves, *Basic Linear Partial Differential Equations*, Academic Press, New York, 1975.

[WF45]   J. A. Wheeler and R. P. Feynman, Interaction with the absorber as the mechanism of radiation, *Rev. Mod. Phys.* **17** (1945), 157-181.

[WF49]   J. A. Wheeler and R. P. Feynman, Classical electrodynamics in terms of direct interparticle action, *Rev. Mod. Phys.* **21** (1949), 425-433.

[Z65]   A. H. Zemanian, *Distribution Theory and Transform Analysis*, McGraw-Hill, New York, 1965.

Gerald Kaiser
Virginia Center for Signals and Waves
1921 Kings Road
Glen Allen, VA 23059, USA
http://www.wavelets.com
E-mail: kaiser@wavelets.com

Received: September 1, 1999; Revised: February 18, 2000

Specific Representations for Members of the Holonomy Group

John Snygg

ABSTRACT Clifford algebra is an ideal medium to express isometry operators. We derive expressions for some members of the holonomy group for n-dimensional spaces with metrics of arbitrary signatures. In particular, we derive expressions for those isometry operators which correspond to coordinate parallelograms that can be continuously shrunk to zero. The isometry operators are expressed in terms of infinite series which are defined by two recursion relations.
Keywords: Holonomy group, Clifford algebra.

1 The 2-dimensional case

When a vector is parallel transported around a closed loop, it undergoes an isometry. For a 2-dimensional space with a Euclidean signature, the angle of rotation is equal to the Gaussian curvature integrated over the area bounded by the loop. In higher dimensions with arbitrary signature, the isometries are more complicated. For a long time (Schouten 1918) and (Pérès 1919), it has been known that, for an infinitesimal loop, the isometry is determined by the Riemann curvature tensor. More elaborate details on the nature of the isometry group and the corresponding algebra for infinitesimal loops were developed by Ambrose and Singer (1953). In this paper we go beyond infinitesimal loops for the special case where the loop is a coordinate parallelogram.

In Clifford algebra, if a vector $\mathbf{v}'$ is the result of applying an isometry operator $\Re$ to a vector $\mathbf{v}$, we may write

$$\mathbf{v}' = \Re \mathbf{v} \Re^{-1}. \tag{1.1}$$

[See (Riesz 1993: 156-173) or (Snygg 1997: 1-10).]

For a coordinate frame, $\mathbf{v} = v_\alpha \gamma^\alpha = v^\alpha \gamma_\alpha$ where

$$\gamma_\alpha \gamma_\beta + \gamma_\beta \gamma_\alpha = 2g_{\alpha\beta}, \ \gamma^\alpha \gamma^\beta + \gamma^\beta \gamma^\alpha = 2g^{\alpha\beta}, \ \gamma_\alpha \gamma^\beta + \gamma^\beta \gamma_\alpha = 2\delta_\alpha^\beta.$$

AMS Subject Classifications: 53C29, 81Q70.

For an orthonormal non-coordinate frame, $\mathbf{v} = v_j \hat{\gamma}^j = v^k \hat{\gamma}_k$ where

$$\hat{\gamma}_j \hat{\gamma}_k + \hat{\gamma}_k \hat{\gamma}_j = 2n_{jk}, \quad \hat{\gamma}^j \hat{\gamma}^k + \hat{\gamma}^k \hat{\gamma}^j = 2n^{jk}, \quad \hat{\gamma}_j \hat{\gamma}^k + \hat{\gamma}^k \hat{\gamma}_j = 2\delta_j^k.$$

We also define an intrinsic differential operator ∇_α :

$$\nabla_\alpha \gamma_\beta = \Gamma_{\alpha\beta}^\nu \gamma_\nu, \quad \nabla_\alpha \gamma^\beta = -\Gamma_{\alpha\nu}^\beta \gamma^\nu,$$

where $\Gamma_{\alpha\beta}^\nu$ is the usual Christoffel symbol. We also define

$$\nabla_\alpha v^\beta = \frac{\partial}{\partial x^\alpha} v^\beta, \quad \nabla_\alpha v_\beta = \frac{\partial}{\partial x^\alpha} v_\beta.$$

This definition for ∇_α is extended to arbitrary Clifford numbers by the usual Leibniz and summation properties. This differential operator is the natural generalization of a derivative applied to vectors tangent to a curved n-dimensional surface embedded in a flat m-dimensional space. However when ∇_α is applied to a tensor, the result is generally not a tensor. In my text, I refer to this as a covariant derivative, but on reflection I see that this is wrong. For further details, see (Snygg 1997: 68-96).

For a non-coordinate basis, an adjustment must be made. This adjustment is most easily made using Fock-Ivanenko 2-vectors [see (Snygg 1997: 90-103)]:

$$\nabla_\alpha \hat{\gamma}_k = -\Gamma_\alpha \hat{\gamma}_k + \hat{\gamma}_k \Gamma_\alpha, \quad \nabla_\alpha \hat{\gamma}^k = -\Gamma_\alpha \hat{\gamma}^k + \hat{\gamma}^k \Gamma_\alpha,$$

where

$$\Gamma_\alpha = \frac{1}{4} \gamma^{\beta\nu} \frac{\partial g_{\alpha\nu}}{\partial x^\beta} - \frac{1}{4} \gamma^\beta \wedge \frac{\partial \gamma_\beta}{\partial x^\alpha}.$$

In general

$$\nabla_\alpha [\ \ ] = \frac{\partial}{\partial x^\alpha} [\ \ ] - \Gamma_\alpha [\ \ ] + [\ \ ] \Gamma_\alpha, \tag{1.2}$$

where any differentiable Clifford number can be inserted into the square brackets.[1]

These Fock-Ivanenko 2-vectors are particularly useful to compute curvature 2-forms. A curvature 2-form is defined by the equation

$$\Re_{\alpha\beta} = \frac{1}{2} R_{\alpha\beta\nu\eta} \gamma^{\nu\eta} = \frac{1}{2} R_{\alpha\beta jk} \hat{\gamma}^{jk},$$

where

$$\gamma^{\nu\eta} = \frac{1}{2}(\gamma^\nu \gamma^\eta - \gamma^\eta \gamma^\nu), \quad \hat{\gamma}^{jk} = \frac{1}{2}(\hat{\gamma}^j \hat{\gamma}^k - \hat{\gamma}^k \hat{\gamma}^j)$$

---

[1]It is understood that the $\hat{\gamma}$'s (but not the γ's) are treated as constants w.r.t. the $\partial/\partial x^\alpha$ operator.

and $R_{\alpha\beta\nu\eta}$ is the usual Riemann tensor.

Note that we are using Greek letters for coordinate indices and Latin letters for non-coordinate indices, so generally $R_{\alpha\beta\nu\eta} \neq R_{\alpha\beta jk}$ even if numerically $\nu = j$ and $\eta = k$.

We also have

$$\frac{1}{2}\Re_{\alpha\beta} = \frac{\partial}{\partial x^\alpha}\Gamma_\beta - \frac{\partial}{\partial x^\beta}\Gamma_\alpha - \Gamma_\alpha\Gamma_\beta + \Gamma_\beta\Gamma_\alpha. \tag{1.3}$$

[See (Snygg 1997: 101-103).] We are now in a position to prove our first theorem.

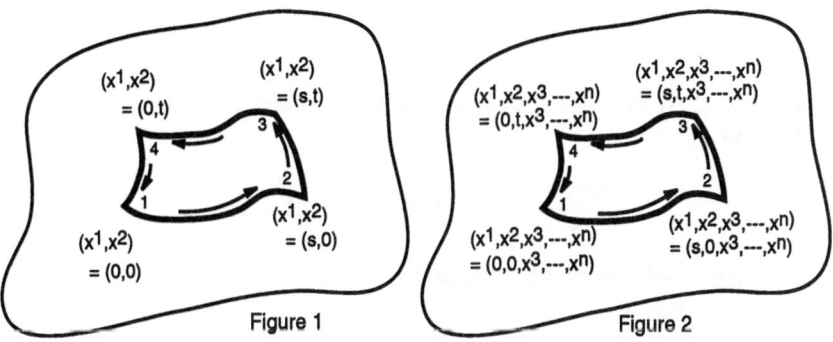

Figure 1   Figure 2

Theorem 1. *If a vector is parallel transported around a coordinate parallelogram in a curved 2-dimensional space, it undergoes an isometry. The isometric operator is*

$$\Re = \exp\left(\frac{1}{2}\int_0^s\int_0^t \Re_{12}(x^1, x^2)dx^1 dx^2\right), \tag{1.4}$$

where $\hat{\gamma}^{12}$ is treated as a constant when the integration is carried out.

Proof. Consider a coordinate parallelogram on a curved 2-dimensional surface. (See Fig. 1). If a vector $\mathbf{v}$ is parallel transported along the coordinate line x^1 while holding $x^2 = 0$, we have

$$\nabla_1\mathbf{v} = \mathbf{0}. \tag{1.5}$$

In this context (1.1) becomes

$$\mathbf{v}(s,0) = \Re(s)\mathbf{v}(0,0)\Re(s)^{-1}. \tag{1.6}$$

It then follows that

$$\frac{\partial\mathbf{v}(s,0)}{\partial s} = \frac{\partial\Re(s)}{\partial s}\mathbf{v}(0,0)\Re(s)^{-1} + \Re(s)\mathbf{v}(0,0)\frac{\partial\Re(s)^{-1}}{\partial s}$$
$$= \frac{\partial\Re(s)}{\partial s}(\Re(s))^{-1}\mathbf{v}(s,0) + \mathbf{v}(s,0)\Re(s)\frac{\partial(\Re(s))^{-1}}{\partial s}. \tag{1.7}$$

Using the fact that $\Re(s)\Re(s)^{-1} = \mathbf{I}$ and thus

$$\Re(s)\frac{\partial\,(\Re(s))^{-1}}{\partial s} = -\frac{\partial\Re(s)}{\partial s}\Re(s)^{-1},$$

it follows that (1.7) becomes

$$\frac{\partial\mathbf{v}(s,0)}{\partial s} = \frac{\partial\Re(s)}{\partial s}\Re(s)^{-1}\mathbf{v}(s,0) - \mathbf{v}(s,0)\frac{\partial\Re(s)}{\partial s}\Re(s)^{-1}. \tag{1.8}$$

Comparing (1.2) and (1.5), we have

$$\frac{\partial\mathbf{v}(s,0)}{\partial s} = \Gamma_1(s,0)\mathbf{v}(s,0) - \mathbf{v}(s,0)\Gamma_1(s,0). \tag{1.9}$$

Comparing (1.8) and (1.9), we see that we are faced with the problem of solving the equation

$$\frac{\partial\Re(s)}{\partial s} = \Gamma_1(s,0)\Re(s), \quad \Re(0) = \mathbf{I}. \tag{1.10}$$

For 2-dimensional spaces, all 2-vectors are scalar multiples of $\hat{\gamma}^{12}$, and thus commute. Therefore,

$$\Re(s) = \exp\int_0^s \Gamma_1(x^1,0)dx^1.$$

This is the isometry operator for the first edge of the parallelogram in Fig. 1. If we combine the isometry operators for all four edges, we get the isometry operator for the complete circuit:

$$\Re = \Re_{4\to1}\Re_{3\to4}\Re_{2\to3}\Re_{1\to2}, \tag{1.11}$$

where

$$\Re_{4\to1} = \exp\left(-\int_0^t \Gamma_2(0,x^2)dx^2\right), \tag{1.12}$$

$$\Re_{3\to4} = \exp\left(-\int_0^s \Gamma_1(x^1,t)dx^1\right), \tag{1.13}$$

$$\Re_{2\to3} = \exp\left(\int_0^t \Gamma_2(s,x^2)dx^2\right), \tag{1.14}$$

$$\Re_{1\to2} = \exp\left(\int_0^s \Gamma_1(x^1,0)dx^1\right). \tag{1.15}$$

Since Γ_1 and Γ_2 commute, (1.3) becomes

$$\frac{1}{2}\Re_{12}(x^1,x^2) = \frac{\partial}{\partial x^1}\Gamma_2(x^1,x^2) - \frac{\partial}{\partial x^2}\Gamma_1(x^1,x^2)$$

and therefore,

$$\frac{1}{2}\int_0^s\int_0^t \Re_{12}(x^1,x^2)dx^1\,dx^2$$

$$= \int_0^t \left[\Gamma_2(s,x^2) - \Gamma_2(0,x^2)\right]dx^2 - \int_0^s \left[\Gamma_1(x^1,t) - \Gamma_1(x^1,0)\right]dx^1.$$

Comparing this equation with (1.11), (1.12), (1.13), (1.14), and (1.15), we arrive at(1.4). □

This can obviously be generalized to any loop in the 2-dimensional surface that can be continuously shrunk to zero. Because of commutation problems, this does not easily generalize to higher dimensions.

Before moving on to the n-dimensional case, consider an alternate form for (1.4). Suppose the Riemann curvature tensor is analytic; then

$$\Re = \exp\left(\sum_{n=1}\sum_{m=1}\frac{s^n}{n!}\frac{t^m}{m!}C_{nm}(0,0)\right),$$

$$\text{where} \quad C_{nm}(s,t) = \frac{\partial^{n+m-2}}{(\partial s)^{n-1}(\partial t)^{m-1}}\frac{1}{2}\Re_{12}(s,t). \tag{1.16}$$

For the n-dimensional case, we will achieve a result similar to (1.16).

2   Maclaurin's series for the n-dimensional case

For the n-dimensional problem, we wish to consider a vector that is parallel transported around a coordinate parallelogram. The coordinate parallelogram is confined to a 2-dimensional coordinate surface, but the vector is not. The isometric operator is the identity operator if the n-dimensional space has zero curvature.

Before proving our next theorem, we need some definitions and a pair of lemmas. Let the operators ∂_s, ∂_t, $\hat{\partial}_s$, and $\hat{\partial}_t$ be defined as follows:

Definition 1.

$$\partial_s\mathbf{X}(s,t) = \frac{\partial}{\partial s}\mathbf{X}(s,t) + \mathbf{X}(s,t)\Gamma_s(s,t), \tag{2.1}$$

$$\partial_t\mathbf{X}(s,t) = \frac{\partial}{\partial t}\mathbf{X}(s,t) + \mathbf{X}(s,t)\Gamma_t(s,t), \tag{2.2}$$

$$\hat{\partial}_s\mathbf{X}(s,t) = \frac{\partial}{\partial s}\mathbf{X}(s,t) - \Gamma_s(s,t)\mathbf{X}(s,t), \tag{2.3}$$

$$\hat{\partial}_t\mathbf{X}(s,t) = \frac{\partial}{\partial t}\mathbf{X}(s,t) - \Gamma_t(s,t)\mathbf{X}(s,t), \tag{2.4}$$

where $\mathbf{X}(s,t)$ *is any differentiable Clifford number.*

Lemma 1.

$$\partial_s^n(\mathbf{X}(s,t)\mathbf{Y}(s,t)) = \sum_{p=0}^{n} \binom{n}{p} \frac{\partial^p \mathbf{X}(s,t)}{\partial s^p} \partial_s^{n-p}\mathbf{Y}(s,t) \qquad (2.5)$$

where $\mathbf{X}$ and $\mathbf{Y}$ are any Clifford numbers that are differentiable sufficiently many times w.r.t. s.

Proof. We first note that, from (2.1),

$$\partial_s(\mathbf{XY}) = \left(\frac{\partial}{\partial_s}\mathbf{X}\right)\mathbf{Y} + \mathbf{X}\left(\frac{\partial}{\partial_s}\mathbf{Y} + \mathbf{Y}\Gamma_s\right) = \left(\frac{\partial}{\partial_s}\mathbf{X}\right)\mathbf{Y} + \mathbf{X}\partial_s\mathbf{Y}.$$

Using the fact that $\binom{n}{p} + \binom{n}{p-1} = \binom{n+1}{p}$, it is not difficult to complete the proof by induction on n. □

A second lemma can be proven in a similar fashion. Namely:

Lemma 2.

$$\hat{\partial}_t^m(\mathbf{X}(s,t)\mathbf{Y}(s,t)) = \sum_{q=0}^{m} \binom{m}{q} \left(\hat{\partial}_t^q \mathbf{X}(s,t)\right)\left(\frac{\partial^{m-q}}{\partial t^{m-q}}\mathbf{Y}(s,t)\right), \qquad (2.6)$$

where X and Y are Clifford numbers differentiable sufficiently many times w.r.t. t.

Our first theorem for the n-dimensional case is:

Theorem 2. *A vector that is parallel transported around a coordinate parallelogram in n-dimensional space undergoes an isometry. Assuming the coordinate parallelogram can be continuously shrunk to zero, the isometric operator is*

$$\Re = \sum_{n=0}^{\infty}\sum_{m=0}^{\infty} \frac{s^n t^m}{n!\,m!}\mathbf{A}_{nm}(0,0), \quad where$$

$$\mathbf{A}_{nm}(s,t) = \sum_{p=0}^{n}\sum_{q=0}^{m} \binom{n}{p}\binom{m}{q}(\hat{\partial}_t^q\hat{\partial}_s^p\mathbf{I})(\partial_s^{n-p}\partial_t^{m-q}\mathbf{I}),$$

and the operator $\mathbf{I}$ is the identity operator.

Proof. To avoid an excess of superscripts, it is convenient to designate the curvature 2-form as follows:

$$\frac{1}{2}\Re_{12}(s,t,x^3,\ldots,x^n) = \frac{1}{4}R_{12\alpha\beta}(s,t,x^3,\ldots,x^n)\gamma^{\alpha\beta} = \frac{1}{2}\Re_{st}(s,t). \qquad (2.7)$$

For the first leg of our journey around the coordinate parallelogram in Fig. 2, we can use essentially the same argument used to derive (1.10). The resulting equation is

$$\frac{\partial}{\partial s}\Re_{1\to2}(s,0) = \Gamma_s(s,0)\Re_{1\to2}(s,0) = (\partial_s\mathbf{I}|_{t=0})\,\Re_{1\to2}(s,0). \qquad (2.8)$$

Because of the lack of commutativity, the solution of this equation for higher dimensional spaces is not as simple as for 2-dimensional spaces, except for special cases. However we can obtain a solution represented by an infinite series. Taking the derivative of (2.8) w.r.t. s, using (2.8), and then (2.5), we have

$$\frac{\partial^2}{\partial s^2}\Re_{1\to2}(s,0) = \frac{\partial}{\partial s}\left((\partial_s\mathbf{I})|_{t=0}\right)\Re_{1\to2}(s,0) + \left((\partial_s\mathbf{I})|_{t=0}\right)\left(\frac{\partial}{\partial s}\Re_{1\to2}(s,0)\right)$$

$$= \left(\frac{\partial}{\partial s}\left((\partial_s\mathbf{I})|_{t=0}\right) + \left((\partial_s\mathbf{I}|_{t=0})^2\right)\right)\Re_{1\to2}(s,0)$$

$$= \left((\partial_s^2\mathbf{I})|_{t=0}\right)\Re_{1\to2}(s,0).$$

Repeating this process, we get

$$\frac{\partial^n}{\partial s^n}\Re_{1\to2}(s,0) = \left((\partial_s^n\mathbf{I})\,\Re_{1\to2}(s,t)\right)|_{t=0}. \qquad (2.9)$$

Applying (2.9) with the assumption that $\Re_{1\to2}(s,0)$ can be represented by a Maclaurin's series in s and noting $\Re_{1\to2}(0,0) = \mathbf{I}$, we have[2]

$$\Re_{1\to2}(s,0) = \sum_{n=0}^{\infty} \frac{s^n}{n!}\left((\partial_s^n\mathbf{I})|_{s,t=0}\right). \qquad (2.10)$$

Similarly, assuming $\Re_{2\to3}(s,t)$ can be represented by a Maclaurin's series in t with s held constant, we have

$$\Re_{2\to3}(s,t) = \sum_{m=0}^{\infty} \frac{t^m}{m!}\left((\partial_t^m\mathbf{I})|_{t=0}\right). \qquad (2.11)$$

If we now assume that the R.H.S. of (2.11) can be expanded as a Maclaurin's series in s, we have

$$\Re_{2\to3}(s,t) = \sum_{p=0}^{\infty}\sum_{m=0}^{\infty} \frac{s^p t^m}{p!\,m!}\left(\left(\frac{\partial^p}{\partial s^p}\partial_t^m\mathbf{I}\right)\Big|_{s,t=0}\right),$$

and thus,

$$\Re_{1\to2\to3} = \Re_{2\to3}\Re_{1\to2}$$

$$= \sum_{p=0}^{\infty}\sum_{m=0}^{\infty} \frac{s^p t^m}{p!\,m!}\left(\left(\frac{\partial^p}{\partial s^p}\partial_t^m\mathbf{I}\right)\Big|_{s,t=0}\right)\sum_{q=0}^{\infty} \frac{s^q}{q!}\left((\partial_s^q\mathbf{I})|_{s,t=0}\right).$$

---

[2]It is understood here that $\partial_s^0\mathbf{X}(s,t) = \mathbf{X}(s,t)$.

Reorganizing the double sum over the exponents of s, we have:

$$\Re_{1\to 2\to 3}$$
$$= \sum_{m=0}^{\infty} \frac{t^m}{m!} \sum_{n=0}^{\infty} \frac{s^n}{n!} \sum_{p=0}^{n} \binom{n}{p} \left(\frac{\partial^p}{\partial s^p} \partial_t^m \mathbf{I} \right)\bigg|_{s,t=0} \left(\partial_s^{n-p} \mathbf{I} \right)\big|_{s,t=0} . \quad (2.12)$$

This expression can be simplified by applying (2.5):

$$\Re_{1\to 2\to 3} = \sum_{n=0}^{\infty} \sum_{m=0}^{\infty} \frac{s^n \, t^m}{n! \, m!} \left(\partial_s^n \left((\partial_t^m \mathbf{I}) \, \mathbf{I} \right)\big|_{s,t=0} \right)$$

$$= \sum_{n=0}^{\infty} \sum_{m=0}^{\infty} \frac{s^n \, t^m}{n! \, m!} \left((\partial_s^n \partial_t^m \mathbf{I})\big|_{s,t=0} \right) . \quad (2.13)$$

To compute the last two legs of our loop, we note that

$$0 = \frac{\partial}{\partial s} \left(\Re_{4\to 3} \Re_{3\to 4} \right) = \frac{\partial \Re_{4\to 3}}{\partial s} \left(\Re_{3\to 4} \right) + \Re_{4\to 3} \frac{\partial \Re_{3\to 4}}{\partial s}$$

$$= \Gamma_s \Re_{4\to 3} \Re_{3\to 4} + \Re_{4\to 3} \frac{\partial \Re_{3\to 4}}{\partial s},$$

and thus,

$$\frac{\partial \Re_{3\to 4}(s,t)}{\partial s} = -\Re_{3\to 4}(s,t) \Gamma_s(s,t).$$

Similarly,

$$\frac{\partial \Re_{4\to 1}(0,t)}{\partial t} = -\Re_{4\to 1}(0,t) \Gamma_t(0,t).$$

Completing computations similar to those for $\Re_{1\to 2\to 3}$, we get

$$\Re_{4\to 1}(0,t) = \sum_{q=0}^{\infty} \frac{t^q}{q!} \left(\hat{\partial}_t^q \mathbf{I}\big|_{s,t=0} \right),$$

$$\Re_{3\to 4}(s,t) = \sum_{n=0}^{\infty} \frac{s^n}{n!} \sum_{p=0}^{\infty} \frac{t^p}{p!} \left(\frac{\partial^p}{\partial t^p} \hat{\partial}_s^n \mathbf{I}\big|_{s,t=0} \right),$$

$$\Re_{3\to 4\to 1}(s,t) = \sum_{n=0}^{\infty} \frac{s^n}{n!} \sum_{m=0}^{\infty} \frac{t^m}{m!} \left(\hat{\partial}_t^m \hat{\partial}_s^n \mathbf{I}\big|_{s,t=0} \right) . \quad (2.14)$$

Combining (2.13) and (2.14), we get

$$\Re = \Re_{3\to 4\to 1} \Re_{1\to 2\to 3} =$$

$$\sum_{n=0}^{\infty} \sum_{m=0}^{\infty} \sum_{p=0}^{\infty} \sum_{q=0}^{\infty} \frac{s^n \, t^m \, s^p \, t^q}{n! \, m! \, p! \, q!} \left((\hat{\partial}_t^m \hat{\partial}_s^n \mathbf{I})(\partial_s^p \partial_t^q \mathbf{I})\big|_{s,t=0} \right) =$$

$$\sum_{n=0}^{\infty} \sum_{m=0}^{\infty} \frac{s^n \, t^m}{n! \, m!} \sum_{p=0}^{n} \sum_{q=0}^{m} \binom{n}{p} \binom{m}{q} \left((\hat{\partial}_t^q \hat{\partial}_s^p \mathbf{I})(\partial_s^{n-p} \partial_t^{m-q} \mathbf{I})\big|_{s,t=0} \right) . \quad (2.15)$$

Thus,

$$\Re = \sum_{n=0}^{\infty} \sum_{m=0}^{\infty} \frac{s^n}{n!} \frac{t^m}{m!} \mathbf{A}_{nm}(0,0), \tag{2.16}$$

where

$$\mathbf{A}_{nm}(s,t) = \sum_{p=0}^{n} \sum_{q=0}^{m} \binom{n}{p} \binom{m}{q} (\hat{\partial}_t^q \hat{\partial}_s^p \mathbf{I})(\partial_s^{n-p} \partial_t^{m-q} \mathbf{I}). \tag{2.17}$$

□

Because of definitions of the operators in (2.17), the formula for $\mathbf{A}_{nm}(s,t)$ appears to depend on the choice of the non-coordinate basis that determines the Fock-Ivanenko 2-vectors. However, it is possible to show that each $\mathbf{A}_{nm}$ is dependent only on the intrinsic derivatives of the curvature 2-form $\Re_{st}$. To prove this, we first need to prove another lemma:

Lemma 3.

$$\nabla_s^n (\mathbf{XY}) = \sum_{p=0}^{n} \binom{n}{p} (\hat{\partial}_s^p \mathbf{X})(\partial_s^{n-p} \mathbf{Y}). \tag{2.18}$$

Proof. It is not difficult to prove this formula by induction. Perhaps the most difficult part of the induction is probably the first step. Namely:

$$\nabla_s (\mathbf{XY}) = \frac{\partial \mathbf{X}}{\partial s} \mathbf{Y} + \mathbf{X} \frac{\partial \mathbf{Y}}{\partial s} - \Gamma_s \mathbf{XY} + \mathbf{XY}\Gamma_s$$

$$= \left(\frac{\partial \mathbf{X}}{\partial s} - \Gamma_s \mathbf{X} \right) \mathbf{Y} + \mathbf{X} \left(\frac{\partial \mathbf{Y}}{\partial s} + \mathbf{Y}\Gamma_s \right)$$

$$= \left(\hat{\partial}_s \mathbf{X} \right) \mathbf{Y} + \mathbf{X} \left(\partial_s \mathbf{Y} \right). \tag{2.19}$$

The rest of the induction is reasonably straightforward.   □

We now have the tools to prove the following theorem:

Theorem 3. *For* $\mathbf{A}_{nm}(s,t)$ *defined by (2.17) and* $n \geq 1$, $m \geq 1$,

$$\mathbf{A}_{00}(s,t) = \mathbf{I}, \quad \mathbf{A}_{n0}(s,t) = 0, \quad \mathbf{A}_{0m}(s,t) = 0, \tag{2.20}$$

$$\mathbf{A}_{n1}(s,t) = \nabla_s^{n-1} \frac{1}{2} \Re_{st}(s,t), \quad \mathbf{A}_{1m}(s,t) = \nabla_t^{m-1} \frac{1}{2} \Re_{st}(s,t) \tag{2.21}$$

$$\mathbf{A}_{nm+1}(s,t) = \nabla_t \mathbf{A}_{nm}(s,t) + \sum_{j=1}^{n} \binom{n}{j} \mathbf{A}_{n-jm}(s,t)\mathbf{A}_{j1}(s,t), \tag{2.22}$$

$$\mathbf{A}_{n+1m}(s,t) = \nabla_s \mathbf{A}_{nm}(s,t) + \sum_{j=1}^{m} \binom{m}{j} \mathbf{A}_{1j}(s,t)\mathbf{A}_{nm-j}(s,t). \tag{2.23}$$

Proof. From (2.17), we have

$$A_{n0}(s,t) = \sum_{p=0}^{n} \binom{n}{p}(\hat{\partial}_s^p I)(\partial_s^{n-p} I). \tag{2.24}$$

Applying (2.18) to (2.24), we get

$$A_{n0}(s,t) = \sum_{p=0}^{n} \binom{n}{p}(\hat{\partial}_s^p I)(\partial_s^{n-p} I) = \nabla_s^n (I \cdot I) = \nabla_s^n (I).$$

Thus,

$$A_{00}(s,t) = I, \quad A_{n0}(s,t) = 0, \quad n \geq 1. \tag{2.25}$$

Similarly,

$$A_{0m}(s,t) = 0, \quad m \geq 1. \tag{2.26}$$

We now set out to establish our two recursion relations for the rest of the A_{nm}. To do this, we first observe that, applying (2.19) to (2.17), we find

$$\nabla_t A_{nm}(s,t) = \sum_{p=0}^{n}\sum_{q=0}^{m} \binom{n}{p}\binom{m}{q}(\hat{\partial}_t^{q+1}\hat{\partial}_s^p I)(\partial_s^{n-p}\partial_t^{m-q} I)$$

$$+ \sum_{p=0}^{n}\sum_{q=0}^{m} \binom{n}{p}\binom{m}{q}(\hat{\partial}_t^q \hat{\partial}_s^p I)(\partial_t \partial_s^{n-p}\partial_t^{m-q} I). \tag{2.27}$$

To deal with the second sum on the R.H.S. of (2.27), we need a formula which follows from a straightforward computation, namely,

$$(\partial_t \partial_s - \partial_s \partial_t) X = -\frac{1}{2} X \Re_{st}, \tag{2.28}$$

where X is any twice differentiable Clifford number. Using the telescopic property and (2.28), we see that for $n - p \geq 1$,

$$\partial_t \partial_s^{n-p} X = \sum_{k=0}^{n-p-1} \partial_s^k (\partial_t \partial_s - \partial_s \partial_t) \partial_s^{n-p-1-k} X + \partial_s^{n-p}\partial_t X$$

$$= \partial_s^{n-p}\partial_t X - \sum_{k=0}^{n-p-1} \partial_s^k \left((\partial_s^{n-p-1-k} X) \frac{1}{2}\Re_{st} \right). \tag{2.29}$$

One can show by induction on k that

$$\partial_s^k (WY) = \sum_{j=0}^{k} \binom{k}{j} (\partial_s^{k-j} W)(\nabla_s^j Y). \tag{2.30}$$

If we let $\mathbf{W} = \partial_s^{n-p-1-k}\mathbf{X}$ and $\mathbf{Y} = \frac{1}{2}\Re_{st}$, then (2.29) becomes

$$\partial_t\partial_s^{n-p}\mathbf{X} = \partial_s^{n-p}\partial_t\mathbf{X} - \frac{1}{2}\sum_{k=0}^{n-p-1}\sum_{j=0}^{k}\binom{k}{j}(\partial_s^{n-p-1-j}\mathbf{X})\,\nabla_s^j\Re_{st}.$$

Reversing the order of summation in this last equation, we get

$$\partial_t\partial_s^{n-p}\mathbf{X} = \partial_s^{n-p}\partial_t\mathbf{X} - \frac{1}{2}\sum_{j=0}^{n-p-1}\sum_{k=j}^{n-p-1}\binom{k}{j}(\partial_s^{n-p-1-j}\mathbf{X})\,\nabla_s^j\Re_{st}$$

$$= \partial_s^{n-p}\partial_t\mathbf{X} - \frac{1}{2}\sum_{j=0}^{n-p-1}\binom{n-p}{j+1}(\partial_s^{n-p-1-j}\mathbf{X})\,\nabla_s^j\Re_{st}.$$

Separating out the $p = n$ term in the second sum of (2.27) and applying this last equation to the remaining terms of the second sum with $\mathbf{X} = \partial_t^{m-q}\mathbf{I}$, we get

$$\nabla_t\mathbf{A}_{nm}(s,t)$$

$$= \sum_{p=0}^{n}\sum_{q=0}^{m}\binom{n}{p}\binom{m}{q}(\hat{\partial}_t^{q+1}\hat{\partial}_s^p\mathbf{I})(\partial_s^{n-p}\partial_t^{m-q}\mathbf{I})$$

$$+ \sum_{q=0}^{m}\binom{m}{q}(\hat{\partial}_t^q\hat{\partial}_s^n\mathbf{I})(\partial_t^{m+1-q}\mathbf{I})$$

$$+ \sum_{p=0}^{n-1}\sum_{q=0}^{m}\binom{n}{p}\binom{m}{q}(\hat{\partial}_t^q\hat{\partial}_s^p\mathbf{I})(\partial_s^{n-p}\partial_t^{m+1-q}\mathbf{I}) \qquad (2.31)$$

$$- \frac{1}{2}\sum_{p=0}^{n-1}\sum_{q=0}^{m}\binom{n}{p}\binom{m}{q}(\hat{\partial}_t^q\hat{\partial}_s^p\mathbf{I})\sum_{j=0}^{n-p-1}\binom{n-p}{j+1}(\partial_s^{n-p-1-j}\partial_t^{m-q}\mathbf{I})\,(\nabla_s^j\Re_{st}).$$

To simplify the R.H.S. of (2.31), we first replace the dummy index q in the first sum by $q - 1$. We also combine the 2nd and 3rd sums and reverse the p and j summations in the 4th sum. The result for $n \geq 1$ is

$$\nabla_t\mathbf{A}_{nm}(s,t)$$

$$= \sum_{p=0}^{n}\sum_{q=1}^{m+1}\binom{n}{p}\binom{m}{q-1}(\hat{\partial}_t^q\hat{\partial}_s^p\mathbf{I})(\partial_s^{n-p}\partial_t^{m+1-q}\mathbf{I})$$

$$+ \sum_{p=0}^{n}\sum_{q=0}^{m}\binom{n}{p}\binom{m}{q}(\hat{\partial}_t^q\hat{\partial}_s^p\mathbf{I})(\partial_s^{n-p}\partial_t^{m+1-q}\mathbf{I}) \qquad (2.32)$$

$$- \frac{1}{2}\sum_{q=0}^{m}\binom{m}{q}\sum_{j=0}^{n-1}\sum_{p=0}^{n-1-j}\binom{n}{p}\binom{n-p}{j+1}(\hat{\partial}_t^q\hat{\partial}_s^p\mathbf{I})(\partial_s^{n-p-1-j}\partial_t^{m-q}\mathbf{I})\,(\nabla_s^j\Re_{st}).$$

Recognizing that

$$\binom{m}{q-1} + \binom{m}{q} = \binom{m+1}{q}, \quad \binom{m}{m} = \binom{m+1}{m+1}, \quad \binom{m}{0} = \binom{m+1}{0},$$

we see that the 1st two sums in (2.32) combine to give us $\mathbf{A}_{nm+1}(s,t)$. To deal with the last sum, we also note that

$$\binom{n}{p}\binom{n-p}{j+1} = \binom{n}{j+1}\binom{n-1-j}{p}.$$

The result is then

$$\nabla_t \mathbf{A}_{nm}(s,t) - \mathbf{A}_{nm+1}(s,t)$$

$$= -\frac{1}{2}\sum_{j=0}^{n-1}\binom{n}{j+1}\sum_{p=0}^{n-1-j}\sum_{q=0}^{m}\binom{n-1-j}{p}\binom{m}{q}$$

$$\times (\hat{\partial}_t^q \hat{\partial}_s^p \mathbf{I})(\partial_s^{n-1-j-p}\partial_t^{m-q}\mathbf{I})\left(\nabla_s^j \mathfrak{R}_{st}\right)$$

$$= -\frac{1}{2}\sum_{j=0}^{n-1}\binom{n}{j+1}\mathbf{A}_{n-1-jm}(s,t)\nabla_s^j \mathfrak{R}_{st}(s,t)$$

or

$$\nabla_t \mathbf{A}_{nm}(s,t) - \mathbf{A}_{nm+1}(s,t)$$

$$= -\frac{1}{2}\sum_{j=1}^{n}\binom{n}{j}\mathbf{A}_{n-jm}(s,t)\nabla_s^{j-1}\mathfrak{R}_{st}(s,t) \quad (2.33)$$

for $n \geq 1$. After substituting $m = 0$ and applying (2.25), (2.33) becomes

$$-\mathbf{A}_{n1}(s,t) = -\frac{1}{2}\binom{n}{n}\mathbf{A}_{00}(s,t)\nabla_s^{n-1}\mathfrak{R}_{st},$$

or

$$\mathbf{A}_{j1}(s,t) = \frac{1}{2}\nabla_s^{j-1}\mathfrak{R}_{st}. \quad (2.34)$$

Substituting this result back into (2.33) gives us (2.22) which is our first recursion relation. A similar calculation yields (2.23) which is our second recursion relation.

This concludes the proof of our theorem except that it probably should be noted that a key step in the derivation of (2.23) is the application of a formula similar to that of (2.30), namely,

$$\hat{\partial}_t^k (\mathbf{WY}) = \sum_{j=0}^{k}\binom{k}{j}\left(\nabla_t^j \mathbf{W}\right)\left(\hat{\partial}_t^{k-j}\mathbf{Y}\right).$$

$\square$

Applying these recursion relations, we find that the $\mathbf{A}_{nm}$ rapidly become quite complicated when the values of n and m increase.

3   The exponential form of the isometry operator

From the discussion at the end of Section I, it might be thought that putting $\Re$ in exponential form would simplify the situation. With this thought in mind, we write

$$\Re = \exp\left(\sum_{n=1}^{\infty}\sum_{m=1}^{\infty}\frac{s^n\, t^m}{n!\, m!}\mathbf{C}_{nm}\right) = \sum_{n=0}^{\infty}\sum_{m=0}^{\infty}\frac{s^n\, t^m}{n!\, m!}\mathbf{A}_{nm}.$$

Since $\mathbf{A}_{00} = \mathbf{I}$ and $\mathbf{A}_{n0} = \mathbf{A}_{0m} = 0$ for $n \neq 0$ and $m \neq 0$, we also have

$$\exp\left(\sum_{n=1}^{\infty}\sum_{m=1}^{\infty}\frac{s^n\, t^m}{n!\, m!}\mathbf{C}_{nm}\right) = \mathbf{I} + \sum_{n=1}^{\infty}\sum_{m=1}^{\infty}\frac{s^n\, t^m}{n!\, m!}\mathbf{A}_{nm}. \tag{3.1}$$

Taking the log of both sides, we get

$$\sum_{n=1}^{\infty}\sum_{m=1}^{\infty}\frac{s^n\, t^m}{n!\, m!}\mathbf{C}_{nm} = \log\left(\mathbf{I} + \mathbf{X}\right) = \sum_{N=1}^{\infty}\frac{(-1)^{N+1}}{N}\mathbf{X}^N \tag{3.2}$$

where

$$\mathbf{X} = \sum_{n=1}^{\infty}\sum_{m=1}^{\infty}\frac{s^n\, t^m}{n!\, m!}\mathbf{A}_{nm},$$

and

$$\mathbf{X}^N =$$

$$\sum_{n_j \geq 1}\sum_{m_k \geq 1}\frac{s^{n_1+n_2+\cdots+n_N}\, t^{m_1+m_2+\cdots+m_N}}{n_1!n_2!\cdots n_N!\; m_1!m_2!\cdots m_N!}\mathbf{A}_{n_1 m_1}\mathbf{A}_{n_2 m_2}\cdots\mathbf{A}_{n_N m_N}.$$

Equating the coefficients of both sides of (3.2), we get

$$\mathbf{C}_{nm} = \sum_{j=1}^{\infty}\frac{(-1)^{j+1}}{j}\sum_{n_p \geq 1}\sum_{m_p \geq 1}\binom{n}{n_1,\ldots,n_j}\binom{m}{m_1,\ldots,m_j}$$

$$\times\mathbf{A}_{n_1 m_1}\mathbf{A}_{n_2 m_2}\cdots\mathbf{A}_{n_j m_j}, \tag{3.3}$$

where

$$\binom{n}{n_1,\ldots} = \begin{cases} \frac{n!}{n_1!n_2!\ldots n_j!} & \text{if } \sum_{k=1}^{j}n_k = n, \\ 0 & \text{if } \sum_{k=1}^{j}n_k \neq n. \end{cases}$$

Note that in the following derivation of the recursion relations for $\mathbf{C}_{nm}$, we apply the above definition to $\binom{n}{n_1}$, which means that[3]

$$\binom{n}{n_1} \neq \frac{n!}{n_1!(n-n_1)!} \quad \text{unless} \quad n_1 = n.$$

It is understood that for (3.3) and for similar sums below, the summation is carried out over all subscripted m's and n's. Observe that the sum in (3.3) is finite since the terms in the sum are zero unless j is less than or equal to $\min(n, m)$.

We can also obtain the inverse relation of (3.3). From (3.1), we see that

$$\sum_{n=1}^{\infty} \sum_{m=1}^{\infty} \frac{s^n\, t^m}{n!\, m!} \mathbf{A}_{nm} = \sum_{N=1}^{\infty} \frac{1}{N!} \mathbf{Z}^N \quad \text{where} \quad \mathbf{Z} = \sum_{n=1}^{\infty} \sum_{m=1}^{\infty} \frac{s^n\, t^m}{n!\, m!} \mathbf{C}_{nm}.$$

Therefore,

$$\mathbf{A}_{nm} = \sum_{N=1}^{\infty} \frac{1}{N!} \sum_{\substack{n_j \geq 1 \\ m_k \geq 1}} \binom{n}{n_1, \dots, n_N} \binom{m}{m_1, \dots, m_N}$$
$$\times \mathbf{C}_{n_1 m_1} \mathbf{C}_{n_2 m_2} \cdots \mathbf{C}_{n_N m_N}. \tag{3.4}$$

To derive a recursion relation for $\mathbf{C}_{nm}$, we begin by applying the operator ∇_s to both sides of (3.3) and apply one of the recursion relations for the $\mathbf{A}_{nm}$ ((2.23)). The result is

$$\nabla_s \mathbf{C}_{nm}$$
$$= \sum_{j=1}^{\infty} \frac{(-1)^{j+1}}{j} \sum_{\substack{n_p \geq 1 \\ m_p \geq 1}} \binom{n}{n_1, \dots, n_j} \binom{m}{m_1, \dots, m_j} \sum_{k=1}^{j} (\mathbf{X}_k - \mathbf{Y}_k), \tag{3.5}$$

where

$$\mathbf{X}_k = \mathbf{A}_{n_1 m_1} \mathbf{A}_{n_2 m_2} \cdots \mathbf{A}_{n_{k-1} m_{k-1}} \mathbf{A}_{n_k + 1 m_k} \mathbf{A}_{n_{k+1} m_{k+1}} \cdots \mathbf{A}_{n_j m_j},$$
$$\mathbf{Y}_k = \sum_{q_p \geq 1} \mathbf{A}_{n_1 m_1} \mathbf{A}_{n_2 m_2} \cdots \mathbf{A}_{n_{k-1} m_{k-1}} \binom{m_k}{q_1, q_2} \times$$
$$\mathbf{A}_{1 q_1} \mathbf{A}_{n_k q_2} \mathbf{A}_{n_{k+1} m_{k+1}} \cdots \mathbf{A}_{n_j m_j}.$$

---

[3]This convention does not apply to binomial coefficients which do not contain subscripted indices.

Let us now examine the sum over the $\mathbf{X}_k$'s when $j = 2$.

$$\sum_{n_p \geq 1} \binom{n}{n_1, n_2} \sum_{k=1}^{2} \mathbf{X}_k$$

$$= \sum_{n_p \geq 1} \binom{n}{n_1, n_2} (\mathbf{A}_{n_1+1 m_1} \mathbf{A}_{n_2 m_2} + \mathbf{A}_{n_1 m_1} \mathbf{A}_{n_2+1 m_2})$$

$$= \sum_{n_p \geq 1} \left(\binom{n}{n_1 - 1, n_2} + \binom{n}{n_1, n_2 - 1} \right) \mathbf{A}_{n_1 m_1} \mathbf{A}_{n_2 m_2}$$

$$- \sum_{n_2 \geq 1} \binom{n}{0, n_2} \mathbf{A}_{1 m_1} \mathbf{A}_{n_2 m_2} - \sum_{n_1 \geq 1} \binom{n}{n_1, 0} \mathbf{A}_{n_1 m_1} \mathbf{A}_{1 m_2}$$

$$= \sum_{n_p \geq 1} \binom{n+1}{n_1, n_2} \mathbf{A}_{n_1 m_1} \mathbf{A}_{n_2 m_2} - \sum_{n_2 \geq 1} \binom{n}{0, n_2} \mathbf{A}_{1 m_1} \mathbf{A}_{n_2 m_2}$$

$$- \sum_{n_1 \geq 1} \binom{n}{n_1, 0} \mathbf{A}_{n_1 m_1} \mathbf{A}_{1 m_2}.$$

More generally,

$$\sum_{n_p \geq 1} \binom{n}{n_1, \ldots, n_j} \sum_{k=1}^{j} \mathbf{X}_k$$

$$= \sum_{n_p \geq 1} \binom{n+1}{n_1, \ldots n_j} \mathbf{A}_{n_1 m_1} \mathbf{A}_{n_2 m_2} \cdots \mathbf{A}_{n_j m_j}$$

$$- \sum_{n_p \geq 1} \sum_{k=1}^{j} \binom{n}{n_1, \ldots \hat{n}_k, \ldots, n_j} \mathbf{Z}_k(j), \quad (3.6)$$

where the cap on the $\hat{n}_k$ signifies a missing index, and

$$\mathbf{Z}_k(j) = \mathbf{A}_{n_1 m_1} \mathbf{A}_{n_2 m_2} \cdots \mathbf{A}_{n_{k-1} m_{k-1}} (\mathbf{A}_{1 m_k}) \mathbf{A}_{n_{k+1} m_{k+1}} \cdots \mathbf{A}_{n_j m_j}. \quad (3.7)$$

We should note that the second sum on the R.H.S. of (3.6) is zero when $j = 1$. To deal with the $\mathbf{Y}_k$ sums, we first note that

$$\binom{m}{m_1, \ldots, m_k, \ldots, m_j} \binom{m_k}{q_1, q_2} = \binom{m}{m_1, \ldots, m_{k-1}, q_1, q_2, m_{k+1}, \ldots, m_j}.$$

Thus,

$$\sum_{m_p \geq 1} \binom{m}{m_1, \ldots, m_j} \mathbf{Y}_k = \sum_{m_p \geq 1} \sum_{q_p \geq 1} \binom{m}{m_1, \ldots, q_1, q_2, \ldots, m_j}$$

$$\times \mathbf{A}_{n_1 m_1} \cdots (\mathbf{A}_{1 q_1}) \mathbf{A}_{n_k q_2} \cdots \mathbf{A}_{n_j m_j}.$$

If we relabel some of the dummy indices, this sum becomes

$$\sum_{m_p \geq 1} \binom{m}{m_1, \cdots, m_{j+1}} \mathbf{A}_{n_1 m_1} \cdots \mathbf{A}_{n_{k-1} m_{k-1}} (\mathbf{A}_{1 m_k})$$
$$\times \mathbf{A}_{n_k m_{k+1}} \cdots \mathbf{A}_{n_j m_{j+1}}.$$

To increase the compatibility of the terms arising from the $\mathbf{X}_k$ and the $\mathbf{Y}_k$ sums, it is useful to relabel some of the n_p indices also, so we then have

$$\sum_{n_p \geq 1} \sum_{m_p \geq 1} \binom{n}{n_1, \ldots, n_j} \binom{m}{m_1, \ldots, m_j} \sum_{k=1}^{j} \mathbf{Y}_k$$
$$= \sum_{n_p \geq 1} \sum_{m_p \geq 1} \binom{n}{n_1, \ldots, \hat{n}_k, \ldots, n_{j+1}} \binom{m}{m_1, \ldots, m_{j+1}} \mathbf{Z}_k(j+1). \quad (3.8)$$

It is important to note that, in the expressions arising from the $\mathbf{X}_k$ sums, the $\mathbf{A}_{1 m_k}$ term may appear in any position of a product including the extreme right position. In the expressions arising from the $\mathbf{Y}_k$ sums, the $\mathbf{A}_{1 m_k}$ term may appear in any position except the extreme right. Before shoving our results back into (3.5), it is useful to note that one of the expressions arising from the $\mathbf{X}_k$ sums results in $\mathbf{C}_{n+1 m}$, namely,

$$\mathbf{C}_{n+1m} = \sum_{j=1} \frac{(-1)^{j+1}}{j} \sum_{n_p \geq 1} \sum_{m_p \geq 1} \binom{n+1}{n_1, \ldots, n_j} \binom{m}{m_1, \ldots, m_j}$$
$$\times \mathbf{A}_{n_1 m_1} \mathbf{A}_{n_2 m_2} \cdots \mathbf{A}_{n_j m_j}.$$

With this thought, (3.5) becomes

$$\nabla_s \mathbf{C}_{nm} - \mathbf{C}_{n+1m}$$
$$= -\sum_{j=2} \frac{(-1)^{j+1}}{j} \sum_{\substack{n_p \geq 1 \\ m_p \geq 1}} \sum_{k=1}^{j} \binom{n}{n_1, \ldots, \hat{n}_k, \ldots, n_j} \binom{m}{m_1, \ldots, m_j} \mathbf{Z}_k(j) -$$
$$\sum_{j=1} \frac{(-1)^{j+1}}{j} \sum_{\substack{n_p \geq 1 \\ m_p \geq 1}} \sum_{k=1}^{j} \binom{n}{n_1, \ldots, \hat{n}_k, \ldots, n_{j+1}} \binom{m}{m_1, \ldots, m_{j+1}} \mathbf{Z}_k(j+1).$$

Replacing j by $j-1$ in the second sum, we get

$$
\nabla_s C_{nm} - C_{n+1m}
$$

$$
= -\sum_{j=2} \frac{(-1)^{j+1}}{j} \sum_{\substack{n_p \geq 1 \\ m_p \geq 1}} \sum_{k=1}^{j} \binom{n}{n_1, \ldots \hat{n}_k, \ldots, n_j} \binom{m}{m_1, \ldots, m_j} \mathbf{Z}_k(j)
$$

$$
+ \sum_{j=2} \frac{(-1)^{j+1}}{j-1} \sum_{\substack{n_p \geq 1 \\ m_p \geq 1}} \sum_{k=1}^{j-1} \binom{n}{n_1, \ldots, \hat{n}_k, \ldots, n_j} \binom{m}{m_1, \ldots, m_j} \mathbf{Z}_k(j). \quad (3.9)
$$

Thus,

$$
\nabla_s C_{nm} - C_{n+1m}
$$

$$
= \sum_{j=2} \frac{(-1)^{j+1}}{j(j-1)} \sum_{\substack{n_p \geq 1 \\ m_p \geq 1}} \binom{n}{\hat{n}_1, n_2, \ldots, n_j} \binom{m}{m_1, \ldots, m_j} \mathbf{Z}_1(j)
$$

$$
+ \sum_{j=3} \frac{(-1)^{j+1}}{j(j-1)} \sum_{\substack{n_p \geq 1 \\ m_p \geq 1}} \sum_{k=2}^{j-1} \binom{n}{n_1, \ldots \hat{n}_k, \ldots, n_j} \binom{m}{m_1, \ldots, m_j} \mathbf{Z}_k(j)
$$

$$
- \sum_{j=2} \frac{(-1)^{j+1}}{j} \sum_{\substack{n_p \geq 1 \\ m_p \geq 1}} \binom{n}{n_1, \ldots, n_{j-1}, \hat{n}_j} \binom{m}{m_1, \ldots, m_j} \mathbf{Z}_j(j). \quad (3.10)
$$

If we pick out the lowest order terms (powers of $\mathbf{A}_{n_p m_p}$) on the R.H.S. of (3.10), we discover they can be grouped into a sum of commutators:

$$
-\frac{1}{2} \sum_{m_p \geq 1} \binom{m}{m_1, m_2} [\mathbf{A}_{1m_1}, \mathbf{A}_{nm_2}].
$$

With this weak motivation, it is useful to add $\frac{1}{2} \sum_{m_p \geq 1} \binom{m}{m_1, m_2} [\mathbf{C}_{1m_1}, \mathbf{C}_{nm_2}]$ to both sides of (3.10). After some calculation, we find that

$$
\frac{1}{2} \sum_{m_p \geq 1} \binom{m}{m_1, m_2} [\mathbf{C}_{1m_1}, \mathbf{C}_{nm_2}]
$$

$$
= \sum_{j=2} \frac{(-1)^j}{2(j-1)} \sum_{\substack{n_p \geq 1 \\ m_p \geq 1}} \binom{n}{\hat{n}_1, n_2, \ldots, n_j} \binom{m}{m_1, \ldots, m_j} \mathbf{Z}_1(j)
$$

$$
- \sum_{j=2} \frac{(-1)^j}{2(j-1)} \sum_{\substack{n_p \geq 1 \\ m_p \geq 1}} \binom{n}{n_1, \ldots, n_{j-1}, \hat{n}_j} \binom{m}{m_1, \ldots, m_j} \mathbf{Z}_j(j).
$$

Adding this to both sides of (3.10), we have

$$\nabla_s \mathbf{C}_{nm} - \mathbf{C}_{n+1m} + \frac{1}{2} \sum_{m_p \geq 1} \binom{m}{m_1, m_2} [\mathbf{C}_{1m_1}, \mathbf{C}_{nm_2}]$$

$$= \sum_{j=3} \frac{(-1)^j (j-2)}{2j(j-1)} \sum_{\substack{n_p \geq 1 \\ m_p \geq 1}} \binom{n}{\hat{n}_1, n_2, \dots, n_j} \binom{m}{m_1, \dots, m_j} \mathbf{Z}_1(j)$$

$$+ \sum_{j=3} \frac{(-1)^{j+1}}{j(j-1)} \sum_{\substack{n_p \geq 1 \\ m_p \geq 1}} \sum_{k=2}^{j-1} \binom{n}{n_1, \dots \hat{n}_k, \dots, n_j} \binom{m}{m_1, \dots, m_j} \mathbf{Z}_k(j)$$

$$+ \sum_{j=3} \frac{(-1)^j (j-2)}{2j(j-1)} \sum_{\substack{n_p \geq 1 \\ m_p \geq 1}} \binom{n}{n_1, \dots, n_{j-1}, \hat{n}_j} \binom{m}{m_1, \dots, m_j} \mathbf{Z}_j(j). \quad (3.11)$$

We can now prove our major recursion relation:

Theorem 4.

$$\mathbf{C}_{n+1m}$$

$$= \nabla_s \mathbf{C}_{nm} + \frac{1}{2} \sum_{m_k \geq 1} \binom{m}{m_1, m_2} [\mathbf{C}_{1m_1}, \mathbf{C}_{nm_2}]$$

$$+ a_3 \sum_{\substack{n_k \geq 1 \\ m_k \geq 1}} \binom{n}{n_2, n_3} \binom{m}{m_1, m_2, m_3} [[\mathbf{C}_{1m_1}, \mathbf{C}_{n_2 m_2}], \mathbf{C}_{n_3 m_3}]$$

$$+ a_5 \sum_{\substack{n_k \geq 1 \\ m_k \geq 1}} \binom{n}{n_2, n_3, n_4, n_5} \binom{m}{m_1, \dots, m_5} \times$$

$$[[[[\mathbf{C}_{1m_1}, \mathbf{C}_{n_2 m_2}], \mathbf{C}_{n_3 m_3}], \mathbf{C}_{n_4 m_4}], \mathbf{C}_{n_5 m_5}]$$

$$+ a_7 \sum_{\substack{n_k \geq 1 \\ m_k \geq 1}} \dots \quad (3.12)$$

where for $k \geq 2$, $a_{2k} = 0$ and for $N \geq 3$,

$$a_N = \frac{1}{(n-1)!} \sum_{j=3}^N \frac{(-1)^{j+1}(j-2)}{2j(j-1)} \sum_{N_k \geq 1} \binom{n-1}{N_1, N_2, N_3, \dots, N_{j-1}}.$$

Proof. To gain some perspective of our task, we note that it can be shown by induction that

$$\sum_{\substack{n_p \geq 1 \\ m_p \geq 1}} \binom{n}{\hat{n}_1, \dots, n_N} \binom{m}{m_1, \dots, m_N}$$

$$\times \left[[[[\mathbf{C}_{1m_1}, \mathbf{C}_{n_2 m_2}], \mathbf{C}_{n_3 m_3}], \dots], \mathbf{C}_{n_N m_N} \right] = \sum_{q=1}^{N} (-1)^{q+1} \mathbf{P}_q(N), \quad (3.13)$$

where

$$\mathbf{P}_q(N) = \binom{n-1}{q-1} \sum_{\substack{n_p \geq 1 \\ m_p \geq 1}} \binom{n}{n_1, \dots \hat{n}_q, \dots, n_N} \binom{m}{m_1, \dots, m_N} \mathbf{C}_q(N),$$

$$\mathbf{C}_q(N) = \mathbf{C}_{n_1 m_1} \cdots \mathbf{C}_{n_{q-1} m_{q-1}} \left(\mathbf{C}_{1m_q} \right) \mathbf{C}_{n_{q+1} m_{q+1}} \cdots \mathbf{C}_{n_N m_N}.$$

To convert the R.H.S. of (3.11) from an expression in terms of $\mathbf{A}_{n_p m_p}$ to an expression in terms of $\mathbf{C}_{n_p m_p}$, we first note that, from (3.4), it easily follows that $\mathbf{A}_{1m} = \mathbf{C}_{1m}$. Using the same equation, it is also possible with more difficulty to show that

$$\sum_{\substack{n_p \geq 1 \\ m_p \geq 1}} \binom{n}{\hat{n}_1, n_2, \dots, n_j} \binom{m}{m_1, \dots, m_j} \mathbf{Z}_1(j)$$

$$= \sum_{N=2} \frac{1}{(n-1)!} \sum_{M_p \geq 1} \binom{n-1}{M_1, M_2, \dots, M_{j-1}} \mathbf{P}_1(N),$$

$$\sum_{\substack{n_p \geq 1 \\ m_p \geq 1}} \binom{n}{n_1, \dots, n_{j-1} \hat{n}_j} \binom{m}{m_1, \dots, m_j} \mathbf{Z}_j(j)$$

$$= \sum_{N=2} \frac{1}{(n-1)!} \sum_{N_p \geq 1} \binom{n-1}{N_1, N_2, \dots, N_{j-1}} \mathbf{P}_N(N),$$

and, for $j - 1 \geq k \geq 2$,

$$\sum_{\substack{n_p \geq 1 \\ m_p \geq 1}} \binom{n}{n_1, \dots, \hat{n}_k, \dots, n_j} \binom{m}{m_1, \dots, m_j} \mathbf{Z}_k(j)$$

$$= \sum_{N=3} \frac{1}{(n-1)!} \sum_{q=2}^{n-1} \sum_{\substack{N_p \geq 1 \\ M_p \geq 1}} \binom{q-1}{N_1, \dots, N_{k-1}} \binom{n-q}{M_1, \dots, M_{j-k}} \mathbf{P}_q(N).$$

With these results, (3.11) becomes

$$
\nabla_s C_{nm} - C_{n+1m} + \frac{1}{2} \sum_{m_p \geq 1} \binom{m}{m_1, m_2} [C_{1m_1}, C_{nm_2}]
$$

$$
= \sum_{N=3} \frac{1}{(n-1)!} \sum_{j=3}^{N} \frac{(-1)^j (j-2)}{2j(j-1)} \sum_{M_p \geq 1} \binom{n-1}{M_1, \ldots, M_{j-1}} \mathbf{P}_1(N)
$$

$$
+ \sum_{N=3} \frac{1}{(n-1)!} \sum_{q=2}^{n-1} \sum_{j=3}^{N} \frac{(-1)^{j+1}}{j(j-1)} \sum_{k=2}^{j-1} \sum_{\substack{N_p \geq 1 \\ M_p \geq 1}} \binom{q-1}{N_1, \ldots, N_{k-1}}
$$

$$
\times \binom{n-q}{M_1, \ldots, M_{j-k}} \mathbf{P}_q(N)
$$

$$
- \sum_{N=3} \frac{1}{(n-1)!} \sum_{j=3}^{N} \frac{(-1)^j (j-2)}{2j(j-1)} \sum_{N_p \geq 1} \binom{n-1}{N_1, \ldots, N_{j-1}} \mathbf{P}_N(N). \quad (3.14)
$$

In view of (3.13), our chief task is to show that, for fixed N, the coefficients of $(-1)^q \mathbf{P}_q(N)$ are constant. We will first prove that, except for the difference in sign, the coefficient of $\mathbf{P}_1(N)$ equals the coefficient of $\mathbf{P}_2(N)$. When we examine the coefficient of $\mathbf{P}_q(N)$ we see that, because of the definition of $\binom{q-1}{N_1,\ldots,N_{k-1}}$, in the sum over k we have $q \geq k$. Thus, we are faced with the task of showing that

$$
\sum_{j=3}^{N} \frac{(-1)^j (j-2)}{2j(j-1)} \sum_{M_p \geq 1} \binom{n-1}{M_1, \ldots, M_{j-1}}
$$

$$
= \sum_{j=3}^{N} \frac{(-1)^j}{j(j-1)} \sum_{M_p \geq 1} \binom{n-2}{M_1, \ldots, M_{j-2}}. \quad (3.15)
$$

We first observe that

$$
\binom{n-1}{M_1, \ldots, M_{j-1}} = \binom{n-2}{M_1 - 1, M_2, \ldots, M_{j-1}}
$$

$$
+ \binom{n-2}{M_1, M_2 - 1, M_3, \ldots, M_{j-1}}
$$

$$
+ - - - + \binom{n-2}{M_1, \ldots, M_{j-2}, M_{j-1} - 1}. \quad (3.16)
$$

Furthermore,

$$\sum_{M_p \geq 1} \binom{n-2}{M_1 - 1, M_2, \ldots, M_{j-1}}$$

$$= \sum_{M_p \geq 1} \binom{n-2}{M_1, \ldots, M_{j-1}} + \sum_{M_p \geq 1} \binom{n-2}{0, M_2, \ldots, M_{j-1}}$$

$$= \sum_{M_p \geq 1} \binom{n-2}{M_1, \ldots, M_{j-1}} + \sum_{M_p \geq 1} \binom{n-2}{M_1, \ldots, M_{j-2}}.$$

The other terms on the R.H.S. of (3.16) can be dealt with in a similar manner when summed. Also, there are $j-1$ terms on the R.H.S. of (3.16). Thus,

$$\sum_{M_p \geq 1} \binom{n-1}{M_1, \ldots, M_{j-1}} = (j-1) \sum_{M_p \geq 1} \binom{n-2}{M_1, \ldots, M_{j-1}}$$

$$+ (j-1) \sum_{M_p \geq 1} \binom{n-2}{M_1, \ldots, M_{j-2}}. \quad (3.17)$$

We should note that for $j = N$, the first multinomial coefficient on the R.H.S. of (3.17) is zero. Thus

$$\sum_{j=3}^{N} \frac{(-1)^j (j-2)}{2j(j-1)} \sum_{M_p \geq 1} \binom{n-1}{M_1, \ldots, M_{j-1}}$$

$$= \sum_{j=3}^{n-1} \frac{(-1)^j (j-2)}{2j} \sum_{M_p \geq 1} \binom{n-2}{M_1, \ldots, M_{j-1}}$$

$$+ \sum_{j=3}^{N} \frac{(-1)^j (j-2)}{2j} \sum_{M_p \geq 1} \binom{n-2}{M_1, \ldots, M_{j-2}}$$

$$= \sum_{j=4}^{N} \frac{(-1)^{j+1} (j-3)}{2(j-1)} \sum_{M_p \geq 1} \binom{n-2}{M_1, \ldots, M_{j-2}}$$

$$+ \sum_{j=3}^{N} \frac{(-1)^j (j-2)}{2j} \sum_{M_p \geq 1} \binom{n-2}{M_1, \ldots, M_{j-2}}. \quad (3.18)$$

Because of the $(j-3)$ factor, we can begin the summation in the first sum on the R.H.S. of (3.18) at $j = 3$ and then combine the two sums to get

$$\sum_{j=3}^{N} \frac{(-1)^j (j-2)}{2j(j-1)} \sum_{M_p \geq 1} \binom{n-1}{M_1, \ldots, M_{j-1}}$$

$$= \sum_{j=3}^{N} \frac{(-1)^j}{j(j-1)} \sum_{M_p \geq 1} \binom{n-2}{M_1, \ldots, M_{j-2}} \quad (3.19)$$

which is identical to (3.15) and which is what we set out to prove.

Our next task is to show that except for the difference in signs, the coefficient of $\mathbf{P}_q(N)$ equals the coefficient of $\mathbf{P}_{q+1}(N)$ for $n-1 \geq q+1$ and $q \geq 2$. This situation arises whenever $N \geq 4$. That is, we must show that

$$\sum_{j=3}^{N} \frac{(-1)^{j+1}}{j(j-1)} \sum_{k=2}^{j-1} \sum_{\substack{N_p \geq 1 \\ M_p \geq 1}} \binom{q-1}{N_1, \ldots, N_{k-1}} \binom{n-q}{M_1, \ldots, M_{j-k}}$$

$$= \sum_{j=3}^{N} \frac{(-1)^j}{j(j-1)} \sum_{k=2}^{j-1} \sum_{\substack{N_p \geq 1 \\ M_p \geq 1}} \binom{q}{N_1, \ldots, N_{k-1}} \binom{n-q-1}{M_1, \ldots, M_{j-k}}. \quad (3.20)$$

From (3.17)

$$\sum_{M_p \geq 1} \binom{n-q}{M_1, \ldots, M_{j-k}}$$

$$= (j-k) \sum_{M_p \geq 1} \left\{ \binom{n-q-1}{M_1, \ldots, M_{j-k}} + \binom{n-q-1}{M_1, \ldots, M_{j-k-1}} \right\} \quad (3.21)$$

and

$$\sum_{N_p \geq 1} \binom{q}{N_1, \ldots, N_{k-1}}$$

$$= (k-1) \sum_{N_p \geq 1} \left\{ \binom{q-1}{N_1, \ldots, N_{k-1}} + \binom{q-1}{N_1, \ldots, N_{k-2}} \right\}. \quad (3.22)$$

The second multinomial coefficient on the R.H.S. of (3.21) is zero when $k = j-1$, and the second multinomial coefficient on the R.H.S. of (3.22) is zero when $k = 2$. If we designate the L.H.S. of (3.20) by A and the R.H.S.

by B, and use (3.21) and (3.22), we have

$$A = \sum_{j=3}^{N} \frac{(-1)^{j+1}}{j(j-1)} \sum_{k=2}^{j-1} \sum_{\substack{N_p \geq 1 \\ M_p \geq 1}} (j-k) \binom{q-1}{N_1, \ldots, N_{k-1}} \binom{n-q-1}{M_1, \ldots, M_{j-k}}$$

$$+ \sum_{j=4}^{N} \frac{(-1)^{j+1}}{j(j-1)} \sum_{k=2}^{j-2} \sum_{\substack{N_p \geq 1 \\ M_p \geq 1}} (j-k) \binom{q-1}{N_1, \ldots, N_{k-1}} \binom{n-q-1}{M_1, \ldots, M_{j-k-1}}$$

and

$$B = \sum_{j=3}^{N} \frac{(-1)^{j}}{j(j-1)} \sum_{k=2}^{j-1} \sum_{\substack{N_p \geq 1 \\ M_p \geq 1}} (k-1) \binom{q-1}{N_1, \ldots, N_{k-1}} \binom{n-q-1}{M_1, \ldots, M_{j-k}}$$

$$+ \sum_{j=4}^{N} \frac{(-1)^{j}}{j(j-1)} \sum_{k=3}^{j-1} \sum_{\substack{N_p \geq 1 \\ M_p \geq 1}} \binom{q-1}{N_1, \ldots, N_{k-2}} \binom{n-q-1}{M_1, \ldots, M_{j-k}}. \quad (3.23)$$

Note that for both A and B, the summation over j in the second term must start at $j = 4$ as a consequence of the comments following (3.22). In the second term on the R.H.S. of (3.23), we replace the dummy index k by $k+1$. We then have

$$B = \sum_{j=3}^{N} \frac{(-1)^{j}}{j(j-1)} \sum_{k=2}^{j-1} \sum_{\substack{N_p \geq 1 \\ M_p \geq 1}} (k-1) \binom{q-1}{N_1, \ldots, N_{k-1}} \binom{n-q-1}{M_1, \ldots, M_{j-k}}$$

$$+ \sum_{j=3}^{N} \frac{(-1)^{j}}{j(j-1)} \sum_{k=2}^{j-2} \sum_{\substack{N_p \geq 1 \\ M_p \geq 1}} \binom{q-1}{N_1, \ldots, N_{k-1}} \binom{n-q-1}{M_1, \ldots, M_{j-k-1}}.$$

It then follows that

$$A - B = \sum_{j=3}^{N} \frac{(-1)^{j+1}}{j} \sum_{k=2}^{j-1} \sum_{\substack{N_p \geq 1 \\ M_p \geq 1}} \binom{q-1}{N_1, \ldots, N_{k-1}} \binom{n-q-1}{M_1, \ldots, M_{j-k}}$$

$$+ \sum_{j=4}^{N} \frac{(-1)^{j+1}}{(j-1)} \sum_{k=2}^{j-2} \sum_{\substack{N_p \geq 1 \\ M_p \geq 1}} \binom{q-1}{N_1, \ldots, N_{k-1}} \binom{n-q-1}{M_1, \ldots, M_{j-k-1}}.$$

Replacing the dummy index j by $j+1$ in the second sum gives us

$$A - B = \sum_{j=3}^{N} \frac{(-1)^{j+1}}{j} \sum_{k=2}^{j-1} \sum_{N_p \geq 1} \binom{q-1}{N_1, \ldots, N_{k-1}} \binom{n-q-1}{M_1, \ldots, M_{j-k}}$$

$$+ \sum_{j=3}^{n-1} \frac{(-1)^{j}}{j} \sum_{k=2}^{j-1} \sum_{\substack{N_p \geq 1 \\ M_p \geq 1}} \binom{q-1}{N_1, \ldots, N_{k-1}} \binom{n-q-1}{M_1, \ldots, M_{j-k}}$$

$$= \frac{(-1)^{N+1}}{N} \sum_{k=2}^{n-1} \sum_{N_p \geq 1} \binom{q-1}{N_1, \ldots, N_{k-1}} \binom{n-q-1}{M_1, \ldots, M_{n-k}}.$$

However, this last expression is zero because for each value of k, at least one of the multinomial coefficients in each product must be zero. The conditions for the first multinomial coefficient and the second multinomial coefficient to be non-zero are, respectively,

$$q - 1 = \sum_{p=1}^{k-1} N_p \geq k - 1 \quad \text{and} \quad n - q - 1 = \sum_{p=1}^{n-k} M_p \geq n - k.$$

Adding the two inequalities, we find that at least one of the multinomials is zero unless $n - 2 \geq n - 1$, which is impossible.

What is left to show is that the coefficient of $\mathbf{P}_{n-1}(N) = -\mathbf{P}_N(N)$. However, a close examination of symmetries in the sums over the N_p and M_p in (3.14) shows that the coefficient of $\mathbf{P}_q(N)$ is the same as the coefficient of $\mathbf{P}_{n-q+1}(N)$. Thus, proving $\mathbf{P}_{n-1}(N) = -\mathbf{P}_N(N)$ is equivalent to proving $\mathbf{P}_2(N) = -\mathbf{P}_1(N)$, which was already done. The equality of the coefficients of $\mathbf{P}_q(N)$ and $\mathbf{P}_{n-q+1}(N)$ has another useful consequence. It implies that the coefficient of $\mathbf{P}_N(N)$ equals the coefficient of $\mathbf{P}_1(N)$. On the other hand, we have shown that the sign of the coefficient of $\mathbf{P}_q(N)$ changes each time the value of q is increased by 1. This means that the coefficient of $\mathbf{P}_N(N)$ is $(-1)^{n-1}$ times the coefficient of $\mathbf{P}_1(N)$. However, $(-1)^{n-1} = 1$ only if N is odd. When N is even, all of the coefficients are zero.[4]

Reviewing (3.13) and (3.14), it is clear that the a_N mentioned in the statement of our theorem is equal to the coefficient of $(-1)^q \mathbf{P}_q(N)$ in (3.14) for any value of q. For $q = 1$,

$$a_N = \frac{1}{(n-1)!} \sum_{j=3}^{N} \frac{(-1)^{j+1}(j-2)}{2j(j-1)} \sum_{N_k \geq 1} \binom{n-1}{N_1, N_2, \ldots, N_{j-1}}.$$

$\square$

---

[4]All those sums in (3.14) corresponding to even values of N are in fact expressions which reduce to zero.

A similar recursion relation may be derived by a similar calculation. Namely

Theorem 5.

$$
\mathbf{C}_{nm+1}
$$

$$
= \nabla_t \mathbf{C}_{nm} - \frac{1}{2} \sum_{n_k \geq 1} \binom{n}{n_1, n_2} [\mathbf{C}_{n_1 1}, \mathbf{C}_{n_2 m}]
$$

$$
+ a_3 \sum_{n_k \geq 1} \sum_{m_k \geq 1} \binom{n}{n_1, n_2, n_3} \binom{m}{m_2, m_3} [[\mathbf{C}_{n_1 1}, \mathbf{C}_{n_2 m_2}], \mathbf{C}_{n_3 m_3}]
$$

$$
+ a_5 \sum_{n_k \geq 1} \sum_{m_k \geq 1} \binom{n}{n_1, \ldots, n_5} \binom{m}{m_2, m_3, m_4, m_5} \times
$$

$$
[[[[\mathbf{C}_{n_1 1}, \mathbf{C}_{n_2 m_2}], \mathbf{C}_{n_3 m_3}], \mathbf{C}_{n_4 m_4}], \mathbf{C}_{n_5 m_5}]
$$

$$
+ a_7 \sum_{n_k \geq 1} \sum_{m_k \geq 1} - - - \tag{3.24}
$$

where the a_N coefficients assume the same values as they do in Theorem 4.

To evaluate the a_N coefficients of Theorems 4 and 5, the following theorem is useful:

Theorem 6.

$$
\sum_{N_k \geq 1} \binom{N}{N_1, N_2, \ldots, N_p} = \sum_{m=1}^{p} \binom{p}{m} (-1)^{p+m} m^N \quad for \quad p \geq 1.
$$

Proof. The equation can be proven by induction on p where the critical step is

$$
\sum_{N_k \geq 1} \binom{N}{N_1, N_2, \ldots, N_{p+1}}
$$

$$
= \sum_{N_k \geq 0} \binom{N}{N_1, \ldots, N_{p+1}} - \sum_{k=1}^{p} \binom{p+1}{k} \sum_{N_j \geq 1} \binom{N}{N_1, \ldots, N_k, 0, 0, \ldots, 0}
$$

$$
= (p+1)^N - \sum_{k=1}^{p} \binom{p+1}{k} \sum_{m=1}^{k} \binom{k}{m} (-1)^{k+m} m^N
$$

$$
= (p+1)^N - \sum_{m=1}^{p} \binom{p+1}{m} (-1)^m m^N \sum_{k=m}^{p} \binom{p+1-m}{k-m} (-1)^k.
$$

$\square$

Using Theorem 6, the a_N coefficients of Theorems 4 and 5 become

$$a_N = \frac{1}{(n-1)!} \sum_{j=3}^{N} \frac{(j-2)}{2j(j-1)} \sum_{m=1}^{j-1} \binom{j-1}{m} (-1)^m m^{n-1}.$$

A few of the non-zero values of a_N are

$$a_3 = \frac{1}{12}, \ a_5 = \frac{-1}{720}, \ a_7 = \frac{1}{30240} = \frac{1}{42(6!)}, \ a_9 = \frac{-1}{30(8!)}, \ a_{11} = \frac{5}{66(10!)}.$$

4   A closing observation

At the end of Section I ((1.16)), we noted that for the 2-dimensional case

$$C_{nm}(s,t) = \frac{\partial^{n+m-2}}{(\partial s)^{n-1}(\partial t)^{m-1}} \frac{1}{2} \Re_{12}(s,t). \tag{4.1}$$

What modification of this formula carries over to higher dimensions? From (3.3) and (2.21)

$$C_{n1}(s,t) = \nabla_s^{n-1} \frac{1}{2} \Re_{st}(s,t) \ \text{ and } \ C_{1m}(s,t) = \nabla_t^{m-1} \frac{1}{2} \Re_{st}(s,t). \tag{4.2}$$

From (3.12), (3.24), and (4.2), it can be shown that

$$C_{n2}(s,t) = \nabla_t \nabla_s^{n-1} \frac{1}{2} \Re_{st}(s,t), \tag{4.3}$$

$$C_{2m}(s,t) = \nabla_s \nabla_t^{m-1} \frac{1}{2} \Re_{st}(s,t). \tag{4.4}$$

For higher values of n and m, the commutator brackets of Theorems 4 and 5 can be replaced by intrinsic differential operators of the same order as that in (4.1). For example, from Theorem 4,

$$C_{33}(s,t) = \nabla_s C_{23} + \frac{3}{2} [C_{11}, C_{22}] + \frac{3}{2} [C_{12}, C_{21}]. \tag{4.5}$$

Also,

$$[C_{12}, C_{21}] = [\nabla_t C_{11}, \nabla_s C_{11}]$$
$$= \nabla_t [C_{11}, \nabla_s C_{11}] - [C_{11}, \nabla_t \nabla_s C_{11}]. \tag{4.6}$$

To carry these computations further, we note that since

$$C_{11}(s,t) = \frac{1}{2} \Re_{st}(s,t),$$

then

$$[\mathbf{C}_{11}, \mathbf{X}] = (\nabla_t \nabla_s - \nabla_s \nabla_t)\,\mathbf{X}. \tag{4.7}$$

[See (Snygg 1997: 101-103).] From (4.4), (4.5), (4.6), and (4.7), we get

$$\mathbf{C}_{33}(s,t) = (\nabla_s^2 \nabla_t^2 - \frac{3}{2}\nabla_s \nabla_t \nabla_s \nabla_t + \frac{3}{2}\nabla_s \nabla_t^2 \nabla_s$$
$$- \frac{3}{2}\nabla_t \nabla_s \nabla_t \nabla_s + \frac{3}{2}\nabla_t^2 \nabla_s^2)\frac{1}{2}\Re_{st}(s,t). \tag{4.8}$$

This representation is not unique because the conversion from commutator brackets to differential operators is not unique. For example, from (4.6) and (4.7),

$$[\mathbf{C}_{12}, \mathbf{C}_{21}] = (\nabla_t^2 \nabla_s^2 - 2\nabla_t \nabla_s \nabla_t \nabla_s + \nabla_s \nabla_t^2 \nabla_s)\frac{1}{2}\Re_{st}(s,t),$$

but it is also true that

$$[\mathbf{C}_{12}, \mathbf{C}_{21}] = [\nabla_t \mathbf{C}_{11}, \nabla_s \mathbf{C}_{11}]$$
$$= \nabla_s\,[\nabla_t \mathbf{C}_{11}, \mathbf{C}_{11}] - [\nabla_s \nabla_t \mathbf{C}_{11}, \mathbf{C}_{11}]$$
$$= (\nabla_s^2 \nabla_t^2 - 2\nabla_s \nabla_t \nabla_s \nabla_t + \nabla_t \nabla_s^2 \nabla_t)\frac{1}{2}\Re_{st}(s,t).$$

For our particular example:

$$\mathbf{C}_{33}(s,t) = \left(\left(\frac{5}{4}+\alpha\right)\nabla_s^2 \nabla_t^2 + \left(-\frac{3}{8}-\frac{3\alpha}{2}+\beta\right)\nabla_s \nabla_t \nabla_s \nabla_t\right)\frac{1}{2}\Re_{st}(s,t)$$
$$+ \left(-\frac{3}{8}-\frac{3\alpha}{2}-\beta\right)\nabla_s \nabla_t^2 \nabla_s\,\frac{1}{2}\Re_{st}(s,t)$$
$$+ \left(-\frac{3}{8}+\frac{3\alpha}{2}-\gamma\right)\nabla_t \nabla_s^2 \nabla_t\,\frac{1}{2}\Re_{st}(s,t)$$
$$+ \left(-\frac{3}{8}+\frac{3\alpha}{2}+\gamma\right)\nabla_t \nabla_s \nabla_t \nabla_s\,\frac{1}{2}\Re_{st}(s,t)$$
$$+ \left(\frac{5}{4}-\alpha\right)\nabla_t^2 \nabla_s^2\,\frac{1}{2}\Re_{st}(s,t),$$

where α, β, and γ are arbitrary real numbers.

Thus, we see that the formula for 2-dimensions can be generalized to higher dimensions, but not in a simple or obvious way.

Acknowledgment

I wish to thank Daniel H. Gottlieb and Ted Shifrin for reading portions of earlier drafts and making suggestions which hopefully made this paper more readable than it otherwise would have been.

REFERENCES

[1] W. Ambrose and I. M. Singer, A theorem on holonomy, *Trans. Am. Math. Soc.* **75** (1953), 428–443.

[2] J. Pérès, Le parallelisme de M. Levi-Civita et la courbure Riemanniene, *Rendiconti dei Lincei,* ser. 5, Vol. **28**, (1919), 425–428.

[3] M. Riesz, *Clifford Numbers and Spinors*, E. Folke Bolinder and Pertti Lounesto, eds., Dordrecht, The Netherlands, Kluwer Academic Publishers, 1993.

[4] J. A. Schouten, Die direkte analysis zur neueren relativitätstheorie, *Verhandelingen Kon. Akad. Amsterdam, Vol. 12*, No. **6** (1918).

[5] J. Snygg, *Clifford Algebra — A Computational Tool for Physicists*, New York, Oxford University Press, 1997.

John Snygg
433 Prospect Street
East Orange, NJ 07017, U.S.A.
E-mail: jsnygg@idt.net

Received: October 7, 1999; Revised: March 10, 2000

An Extension of Clifford Analysis Towards Super-symmetry

Frank Sommen[*]

ABSTRACT In this paper we present an extension of Clifford analysis using commuting as well as anti-commuting variables, thus following the lines of thinking of super-symmetry. However, it turns out that on the level of abstract vector variables, the calculus remains the same.

Keywords: Clifford algebra, geometric calculus, q-deformation.

1 Introduction

Clifford analysis in its most basic form deals with monogenic polynomials, also called spherical monogenics. Let $e_1, \ldots, e_m$ be an orthonormal basis of Euclidean space and consider the Clifford algebra R_m with defining relations $e_i e_j + e_j e_i = -2\delta_{ij}$; then we consider the Dirac operator $\partial_{\underline{x}} = \sum e_j \partial_{x_j}$ and define monogenic functions to be R_m-valued solutions to $\partial_{\underline{x}} f(\underline{x}) = 0$. By restricting attention to polynomials, one is in fact working within the algebra generated by the real variables $x_1, \ldots, x_m$, the corresponding derivatives $\partial_{x_1}, \ldots, \partial_{x_m}$ and the Clifford basis $e_1, \ldots, e_m$. This is the most basic "Clifford analysis toolkit" and any extension of it leads to an extension of Clifford analysis. On this algebra the anti-involution is given by

$$x_j^+ = \partial_{x_j}, \quad \partial_{x_j}^+ = x_j, \quad e_j^+ = -e_j, \quad (ab)^+ = b^+ a^+,$$

leading to the Fischer inner product for Clifford polynomials

$$(R(\underline{x}), S(\underline{x})) = [(R(\underline{x}))^+ S(\underline{x})]_0 = [\bar{R}(\partial_{\underline{x}}) S(\underline{x})]_0,$$

whereby $a \to \bar{a}$ denotes the anti-involution restricted to R_m and $[a]_0$ denotes the scalar part of $a \in R_m$.

---

[*] Senior Research Associate F.W.O.
AMS Subject Classification: 30G35, 81T60.

For this Fischer inner product, every Clifford polynomial $R(\underline{x})$ admits an orthogonal decomposition of the form

$$R(\underline{x}) = P(\underline{x}) + \underline{x}\,S(\underline{x}), \quad \partial_{\underline{x}} P(\underline{x}) = 0,$$

called the Fischer decomposition and $P(\underline{x}) = M(R(\underline{x}))$ is called the mono-genic part of $R(\underline{x})$. In particular, the monogenic part of the basic repro-ducing polynomial $\frac{1}{k!} < \underline{x}, \underline{u} >^k$ has the form

$$\frac{1}{k!} < \underline{x}, \underline{u} >^k + a_1 \underline{x} < \underline{x}, \underline{u} >^{k-1} \underline{u} + \ldots + a_k \, \underline{x}^k \underline{u}^k$$

whereby the coefficients $a_1, \ldots, a_k$ depend only on k and the dimension m. The above polynomial is called the zonal spherical monogenic and may be expressed in terms of Gegenbauer polynomials (see also [4]).

Important for our discussion are the following observations:

(i) zonal polynomials belong to the algebra generated by two vector vari-ables $\underline{x}$ and $\underline{u}$, which is a subalgebra of the algebra of Clifford poly-nomials in two vector variables,

(ii) the zonal functions are obtained by letting $\partial_{\underline{x}}$ act several times on the Fischer decomposition formula, so that the dimension m enters the formulae only through the evaluation $\partial_{\underline{x}}[\underline{x}] = -m$.

These observations lead to the idea of defining an algebra of abstract vector variables $x, u, \ldots$ and to reintroduce the Dirac operator axiomatically as an endomorphism on this algebra which may then be called "vector deriva-tive" after [6]. We developed this calculus to some extent in [9] and called it algebra of abstract vector variables or "radial algebra." It also provides a solid axiomatic background for "geometric calculus" in the sense of [6].

To define it, we start from a set $S = x, u, \ldots$ of so-called "abstract vector variables" as opposed to the true vector variables $\underline{x} = \sum e_j x_j$ and $\underline{u} = \sum u_j e_j$, and we consider the associative algebra $R(S)$ generated by S together with the relations

$$\{x, y\}z = z\{x, y\},$$

i.e., the anti-commutator of any two vector variables is a scalar (commu-tative). The algebra $R(S)$ may be represented by an algebra of real vector variables by making the application

$$x \to \underline{x} = \sum e_j x_j,$$

whereby x_j are real coordinates and $e_1, \ldots, e_m$ satisfy general Clifford algebra defining relations $e_i e_j + e_j e_i = -2g_{ij}$.

In case the dimension m exceeds the number of vector variables, this application is injective; in case S is infinite this is never the case.

This however means that the radial algebra $R(S)$ is itself independent of the dimension m and the metric g_{ij}. Moreover, the functions $F(x, u, \dots)$ belonging to it are **Spin**(m)-invariant in the sense that the corresponding Clifford polynomial representations $F(\underline{x}, \underline{u}, \dots)$ satisfy

$$F(\underline{x}, \underline{u}, \dots) = sF(\bar{s}\underline{x}s, \bar{s}\underline{u}s, \dots)\bar{s}.$$

One may also consider certain vector variables as fixed objects. For example if x is variable and u fixed, one needs to consider the actions of **Spin**(m) on the Clifford vector representation

$$H(s)F(x, u) = sF(\bar{s}xs, u)\bar{s}, \quad L(s)F(x, u) = sF(\bar{s}xs, u)$$

and of course polynomials of $R(S)$ need not be invariant in this sense. To arrive at a sufficient toolkit for Clifford analysis we also consider the vector derivative which is defined axiomatically as an endomorphism by

(D1) $\partial_x[fF] = \partial_x[f]F + f\partial_x[F]$,

   $[fF]\partial_x = F\partial_x[f] + f[F]\partial_x$, f, $F \in R(S)$, f scalar,

(D2) $\partial_x[FG] = \partial_x[F]G$,

   $[GF]\partial_x = G[F]\partial_x$ in case $G \in R(S\backslash\{x\})$,

(D3) $[\partial_x F]\partial_y = \partial_x[F\partial_y]$,

(D4) $\partial_x x^2 = x^2\partial_x = 2x$, $\quad \partial_x\{x, y\} = \{x, y\}\partial_x = 2y$, $\quad x \neq y$.

In [9] we proved that, under certain obvious extra conditions, these axioms lead to the property

(D5) $\partial_x x] = [x]\partial_x = $ scalar.

This means that if abstract vector variables $x, \dots$ are represented by true ones $\underline{x}, \dots$, then, under the assumption of the extra constraint

$$\partial_x[x] = m,$$

∂_x is represented by the operator $\underline{x}^+ = -\partial_{\underline{x}}$ rather than $\partial_{\underline{x}}$, so that, on radial algebra we may also define the anti involution by $x^+ = \partial_x, u^+ = \partial_u$. In any case this extra constraint is not needed to do many of the Clifford analysis computations, which leads to the introduction of the "formal dimension" $\partial_x[x] = M$ which may be any type of real or complex scalar. Standard Clifford analysis then corresponds to $M = m = 1, 2, 3, \dots$ while Clifford analysis using anti-commuting variables will lead to the consideration of a formal dimension of the form $M = -2m$.

What we are going to illustrate in this paper is exactly the fact that radial algebra may be represented by using commuting and anti-commuting variables and any of these "super-symmetric representations" leads to the same Clifford analysis calculus on the level of abstract vector variables. But on the level of coordinates it leads to super-vector variables, and the action of both the symplectic and rotation group unified in the super spin group, studied in section one, and to the super Dirac operator studied in section two. Moreover, we also extend the radial algebra with an exterior derivative, leading to abstract differential forms and even more super-symmetric representations. This whole discussion shows that radial algebra is a canonical background that is invariant in the group sense and independent of dimension and quadratic form and no extension of it is needed for super-symmetry.

Our extension of Clifford analysis towards super-symmetry is in accordance with superanalysis as defined in [1, 13]: the Dirac operator is defined on the "super-space" introduced in this paper and all functions considered are super-differentiable. We also refer to the approach towards super-symmetry presented in [3], which is also based on the use of symbolic vectors and to [7] in which a Stokes' type theorem in superanalysis is obtained.

2   Bosonic, fermionic and super-symmetric representation of radial algebra

Let S be a set of abstract labels, and let $R(S)$ be the associated radial algebra generated by S. By making the assignment

$$x \to \underline{x} = \sum_{1}^{m} x_j e_j,$$

whereby x_j are real coordinates and e_j are generators of a Clifford algebra (satisfying defining relations of the form $e_i e_j + e_j e_i = -2g_{ij}$), we obtain the standard representation, which is only injective if the number m of commuting variables x_j exceeds the cardinality l of S and always if one allows infinite dimensional vectors of the form $\underline{x} = \sum e_j x_j$. Of course in that case one needs to be careful about the topology, the discussion of which is beyond the scope of this paper.

As already explained in the introduction, the radial algebra $R(S)$ is independent of the number m of variables (classical dimension) as well as the quadratic form g_{ij}. In other words, the algebra $R(S)$ is highly invariant and nevertheless behaves in much the same way as the algebra of Clifford polynomials because one may fix any number of vector variables to form a frame. In this sense the algebra $R(S)$ is also similar though not identical to the "geometric algebra" as described in [6]; it is the dual picture.

In this section we'll show that the radial algebra is even invariant in the super-symmetric sense, i.e., there exist bosonic, fermionic and mixed representations for it, the above Clifford vector representation being the bosonic one.

To arrive at a fermionic representation for any vector variable x, we introduce $\overset{\backprime}{m}$ fermionic coordinates $x_j^{\backprime}$, i.e., $x_j^{\backprime} x_k^{\backprime} = -x_k^{\backprime} x_j^{\backprime}$, and we introduce also "fermionic Clifford algebra" generators $e_j^{\backprime}$ to form the fermionic vector variable $\underline{x}^{\backprime} = \sum x_j^{\backprime} e_j^{\backprime}$. The condition that $\underline{x}^{\backprime} \underline{y}^{\backprime} + \underline{y}^{\backprime} \underline{x}^{\backprime}$ is a scalar (a commutative object) then automatically leads to

$$\underline{x}^{\backprime} \underline{y}^{\backprime} + \underline{y}^{\backprime} \underline{x}^{\backprime} = \sum (x_j^{\backprime} y_k^{\backprime} e_j^{\backprime} e_k^{\backprime} + y_k^{\backprime} x_j^{\backprime} e_k^{\backprime} e_j^{\backprime})$$

$$= \sum x_j^{\backprime} y_k^{\backprime} (e_j^{\backprime} e_k^{\backprime} - e_k^{\backprime} e_j^{\backprime}) = \text{scalar}$$

and as any $x_j^{\backprime} y_k^{\backprime}$ is scalar (i.e., commutative) we arrive at defining relations of the form (see also [2])

$$e_j^{\backprime} e_k^{\backprime} - e_k^{\backprime} e_j^{\backprime} = f_{jk} = \text{scalar}.$$

Of course, while in the commutative sense $g_{jk} = g_{kj}$ is a symmetric form, in this case, $f_{jk} = -f_{kj}$ will be a symplectic form and the vector variables $\underline{x}^{\backprime} = \sum x_j^{\backprime} e_j^{\backprime}$ are to be expected to be symplectic invariant. Particular examples of non-degenerate symplectic forms may only be defined in even dimensional cases and the standard example is given by

$$f_{2j,2k} = f_{2j-1,2k-1} = 0, \quad f_{2j-1,2k} = -f_{2k,2j-1} = \delta_{jk}, \quad j,k = 1,\dots,n.$$

In this example, the elements $e_j^{\backprime}$ and the algebra generated by them may be represented by polynomial differential operators in n dimensions whereby we introduce commuting variables a_j and the corresponding derivatives ∂_{a_j} and make the assignments

$$e_{2j-1}^{\backprime} \to \partial_{a_j}, \quad e_{2j}^{\backprime} \to a_j.$$

One indeed has the "Weyl algebra defining relations"

$$\partial_{a_j} a_k - a_k \partial_{a_j} = \delta_{jk} \quad \text{as operators on polynomials in} \quad a_1,\dots,a_n.$$

The fermionic vector variable corresponding to an abstract vector variable $\underline{x}^{\backprime}$ may thus be rewritten into the form $\underline{x}^{\backprime} = \sum x_{2j-1}^{\backprime} \partial_{a_j} + \sum x_{2j}^{\backprime} a_j$ whereby the variables $x_j^{\backprime}, j = 1,\dots,2n$ are generators of a Grassmann algebra.

Note that in the bosonic case, the variables x_j form an infinite dimensional polynomial algebra, and the "values e_j" form a finite dimensional Clifford. On the other hand in the fermionic case, the variables $x_j^{\backprime}$ form a finite dimensional Grassmann algebra, and the values a_j, ∂_{a_j} form an infinite dimensional operator algebra.

In the fermionic case, the anti-commutator is given by

$$\underline{x}^{\backprime} \underline{y}^{\backprime} + \underline{y}^{\backprime} \underline{x}^{\backprime} = \sum x_j^{\backprime} y_k^{\backprime} f_{jk} =: 2 < \underline{x}^{\backprime}, \underline{y}^{\backprime} >$$

and thought of as a scalar. But being an element of a Grassmann algebra, the algebra generated by the elements $< \underline{x}^{\backprime}, \underline{y}^{\backprime} >$ is finite dimensional at least if the dimension $2n$ is finite. Hence, in the fermionic case, the representation

$$x \to \grave{\underline{x}}, \quad x \in S$$

never leads to an isomorphism, at least not in the case where dimension n is finite. In the infinite dimensional setting, the above representation also becomes an isomorphism and, using both bosonic and fermionic representations, it is possible to make a complete transition from a bosonic to a fermionic space by means of the "transformation formulae"

$$< \underline{x}, \underline{y} > = < \grave{\underline{x}}, \grave{\underline{y}} >,$$

i.e., there is a transformation "via the radial algebra". Only in case there are infinitely many vector variables (e.g., a sequence of vector variables) one can really speak of a transformation from bosonic to fermionic; infinite dimensionality is a requirement for this. Apart from bosonic and fermionic representations one may of course also consider the mixed "supersymmetric" representations of the form

$$x \to \underline{x} + \grave{\underline{x}} = \sum_{j=1}^{m} e_j x_j + \sum_{j=1}^{2n} \grave{e_j} x_j$$

and the radial algebra condition $x\,y + y\,x = $ scalar together with the obvious commutation relations $x_j \grave{x_k} = \grave{x_k} x_j$ lead back to the above defining relations for e_j and $\grave{e_j}$ together with the commutation relations

$$e_j \grave{e_k} = -\grave{e_k} e_j,$$

so that one cannot really keep the identification

$$\grave{e_{2j-1}} \to \partial_{a_j}, \quad \grave{e_{2j}} \to a_j, \quad j = 1, \dots, n.$$

However, if one introduces an extra Clifford algebra generator e_{m+1} with

$$e_{m+1}^2 = -1 \quad \text{and} \quad e_{m+1} e_j = -e_j e_{m+1},$$

we may make the assignment

$$\grave{e_{2j-1}} \to e_{m+1} \partial_{a_j}, \quad \grave{e_{2j}} \to -e_{m+1} a_j, \quad j = 1, \dots, n,$$

to prove that there exists a non-trivial algebra generated by the whole set $\{\grave{e_1}, \dots, \grave{e_{2n}}; e_1, \dots, e_m\}$ with the above defining relations such that the subalgebras generated by the sets $\{\grave{e_1}, \dots, \grave{e_{2n}}\}$ and $\{e_1, \dots, e_m\}$ remain the same as before.

Although radial algebra in the usual sense includes bosonic as well as fermionic variables, it is also possible to define a fermionic analogue of the radial algebra by requiring the axiom

$$[x, y]z = z[x, y]$$

to be satisfied. One then introduces representations of the form

$$\underline{x} = \sum x_j e_j^{\,\grave{}},$$

and the commutator is given by

$$\underline{x}\,\underline{y} - \underline{y}\,\underline{x} = \sum x_j\, y_k (e_j^{\,\grave{}} e_k^{\,\grave{}} - e_k^{\,\grave{}} e_j^{\,\grave{}}),$$

so that we again arrive at defining relations of the form

$$e_j^{\,\grave{}} e_k^{\,\grave{}} - e_k^{\,\grave{}} e_j^{\,\grave{}} = f_{jk}.$$

The fermionic representation is then given by

$$\underline{x}^{\,\grave{}} = \sum x_j^{\,\grave{}} e_j$$

of which the commutator has the form

$$\underline{x}^{\,\grave{}}\underline{y}^{\,\grave{}} - \underline{y}^{\,\grave{}}\underline{x}^{\,\grave{}} = \sum x_j^{\,\grave{}} y_k^{\,\grave{}} (e_j\, e_k + e_k\, e_j),$$

leading to the defining relations

$$e_j\, e_k + e_k\, e_j = -2g_{jk}.$$

Nevertheless, we are not going to consider this "fermionic version" of radial algebra because the differentials dx, $x \in S$ already form such a fermionic radial algebra. In other words in the extended version of radial algebra including abstract vector differentials dx, $x \in S$, there is no need for further super-symmetrization; the concept of radial algebra already includes super-symmetry.

To finish this section we'll discuss the group invariance whereby one of the variables x is varying and the other vector variables are mere parameters (i.e., they are symbols for fixed objects).

In case of a bosonic vector variable $\underline{x} = \sum e_j\, x_j$, the group action is determined by

$$F(\underline{x}) \rightarrow s\, F(\bar{s}\underline{x}s) \quad \text{or} \quad F(\underline{x}) \rightarrow s\, F(\bar{s}\,\underline{x}\,s)\bar{s}$$

$s \in \mathbf{Spin}(m)$, which is derived from the group actions on vectors and Clifford numbers

$$\underline{x} \rightarrow s\,\underline{x}\bar{s}, \quad a \rightarrow s\,a.$$

The infinitesimal generators of $\mathbf{Spin}(m)$ are of the form

$$s = 1 + e \sum b_{ij}\, e_{ij},$$

e being infinitesimal and $\sum b_{ij}\, e_{ij}$ a standard bivector with commuting coordinates b_{ij}.

In the fermionic case things are quite similar to this (see also [2]): first we make the observation that, putting $e_{jk}\grave{} = e_j\grave{}e_k\grave{}$

$$
\begin{aligned}
e_{jk}\grave{}\,e_l\grave{} - e_l\grave{}\,e_{jk}\grave{} &= e_j\grave{}(e_k\grave{}e_l\grave{} - e_l\grave{}e_k\grave{}) + (e_j\grave{}e_l\grave{} - e_l\grave{}e_j\grave{})e_k\grave{} \\
&= e_j\grave{}\,f_{kl} + f_{jl}\,e_k\grave{}
\end{aligned}
$$

which already means that the commutator of $\flat = \sum b'_{jk}\,e_{jk}\grave{}$, b'_{jk} scalar, with the variable $\underline{x}\grave{}$ is at least again a vector. In other words, if we consider the infinitesimal algebra element

$$
\acute{s} = \exp(\varepsilon\,\flat) = 1 + \varepsilon\,\flat
$$

then we have that

$$
\acute{s}^{-1} = \exp(-\varepsilon\,\flat) = 1 - \varepsilon\,\flat
$$

and we have established that the map

$$
h(\acute{s}) : \underline{x}\grave{} \rightarrow \acute{s}\,\underline{x}\grave{}\,\acute{s}^{-1}
$$

transforms vectors into vectors.

For this map to be symplectic (i.e., to preserve the symplectic structure), no extra conditions are needed because the symplectic form is just the anti-commutator $\underline{x}\grave{}\underline{y}\grave{} + \underline{y}\grave{}\underline{x}\grave{}$, which is preserved by $h(\acute{s})$. But of course the kernel of the representation $\acute{s} \rightarrow h(\acute{s})$ is more than only $\acute{s}$ and $-\acute{s}$; the bivector coordinates b'_{ij} need to satisfy the extra symmetry condition $b'_{ij} = b'_{ji}$ to lead to a double covering and to the symplectic analogue $\mathbf{Spin}\grave{}(n)$ of the spin group which is the metaplectic group (see [2]). On the super-symmetric level, considering the super-vector-variable

$$
x = \sum_{j=1}^{m} x_j\,e_j + \sum_{j=1}^{2n} x_j\grave{}\,e_j\grave{}
$$

we already have the action of the groups $\mathbf{Spin}(m)$ on $\underline{x}$ and $\mathbf{Spin}\grave{}(n)$ on $\underline{x}\grave{}$ corresponding to the exponentials of the bivector $b = \sum b_{ij}\,e_{ij}$, $b_{ij} = -b_{ji}$ and the anti-bivector $\flat = \sum b'_{ij}\,e_{ij}\grave{}$, $b'_{ij} = b'_{ji}$. Moreover, for bivectors we have the commutation relations

$$
e_{ij}\,e_k\grave{} = e_k\grave{}\,e_{ij}, \qquad e_{ij}\grave{}\,e_k = e_k\,e_{ij}\grave{},
$$

so that action of the element $s = \exp b$ (resp. $\acute{s} = \exp \flat$) leaves invariant the variable $\underline{x}\grave{}$ (resp. $\underline{x}$) and so the action of $s \in \mathbf{Spin}(m)$ and $\acute{s} \in \mathbf{Spin}\grave{}(n)$ on the super-vector-variable $x = \underline{x} + \underline{x}\grave{}$ is well defined and leaves the anti-commutator invariant. But no group action seems to cover these two groups and leads to a grand invariance of the super-vector-variable x. There is something which covers both group actions if we consider so-called "super-group" actions which are introduced the easiest as follows:

The mapping $b \to \exp b, b = \sum e_{ij}\, b_{ij}$, covers the spin group $\mathbf{Spin}(m)$ (it is the image) and hence determines this group. Hence, in some sense, the expression $\exp(b)$ is the spin group although this is not to be seen as an equality of sets. Moreover, one may now consider the bivector $b = \sum b_{ij}\, e_{ij}$ as a purely symbolic object whereby the entries $b_{ij} = -b_{ji}$ are merely commuting variables (not elements of the set of real numbers) and one might think of using other types of bivector variables such as anti-commuting ones thus leading to an extension of the notion of a group. We do not need to do this for the bivectors e_{ij} or $e_{ij}\grave{}$, but we do need this for the mixed elements $e_i\, e_j\grave{}$ for which we have the commutation and anti-commutation relations

$$e_i\, e_j\grave{}\, e_k\grave{} + e_k\grave{}\, e_i\, e_j\grave{} = e_i\,(e_j\grave{}\, e_k\grave{} - e_k\grave{}\, e_j\grave{}) = e_i\, f_{jk}$$

$$e_i\, e_j\grave{}\, e_k - e_k\, e_i\, e_j\grave{} = -e_j\grave{}\,(e_i\, e_k + e_k\, e_i) = 2e_j\grave{}\, g_{ik}$$

under which vectors are transformed into vectors. For the bosonic and fermionic variables $\underline{x}$ and $\underline{x}\grave{}$ this means that

$$e_i\, e_j\grave{}\, \underline{x}\grave{} + \underline{x}\grave{}\, e_i\, e_j\grave{} = \text{type } e_i\, X_i\grave{}, \quad X_i\grave{} \text{ anti-commuting}$$

$$e_i\, e_j\grave{}\, \underline{x} - \underline{x}\, e_i\, e_j\grave{} = \text{type } e_j\grave{}\, X_j, \quad X_j \text{ commuting};$$

and so if we consider anti-commuting bivector variables

$$\eth_{ij} \quad i = 1,\dots,m, \quad j = 1,\dots,2n,$$

and produce the super-bivector $B = \sum b_{ij}\grave{}\, e_i\, e_j\grave{}$, then in both cases the commutators

$$B\underline{x}\grave{} - \underline{x}\grave{}B, \quad B\underline{x} - \underline{x}B$$

are of the vector type $\sum e_i X_i$, with X_i commuting, and of the vector type $\sum e_j\grave{} X_j$, with $X_j\grave{}$ anti-commuting, respectively.

Hence, if we introduce the infinitesimal element $S = \exp(\varepsilon B) = 1 + \varepsilon B$, we clearly have that $S^{-1} = \exp(-\varepsilon B) = 1 - \varepsilon B$ and the action of S on vectors and anti-vectors is given by

$$\underline{x} \to S\underline{x}S^{-1}, \quad \underline{x}\grave{} \to S\underline{x}\grave{}S^{-1}$$

and results in a super vector. Hence, the map

$$x \to SxS^{-1}$$

transforms super vectors into super-vectors and leaves invariant the anti-commutator. Also the exponential function $S = \exp B$ makes sense because the element B belongs to the algebra generated by $e_j, e_j\grave{}$ and $\eth_{ij}$ and we hence have three different basic "super-spin group actions"

$$x \to s\, x\, s^{-1}, \quad s = \exp b, \quad b = \sum e_{ij}\, b_{ij},$$

$$x \to \grave{s}\, x\, \grave{s}^{-1}, \quad \grave{s} = \exp \eth, \quad \eth = \sum e_{ij}\grave{}\, b_{ij}',$$

$$x \to SxS^{-1}, \quad S = \exp B, \quad B = \sum b_{ij}\grave{}\, e_i\, e_j\grave{},$$

whereby b_{ij}, b'_{ij} are commuting variables and b_{ij} anti-commuting variables. But this is still not good enough to define something like a super-spin group; one does not only need the above three actions but also all possible compositions of them, which does give rise to the problem that there may be an unlimited collection of bivector variables b_{ij}, b'_{ij} and b_{ij} to be considered. One needs the possibility to substitute variables by fixed values in one way or another to arrive at as "set" of group elements. For b_{ij}, b'_{ij} this is not so difficult because they may assume all real numbers as values and thus lead to the definition of the sets $\mathbf{Spin}(m)$ and $\mathbf{Spin}(n)$ as the image of s resp. $\dot{s}$ whereby b_{ij}, b'_{ij} assume real values.

The problem with anti-commuting variables like x_j or b_{ij} is that they can only assume the real value zero, and the only way to replace anti-commuting variables by fixed objects is by letting them take their values in an algebra which generalizes the real numbers and contains commuting as well as anti-commuting elements such that the product of two anti-commuting elements is commuting. The natural candidate for this (see also [1, 13]) is a Grassmann algebra $L(f_1, \dots, f_k)$ generated by a set of anti-commuting "fixed elements" $\{f_1, \dots, f_k\}$. The algebra admits a Z mod 2 -grading $L = L_+ + L_-$ whereby L_+ is the even subalgebra and L_- the subspace of odd elements, i.e.,

$$L_+ = L_o + L_2 + L_4 + \dots, \quad L_- = L_1 + L_3 + L_5 + \dots,$$

L_j being the subspace of j-vectors inside L. Commuting variables may now take their values in L_+ while anti-commuting variables take values in L_-.

3   The bosonic and fermionic vector derivatives

In case we use a vector variable $\underline{x} = \sum e_j x_j$ of bosonic type, we may consider the Dirac operator (or vector derivative) $-\partial_{\underline{x}} = \sum (-e_j) \partial_{x_j}$ so that, via the representation $x \to \underline{x}$, we arrive at an endomorphism $\partial_x \to -\partial_{\underline{x}}$ for which the following abstract relations are satisfied

(D1) $\partial_x[f\ F] = \partial_x[f]\ F + f\ \partial_x[F]$,
      $[f\ F]\partial_x = F\ \partial_x[f] + f\ [F]\partial_x$, f, $F \in R(S)$, f scalar,

(D2) $\partial_x[F\ G] = \partial_x[F]\ G$,
      $[G\ F]\partial_x = G[F]\partial_x$ in case $G \in R(S\backslash\{x\})$

(D3) $[\partial_x\ F]\partial_y = \partial_x[F\ \partial_y]$,

(D4) $\partial_x x^2 = x^2\partial_x = 2x$,
      $\partial_x\{x, y\} = \{x, y\}\partial_x = 2y, x \neq y$.

And in [So] we have proven that these relations define the abstract endomorphism ∂_x, and it was also shown that the quantity $\partial_x[x]$ is a scalar which does not depend on the variable x. Therefore, we may define the "abstract or formal dimension m" to be equal to

$$(D5) \quad \partial_x[x] = [x]\partial_x = m.$$

The above relations (D1), ... , (D5) are, hence, established on the level of abstract vector variables and were inspired by the actual properties of the Dirac operator $-\partial_{\underline{x}}$. Hence, they have to remain satisfied by any extended version of the Dirac operator, e.g., to the fermionic or super-symmetric situation.

Let us first try to define a good fermionic Dirac operator according to these ideas using the special fermionic representation from the previous section:

$$\underline{x}\grave{} = \sum x_{2j-1}\grave{}\, e_{2j-1}\grave{} + \sum x_{2j}\grave{}\, e_{2j}\grave{}$$

$$e_{2j-1}\grave{} = \partial_{a_j}\, e_{m+1}, \quad e_{2j}\grave{} = -a_j\, e_{m+1}.$$

Next denote by $\partial_{x_j}\grave{}$ the fermionic analogue of the partial derivatives which are determined by

$$\partial_{x_j}\grave{}[F] = 0 \quad \text{and} \quad \partial_{x_j}\grave{}[x_j\grave{}\, F] = F$$

whereby $x_j\grave{}$ does not occur in $F \in \text{Alg}\{x_1\grave{},\dots,x_{2n}\grave{}\}$ (thinking of $x_j\grave{}$ as differential forms dx_j, these derivatives $\partial_{x_j}\grave{}$ may be thought of as contraction operators $\partial_{x_j}|$).

We then define the fermionic analogue of the Dirac operator to be

$$\partial_{\underline{x}}\grave{} = 2\sum \partial_{x_{2j-1}}\grave{}\, e_{2j}\grave{} - 2\sum \partial_{x_{2j}}\grave{}\, e_{2j-1}\grave{}$$

and it is easy to verify these axioms

$$(D1.a) \quad \partial_{\underline{x}}\grave{}[fF] = \partial_{\underline{x}}\grave{}[f]\, F + f\, \partial_{\underline{x}}\grave{}[F]$$

$$(D1.b) \quad [F\, f]\partial_{\underline{x}}\grave{} = F\, [f]\partial_{\underline{x}}\grave{} + [F]\, \partial_{\underline{x}}\grave{}\, f,$$

whereby f is scalar in the abstract vector variable sense (i.e., in the algebra generated by the inner products). (D1) would follow if one would have the extra condition $\partial_{\underline{x}}\grave{}[f] = [f]\partial_{\underline{x}}\grave{}$, but this condition is false unless $\partial_{\underline{x}}\grave{}[f] = 0$.

There is no problem with the validity of (D2) and the similar property for the right vector derivative and also the associativity property (D3) remains valid. As to the axiom (D4), we have that

$$\underline{x}\grave{}^2 = \sum x_{2j-1}\grave{}\, x_{2k}\grave{}\,(e_{2j-1}\grave{}\, e_{2j}\grave{} - e_{2j}\grave{}\, e_{2j-1}\grave{}) = \sum x_{2j-1}\grave{}\, x_{2j}\grave{},$$

so that

$$\partial_{\underline{x}}\grave{}[\underline{x}\grave{}^2] = 2\sum e_{2j}\grave{}\, x_{2j}\grave{} + 2\sum e_{2j-1}\grave{}\, x_{2j-1}\grave{} = 2x - [\underline{x}\grave{}^2]\partial_{\underline{x}}\grave{}$$

$$\partial_{\underline{x}}\grave{}[\underline{x}\grave{}] = 2\sum(e_{2j}\grave{}\, e_{2j-1}\grave{} - e_{2j-1}\grave{}\, e_{2j}\grave{}) = -2n = -[\underline{x}\grave{}]\partial_{\underline{x}}\grave{},$$

and we have that

$$\dot{x}\dot{y} + \dot{y}\dot{x} = \sum x_{2j-1}\, y_{2j}\, (e_{2j-1}\grave{}\, e_{2j}\grave{} - e_{2j}\grave{}\, e_{2j-1}\grave{}) + \text{ibid } (x \leftrightarrow y)$$
$$= \sum (x_{2j-1}\grave{}\, y_{2j}\grave{} + y_{2j-1}\grave{}\, x_{2j}\grave{}),$$

so that

$$\partial_{\underline{x}}\grave{}(\dot{x}\dot{y} + \dot{y}\dot{x}) = 2\,(e_{2j}\grave{}\, y_{2j}\grave{} + e_{2j-1}\grave{}\, y_{2j-1}\grave{}) = 2\,\dot{y} = -(\dot{x}\dot{y} + \dot{y}\dot{x})\partial_{\underline{x}}\grave{}.$$

Hence, if we now reconsider $\dot{x}$ to be a representation of the abstract vector variable $x \in S$, then the abstract vector derivative seems to be well represented by the operator

$$\partial_x[F] \rightarrow \partial_{\underline{x}}\grave{}[F], \quad [F]\partial_x \rightarrow -[F]\partial_{\underline{x}}\grave{}.$$

For all scalar elements in the radial algebra, we have verified the identity

$$\partial_{\underline{x}}\grave{}[f] = -[f]\partial_{\underline{x}}\grave{},$$

while, when seen as a left operator, $\partial_{\underline{x}}\grave{}$ satisfies also (D4) and (D5) with

$$(D5) \quad \partial_{\underline{x}}\grave{}[\dot{x}] = M = -2n.$$

Hence the abstract operator ∂_x corresponds as a left operator to the fermionic vector derivative $\partial_{\underline{x}}\grave{}$ while as a right operator it corresponds to $-\partial_{\underline{x}}\grave{}$ whereby the action of these fermionic operators is restricted to elements F' which correspond in the fermionic picture to actual elements of $R(S)$. Surprising in the fermionic representation is the formula for the formal dimension valid for the representation of ∂_x :

$$(D5) \quad \partial_x[x] = -2n = [x]\partial_x,$$

which is needed to choose a specific value for $\partial_x[x]$ allowing complete evaluation of ∂_x. In other words, in the fermionic picture, on the level of the radial algebra $R(S)$, everything is exactly the same as in the bosonic picture, except for the fact that the formal dimension $\partial_x[x]$ is the negative even integer $-2n$, $2n$ being the number of basic fermionic variables. Finally let us consider the super-vector variable

$$x = \dot{x} + \underline{x} = \sum_{j=1}^{n}(x_{2j-1}\grave{}\, e_{2j-1}\grave{} + x_{2j}\grave{}\, e_{2j}\grave{}) + \sum_{j=1}^{m} x_j\, e_j$$

then using the already made evaluations it seems that one has to consider the following representations for the abstract vector derivative ∂_x acting on the algebra $R(S)$, here represented by the algebra generated by the elements $\dot{x} + \underline{x},\ x \in S$:

$$\partial_x[F] \rightarrow (\partial_{\underline{x}}\grave{} - \partial_{\underline{x}})[F], \quad [F]\partial_x \rightarrow [F](-\partial_{\underline{x}}\grave{} - \partial_{\underline{x}}).$$

It indeed seems that the computations already made lead to the relations (D1), (D2), (D3), (D4) for ∂_x together with the "formal dimension formula"

$$(D5) \quad \partial_x[x] = [x]\partial_x = M = m - 2n.$$

Note that in the very exceptional case $m = 2n$ one has that

$$\partial_x[x] = [x]\partial_x = 0,$$

i.e., the vector variable x is itself left and right monogenic.

The definition of fermionic Dirac operator given here corresponds to the definition of the Dirac operator on symplectic spinors given in [5] except for the fact that the coordinates x_j are here anti-commutative. In [5] the Dirac operator is indeed given by $\partial_x = \sum e_{2j} \partial_{x_{2j-1}} - \sum e_{2j-1} \partial_{x_{2j}}$; it will correspond to the contraction operator $\partial_{\underline{x}}|$ in our approach. Writing down a super-Dirac operator is a purely formal algebraic matter; in the infinite dimensional setting there is complete symmetry between the bosonic and fermionic cases. But when solving equations, one has to employ suitable function spaces, and infinite dimensionality gives rise to extra problems which we won't discuss here. Hence, due to the analysis aspect there is a substantial difference between the fermionic and bosonic cases. Assuming the reader to be familiar with the standard (bosonic) setting, we'll first discuss the fermionic theory.

Hereby (up to a factor $\pm 2e_{m+1}$) the Dirac may be written in the more suitable form

$$p_{\underline{x}} = \sum a_j \partial_{x_{2j-1}} - \sum \partial_{a_j} \partial_{x_{2j}},$$

i.e., the variables are the anti-commuting objects $x_j, j = 1, \ldots, 2n$ and the algebra of values is generated by a_j, ∂_{a_j}. In other words, as opposed to the bosonic case, the algebra of variables is finite dimensional and the algebra of values infinite dimensional. Hence, from the analysis point-of-view it seems more natural to consider solutions to be functional or distributional objects of the form $T(a_j, b_j)$ with values in the Grassmann algebra generated by $x_1, \ldots, x_{2n}$ (whereby b_j stands for the symbol of ∂_{a_j}) rather than functions of the form $F(x_1, \ldots, x_{2n})$. But the consideration of infinite order differential operators is still complicated and it is in first instance more natural to consider $\mathrm{Alg}\{x_1, \ldots, x_{2n}\}$-valued function like objects $T(a_1, \ldots, a_n)$ on which a_j and ∂_{a_j} act like endomorphisms. The same happens in the bosonic case where Clifford numbers are interpreted as endomorphisms on spinor spaces. In the fermionic case, the analogue of the spinor space could be the function space $S(R^n)$ or distribution space $S'(R^n)$ or even $L_2(R^n)$ (compare with [5]).

As an example, we'll work out the analogue of the Cauchy-Kovalewski theorem in the fermionic setting. Roughly speaking the C.-K. extension theorem has to do with a splitting of R^m as $R^p + R^q$, the standard case is $p = 1$ and $q = m - 1$, which leads to a splitting of the Dirac operator. In the fermionic setting, it seems most natural to consider the splitting

$$\partial_{\underline{x}}^{\grave{}} = a_1 \partial_{x_1}^{\grave{}} - \partial_{a_1} \partial_{x_2}^{\grave{}} + \partial_{\underline{x}},$$

"$\partial_{\underline{x}}$" being the remaining part.

Now in the bosonic case one would write the solution as a power series in the first coordinate x_1; in the fermionic case we separated the coordinates $x_1^{\grave{}}$, $x_2^{\grave{}}$ and the power series becomes simply a sum of four terms

$$T = T_0 + x_1^{\grave{}} T_1 + x_2^{\grave{}} T_2 + x_1^{\grave{}} x_2^{\grave{}} T_3$$

whereby $x_1^{\grave{}}$, $x_2^{\grave{}}$ no longer appears in T_0, T_1, T_2, T_3. Hence, the monogenic system $\partial_{\underline{x}}^{\grave{}} T = 0$ may be rewritten as

$$\partial_{\underline{x}} " T_3 = 0,$$
$$-\partial_{\underline{x}} " T_2 + a_1 T_3 = 0,$$
$$-\partial_{\underline{x}} " T_1 + \partial_{a_1} T_3 = 0,$$
$$\partial_{\underline{x}} " T_0 + a_1 T_1 - \partial_{a_1} T_2 = 0,$$

system which –like in the bosonic case– may be solved recursively, but this time there are only 3 steps:

 (i) to choose a solution T_3 of the reduced Dirac equation

 (ii) using this T_3, to chose solutions T_1 and T_2 of the following two restricted inhomogeneous Dirac equations

 (iii) using these T_1, T_2 to solve the fourth restricted inhomogeneous Dirac equation for T_0.

Note that both the bosonic and fermionic C.-K. theorems readily generalize to the super-symmetric situation.

Another fundamental result is the Fischer decomposition which in this case is the writing of a general distribution $T(a) \in S'(R^n)$ into the form

$$T(a) = M(a) + \left(\sum \partial_{a_j} x_{2j-1}^{\grave{}} + \sum a_j x_{2j}^{\grave{}} \right) [T'(a)]$$

whereby $\partial_{\underline{x}}^{\grave{}} M(a) = 0$ and $T'(a)$ is another distribution in $S'(R^n)$ with values in the Grassmann algebra $\mathrm{Alg}\{x_1^{\grave{}}, \dots, x_{2n}^{\grave{}}\}$, and this splitting is unique. The proof of this identity is left as a very good exercise to the interested reader (hint: split the Grassmann algebra into homogeneous subspaces as is done in the bosonic case for polynomials in $x_1, \dots, x_m$).

4 Exterior differentiation on the level of abstract vector variables

When starting with a set S of abstract vector variables, one may define the radial algebra $R(S)$, leading to a canonical form of Clifford analysis including abstract Dirac operators. But this formalism includes no differential forms and no objects like the fundamental form $d\underline{x} = \sum dx_j e_j$ or fundamental contraction operator $\partial_{\underline{x}}| = \sum \partial_{x_j}|e_j$ whereby $\partial_{x_j}|$ are the classical contraction operators. These are used in connection with monogenic differential forms (see also [11, 12, 10, 4]). What we in fact want is an abstract version of the following: suppose that $x_1, \ldots, x_m$ are real variables; then we can define an operator d of exterior differentiation with the following properties

$$d\,d = 0, \quad d(f\,g) = df\,g + f\,dg, \quad f \text{ being scalar (zero forms).}$$

From this it is possible to define the elements dx_j and derive the relation $dx_j dx_k = -dx_k dx_j$ from $d(d(x_j x_k)) = 0$, and we may consider the algebra of differential forms with polynomial coefficients to be the algebra generated by the sets $\{x_1, \ldots, x_m,\ dx_1, \ldots, dx_m\}$. On this algebra the exterior derivative is fully defined by

(i) $d\,d = 0$

(ii) $d(f\,F) = df\,F + f\,dF, \quad f \in \text{Alg}\{x_1, \ldots, x_m\}$,

and one may easily prove from this the property

(iii) $d(dx_j\,F) = -dx_j\,d\,F$

because it suffices to prove this for elements of the form $F = f\,G$ with $dG = 0$ and one has that

$$d(dx_j\,f\,G) = df\,dx_j\,G + f\,d(dx_j\,G)$$

whereby

$$df\,dx_j = -dx_j\,df \quad \text{and} \quad d(dx_j\,G) = d(d(x_j\,G)) = 0.$$

Next, one may consider instead of scalar variables, vector variables like $\underline{x}_j = \sum x_{jk} e_k$ and the corresponding differentials $d\underline{x}_j = \sum dx_{jk} e_k$ and consider also the algebra $\text{Alg}\{\underline{x}_1, \ldots, \underline{x}_l; d\underline{x}_1, \ldots, d\underline{x}_l\}$. Then using the above defined operator d on scalar valued differential forms with polynomial coefficients, one may extend the operator d to Clifford algebra valued forms and in that way define the operator d as an endomorphism on the algebra $\text{Alg}\{\underline{x}_1, \ldots, x_l;\ d\underline{x}_1, \ldots, d\underline{x}_l\}$.

What we want is an abstract version of this operator d acting on an algebra of the form $\text{Alg}(S \cup dS)$ whereby S is the set of abstract vector variables $x \in S$, and dS is the corresponding set of elements dx, $x \in S$. But there is no direct way of defining the object dx; the operator d is to be defined also through axioms.

First we assume the unpronounced axioms

$$(U1) \quad d : S \to dS \text{ is a bijection,}$$
$$(U2) \quad S \cap dS = \text{ empty.}$$

Now, using our experience with real vector variables it seems we have to assume as basic properties

$$(P1) \quad d\,d = 0,$$
$$(P2) \quad d(x\,F) = dx\,F + x\,dF \text{ for } x \in S, \quad F \in \text{Alg}(S \cup dS),$$
$$(P3) \quad d(dx\,F) = -dx\,dF, \quad dx \in dS, \quad F \in \text{Alg}(S \cup dS).$$

Knowing that every element of $\text{Alg}(S \cup dS)$ can be written as a sum of products of the form $a_1 \ldots a_k$ with $a_j \in S \cup dS$, it is clear that (P2) and (P3) are sufficient to define d as an endomorphism on $\text{Alg}(S \cup dS)$. Hence (P1) is a consequence of (P2) and (P3) which is easy to verify (one has to assume the unpronounced axiom $d1 = 0$). One may be surprised by the fact that we here do not present (P3) as a consequence of (P1) and (P2), but at the present level of development this isn't possible.

What one still needs are properties which define the multiplication within the algebra $\text{Alg}(S \cup dS)$ (the above claims are true for any algebra generated by $S \cup dS$ as long as (P2) and (P3) are consistent with the product structure). First consider $\text{Alg}(dS)$; as the differentials dx, dy are to behave like true vector differentials $\sum e_j\,dx_j$, $\sum e_j\,dy_j$, the commutator $[dx, dy]$ commutes with everything; one should, hence, assume that

$$(P4) \quad [F, [dx, dy]] = 0 \text{ for } F \in \text{Alg}(dS).$$

For real vector differentials one also has that $dx^{m+1} = 0$, but as abstract vector variables are already represented by true vector variables in any dimension, no abstract version of this property is specified. It seems that property (P4) is sufficient for the abstract definition of $\text{Alg}(dS)$, which shows that this is exactly the "fermionic version" of the algebra $R(S)$.

Finally for the algebra $\text{Alg}(S \cup dS)$ we need to extend (P4) as well as the basic axiom of radial algebra $[F, \{x, y\}] = 0$, $F \in R(S)$ to

$$(P5) \quad [F, \{x, y\}] = 0, \ F \in \text{Alg}(S \cup dS),$$
$$(P6) \quad [F, [dx, dy]] = 0, \ F \in \text{Alg}(S \cup dS),$$
$$(P7) \quad [F, \{x, dy\}] = 0 \text{ for } F \in \text{Alg}(S),$$
$$(P8) \quad \{dz, \{x, dy\}\} = 0 \text{ for } x, y, z \in S.$$

The next problem is to rearrange these properties into independent axioms for the algebra $\text{Alg}(S \cup dS)$. We already mentioned (P1) as a consequence of (P2) and (P3), which leads to the defining axioms for d :

$$(R1) \quad d(x\,F) = dx\,F + x\,dF, \;\; F \in \mathrm{Alg}(S \cup dS),$$
$$(R2) \quad d(dx\,F) = -dx\,dF, \;\; F \in \mathrm{Alg}(S \cup dS).$$

The basic axiom for the product in $R(S)$ is

$$(R3) \quad [z, \{x, y\}] = 0, \;\; x, y, z \in S.$$

Differentiation of this identity leads to $[dz, \{x, y\}] + [z, d\{x, y\}] = 0$ though clearly not to

$$(R4) \quad [dz, \{x, y\}] = 0, \;\; x, y, z \in S.$$

Taken together, (R1) up to (R4) lead to the evaluation of $d\{x, y\}$ and hence, to the identity $[z, \{dx, y\}] + [z, \{x, dy\}]$ though not to

$$(R5) \quad [z, \{dx, y\}] = 0, \;\; x, y, z \in S.$$

Differentiation of (R4) resp. (R5) leads to the identities

$$\{dz, d\{x, y\}\} = 0, \quad \{dz, \{dx, y\}\} = [dz, [dx, dy]] = 0,$$

though again not to the identities

$$(R6) \quad \{dz, \{dx, y\}\} = 0, \;\; x, y, z \in S,$$
$$(R7) \quad [z, [dx, dy]] = 0, \;\; x, y, z \in S.$$

Note that differentiation of either (R6) or (R7) leads to the identity

$$[dz, [dx, dy]] = 0,$$

i.e., (P4) is a consequence of the other axioms.

We thus have seven axioms or basic rules (R1)–(R7) which define what one could call "super-radial" algebra although we prefer the name radial algebra with differentiation. The next problem is to examine to what extent an algebra generated by $S \cup dS$ and the rules (R1)–(R7) is close to an algebra generated by true vector variables like $\underline{x} = \sum e_j\,x_j$ and true vector differentials $d\underline{x} = \sum e_j\,dx_j$. First, we define the "scalar subalgebra" by

Definition 1. *By $S\,Alg(S \cup dS)$ we denote the algebra of scalar valued objects in $R(S \cup dS)$, which is the algebra generated by $\{x, y\}$, $[dx, dy]$ and $\{dx, y\}$.*

The objects $\{x, y\}$ and $[dx, dy]$ are commutative and it is, hence, acceptable that these are scalars. But $\{dx, y\}$ is no commutative object; the reason why $\{dx, y\}$ is called scalar is because for true vector variables $\underline{x}, \underline{y}$ the object $\{d\underline{x}, \underline{y}\} = -2\sum dx_j\,y_j$ is a scalar valued (a real valued) as opposed to a general Clifford algebra valued form. Note that $S\mathrm{Alg}(S \cup dS)$

is also closed under differentiation and this in view of the relation $d^2 = 0$ together with

$$d\{x, y\} = \{dx, y\} + \{x, dy\}, \quad d\{dx, y\} = -[dx, dy],$$

i.e., the operator d is itself scalar valued. Moreover, the elements $\{x, y\}$ and $[dx, dy]$ are central while the forms $\{dx, y\}$ mutually anti-commute.

Next we are going to define a wedge product in the algebra $\text{Alg}(S \cup dS)$ which is based on the product in $\text{Alg}(S \cup dS)$ and which for true vector variables and vector differentials

$$\underline{v}_1 = \underline{x}_1, \dots, \underline{v}_k = \underline{x}_k, \quad \underline{v}_{k+1} = d\underline{y}_1, \dots, \underline{v}_{k+l} = d\underline{y}_l$$

corresponds to taking the $(k + l)$−vector part in the Clifford algebra, i.e.,

$$\underline{v}_{\pi(1)} \wedge \dots \wedge \underline{v}_{\pi(k+l)} = [\underline{v}_{\pi(1)} \cdots \underline{v}_{\pi(k+l)}]_{k+l}$$

whereby $a \to [a]_k$ denotes the projection of $a \in R_m$ on the k-vectors $R_{m;k}$. In other words the symbol "$\wedge$" for the wedge product refers to the wedge product in the Clifford algebra R_m in which the vector variables and differentials

$$\underline{x}_i = \sum x_{ik} \, e_k, \quad d\underline{y}_j = \sum dy_{jk} \, e_k$$

take their values. In particular the wedge product is associative and one has the commutation relations

$$\underline{x}_i \wedge \underline{x}_j = -\underline{x}_j \wedge \underline{x}_i, \quad \underline{x}_i \wedge d\underline{y}_j = -d\underline{y}_j \wedge \underline{x}_i, \quad d\underline{y}_i \wedge d\underline{y}_j = d\underline{y}_j \wedge d\underline{y}_i.$$

The problem with the algebra $\text{Alg}\{S \cup dS\}$ is that the variables $x \in S$ and differentials $dx \in dS$ do not take values in a Clifford algebra; one has to define the wedge product abstractly. This was done for $R(S)$ in [9] by directly putting

$$x_1 \wedge \dots \wedge x_k = \frac{1}{k!} \sum \text{sgn}(\pi) x_{\pi(1)} \cdots x_{\pi(k)}.$$

The same can be repeated here for $v_1, \dots, v_k \in S \cup dS$ in the sense that one has a formula of the form

$$v_1 \wedge \dots \wedge v_k = \frac{1}{k!} \sum c(\pi) v_{\pi(1)} \cdots v_{\pi(k)},$$

where this time the signum $c(\pi) \in \{1, -1\}$ is determined by the extra conditions

(i) the wedge product is associative

(ii) $x \wedge y = -y \wedge x$, $x \wedge dy = -dy \wedge x$, $dx \wedge dy = dy \wedge dx$, $x, \, y, \in S$

(iii) $v_1 \wedge \ldots \wedge v_k = c(\pi) \, v_{\pi(1)} \wedge \ldots \wedge v_{\pi(k)}$.

The notion of a wedge product allows the following splitting of elements in $Alg(S \cup dS)$, which corresponds to the splitting of a Clifford number into k-vector parts.

Theorem 1. *Every $F \in Alg(S \cup dS)$ may be written in the canonical form*

$$F = \sum f v_1 \wedge \ldots \wedge v_l, \ v_j \in S \cup dS, \ f \in SAlg(S \cup dS).$$

Hereby the sum runs over all distinct wedge products $v_1 \wedge \ldots \wedge v_l$, $v_j \in S \cup dS$ and the sum is finite.

Proof. It clearly suffices to consider F to be a product $F = v_1 \ldots v_l$ with $v_j \in S \cup dS$. Using the axioms (R1) up to (R7) there is a canonical way of writing

$$v_1 \ldots v_l = c(\pi) v_{pi(1)} \ldots v_{pi(l)} + \sum GH$$

whereby $c(\pi)$ is the same signum as used in the definition of the wedge product and whereby $G \in SAlg(S \cup dS)$ and H is a product of elements in $S \cup dS$ of order less than l. (This is easy to check for transpositions and, hence, follows for general permutations.) From this it readily follows that $v_1 \ldots v_l = v_1 \wedge \ldots \wedge v_l + \sum GH$. $\square$

Remark 1. *Note that the signature coefficients $c(\pi)$ and, hence, also the definition of the wedge product follow directly from the axioms (R1)–(R7), and, hence, the above conditions (i), (ii), (iii) are not axioms but properties.*

The next problem is to examine whether the algebra $Alg(S \cup dS)$ is representable by an algebra consisting of true vector variables and differentials. That such a representation exists is clear: choose a certain dimension m and consider the Clifford algebra R_m determined by $e_i e_j + e_j e_i = -2\delta_{ij}$; then we may make the assignments

$$x \in S \rightarrow \underline{x} = \sum x_j \, e_j, \quad dx \in dS \rightarrow d\underline{x} = \sum dx_j \, e_j,$$

introducing sets of vector variables and differentials $\underline{S} = \{\underline{x} : x \in S\}$, $d\underline{S} = \{d\underline{x} : x \in S\}$ whereby d is the standard exterior derivative. It is readily seen that we have a map

$$\underline{\ \ } : Alg(S \cup dS) \rightarrow Alg(\underline{S} \cup d\underline{S})$$

which is an algebra representation which carries over the operator d as well as the wedge product. But it is never an isomorphism because $d\underline{x}^{m+1} = 0$ for every vector differential $d\underline{x} = \sum dx_j \, e_j$ while in $Alg(S \cup dS)$ no power dx^k of dx vanishes. This need not be a surprise because the above representations exist for all $m \in N$ and $d\underline{x}^m$ does not vanish. What we have is the following:

Theorem 2. *The totality of all Clifford algebra representations $\cdot$ is injective, i.e., when formally introducing an infinite dimensional Clifford algebra representation one obtains an isomorphism.*

Proof. We are going to prove injectivity of the map $\cdot$ for certain special subspaces of $\text{Alg}(S \cup dS)$. First we consider only finitely many vector variables $x_1, \ldots, x_K \in S$ and the corresponding differentials $dx_1, \ldots, dx_K$. Let Alg denote the radial algebra generated by these; then we perform the canonical decomposition from Theorem 1 and one may also perform the corresponding decomposition in any representation $\cdot$ [Alg]. Hereby elements of the scalar subalgebra $S\text{Alg}$ of Alg are mapped onto the corresponding scalar elements in $\cdot$ [Alg]. Moreover, if one puts an upper bound on the degree of wedge products, thus leading to a subspace V of Alg, one may always find a dimension m which is high enough so that the number of terms in the decomposition of Theorem 1 is the same for both V and $\cdot[V]$. Hence our next task is to check whether injectivity may be achieved for $\cdot$ acting on the scalar subalgebra $S\text{Alg}$ of Alg, which is generated by the variables $\{x_i, x_j\}$, $[dx_i, dx_j]$, $\{dx_i, x_j\}$.

Now let us for $i, j \in \{1, \ldots, K\}$ introduce commuting variables B_{ij} and anti-commuting variables F_{ij}; then the set $\{B_{ij}, F_{ij}\}$ generated the algebra $P(K^2) \otimes L(K^2)$ whereby $P(K^2)$ is the polynomial algebra generated by the K^2 commuting variables B_{ij} while $L(K^2)$ is the Grassmann algebra generated by the K^2 anti-commuting variables F_{ij}. Moreover, by making the substitution

$$B_{ij} = \{x_i, x_j\} - [dx_i, dx_j], \quad F_{ij} = \{dx_i, x_j\},$$

it is clear that the scalar subalgebra $S\text{Alg}$ is a representation of $P(K^2) \otimes L(K^2)$, for it also follows that

$$B_{ij} + B_{ji} = 2\{x_i, x_j\}, \quad B_{ij} - B_{ji} = -2[dx_i, dx_j].$$

Hence, the representations $\cdot$ lift to representations from $P(K^2) \otimes L(K^2)$ onto $\cdot$ [$S\text{Alg}$] and it suffices to check whether this lifting is injective provided the dimension is high enough. Now we can write

$$\underline{x}_j = \sum b_{jl}\, e_l, \quad d\underline{x}_j = \sum f_{jl}\, e_l,$$

whereby b_{jl} resp. f_{jl}, $j = 1, \ldots, K$, $l = 1, \ldots, m$ are new commuting resp. anti-commuting variables and the above lifted representation corresponds in fact to the quadratic "super-transformation"

$$B_{ij} = \sum b_{il}\, b_{jl} - \sum f_{il}\, f_{jl}, \quad F_{ij} = \sum b_{il} f_{jl}$$

which defines a representation from $P(K^2) \otimes L(K^2)$ to $P(Km) \otimes L(Km)$. Now the variables B_{ij} are completely independent scalar variables and may be written as $B_{ij} = S_{ij} + A_{ij}$, $A_{ij} = \frac{1}{2}(B_{ij} - B_{ji})$ being an anti-symmetric matrix of commuting variables. Hence, the algebra generated by A_{ij} is an infinite dimensional polynomial algebra, while (according to the above transformation $A_{ij} = -\sum f_{il} f_{jl}$) the image is finite dimensional for all values of m. This means that injectivity can only be achieved for $m =$ infinite, or for $m =$ finite, provided that one restricts the map $\colon$ to subspaces of SAlg obtained by putting a limit on the degree of elements in SAlg leading to finite dimensional subspaces SAlg$_p$ of SAlg, of which the elements are polynomials $F(S_{ij}; A_{ij})$ with values in the Grassmann algebra generated by F_{ij}, $i,j = 1, \dots, K$.

Now we can assume that $m > K$ and that for $l \leq K$, $b_{il} = \delta_{il}$ so that we already get an injective representation $F_{ij} = f_{ij}$, $i,j = 1, \dots, k$ for the Grassmann values. This means that we have to investigate substitutions of the form

$$S_{ij} = \sum_1^K \delta_{il} \delta_{jl} + \sum_{K+1}^m b_{il} b_{jl} = \delta_{ij} + \sum_{K+1}^m b_{il} b_{jl}$$

and by taking $m \geq 2K$ it is readily seen that this substitution leads to an injective map for all polynomials of the variables S_{ij} (see also [9]). But we still have to investigate the substitutions

$$A_{ij} = -\sum_1^K F_{il} F_{jl} - \sum f_{il} f_{jl}$$

and, depending on the upper bound p imposed on the degree, one may find a value P such that for $m \geq P$ the above substitution is injective on the polynomials of the variables A_{ij} of degree at most p. To obtain injectivity on SAlg$_p$, it hence suffices to choose $m \geq \max\{P, 2K\}$. $\square$

Remark 2. *The algebra generated by $S \cup dS$ together with (R1)–(R7) is clearly only representable by a Clifford differential calculus in an infinite dimensional setting, but this is also possible, i.e., by restrictions on the degrees one truly obtains injective representations using standard vector variables and vector differentials.*

Remark 3. *Using the notation*

$$B_{ij} = \{x_i, x_j\} - [dx_i, dx_j], \quad F_{ij} = \{dx_i, x_j\},$$

it also follows that the variables B_{ij} resp. F_{ij} are completely indepen-dent commuting resp. anti-commuting variables among which no extra con-straints exist showing that the scalar subalgebra S $Alg(S \cup dS)$ is nothing but $P(K^2) \otimes L(K^2)$. Moreover, it is possible to redefine the operator d on this algebra simply by putting

(i) $d(B_{ij}G) = (F_{ij} + F_{ji})G + B_{ij}dG$

(ii) $d(F_{ij}G) = \frac{1}{2}(B_{ij} - B_{ji})G - F_{ij}dG.$

This suggests the introduction of matrices of commuting variables $S_{ij} = S_{ji}$ and anti-commuting variables $A_{ij} = -A_{ji}$ and to consider the transformation

$$B_{ij} = S_{ij} + dA_{ij}, \quad F_{ij} = A_{ij} + \frac{1}{2}dS_{ij},$$

leading back to the required relations

$$dB_{ij} = dS_{ij} = F_{ij} + F_{ji}, \quad dF_{ij} = dA_{ij} = \frac{1}{2}(B_{ij} - B_{ji})$$

so that the rules (i), (ii) may be rewritten into a more standard form.

(i) $d(S_{ij}\,G) = dS_{ij}\,G + S_{ij}\,dG, \quad d(dS_{ij}\,G) = -dS_{ij}\,dG,$

(ii) $d(A_{ij}\,G) = dA_{ij}\,G - A_{ij}\,dG, \quad d(dA_{ij}\,G) = dA_{ij}\,dG.$

The above rule (i) is identical to the standard exterior differentiation defined for commuting variables. The rule (ii) introduces a definition of exterior differentiation for anti-commuting variables A_{ij}, which shows that ideas from super-symmetry are already included in differential calculus on abstract vector variables; also in case these abstract vector variables are represented by standard Clifford-vector variables with commuting coordinates even before any "fermionic representation"

$$x \to \sum x_j\grave{}\, e_j\grave{}, \quad e_j\grave{}\, e_k\grave{} - e_k\grave{}\, e_j\grave{} = f_{jk}$$

is considered.

The above calculations also suggest that the differential $dx_j\grave{}$ of an anti-commuting variable should be a commutative object; even more, the set of objects $dx_1\grave{},\dots,dx_{2n}\grave{}$ should behave like the above objects dA_{ij} which generate an infinite dimensional polynomial algebra in which dA_{ij} are independent commuting variables. The same should hold for the elements $dx_j\grave{}$ and on the algebra generated by the elements $\{x_1\grave{},\dots,x_{2n}\grave{}; dx_1\grave{},\dots,dx_{2n}\grave{}\}.$ One should expect as rules for differentiation:

(i) $x_j\grave{}\, dx_k\grave{} = dx_k\grave{}\, x_j\grave{}$

(ii) $d(x_j\grave{}\,G) = dx_j\grave{}\,G - x_j\grave{}\,dG$

(iii) $d(dx_j\grave{}\,G) = dx_j\grave{}\,dG.$

Moreover, it seems obvious to assume that $e_j\grave{}$ commutes with all $x_j\grave{}, dx_j\grave{}$, and that $d(e_j\grave{}\,G) = e_j\grave{}\,dG$, i.e., the elements $e_j\grave{}$ are "values" and not variables. We now have to check whether all this is consistent with the rules (R1)–(R7). But already the first rules (R1), (R2) seem to be violated because the above assumptions directly lead to

$$d(\underline{x}\grave{}\,G) = d\underline{x}\grave{}\,G - \underline{x}\grave{}\,dG, \quad d(d\underline{x}\grave{}\,G) = d\underline{x}\grave{}\,dG,$$

and to correct the situation it seems we have to assume the controversial rules

(iv) $e_j\grave{}\,x_j\grave{} = x_j\grave{}\,e_j\grave{}$ (as before)

(v) $dx_j\grave{}\,e_j\grave{} = -e_j\grave{}\,dx_j\grave{}$

(vi) $d(e_j\grave{}\,G) = -e_j\grave{}\,dG,$

leading to the correct relations for $d\underline{\grave{x}} = \sum dx_j\grave{}\,e_j\grave{}$

$$(R1) \quad d(\underline{\grave{x}}\,G) = d\underline{\grave{x}}\,G + \underline{\grave{x}}\,dG,$$
$$(R2) \quad d(d\underline{\grave{x}}\,G) = -d\underline{\grave{x}}\,dG.$$

Condition (R3) was already valid for the vector variables $\underline{\grave{x}}$, $x \in S$ and need not be rechecked. Moreover, as $\{\underline{\grave{x}},\ \underline{\grave{y}}\}$ is a quadratic expression in anti-commuting coordinates $x_j\grave{}$, $y_j\grave{}$, (R4) follows from (i) and (iv).

For (R5) and (R6) we have that by (i), (iv), (v)

$$\{d\underline{\grave{x}},\underline{\grave{y}}\} = d\underline{\grave{x}}\,\underline{\grave{y}} + \underline{\grave{y}}\,d\underline{\grave{x}} = \sum dx_j\grave{}\,e_j\grave{}\,y_k\grave{}\,e_k\grave{} + \sum y_k\grave{}\,e_k\grave{}\,dx_j\grave{}\,e_j\grave{}$$
$$= \sum dx_j\grave{}\,y_k\grave{}\,(e_j\grave{}\,e_k\grave{} - e_k\grave{}\,e_j\grave{}) = \sum dx_j\grave{}\,y_k\grave{}\,f_{jk},$$

and from (i), (v) and the previously assumed relations $z_j\grave{}\,x_k\grave{} = -x_k\grave{}\,z_j\grave{}$, it follows that $\underline{\grave{z}}$ commutes while $d\underline{\grave{z}}$ anti-commutes with $\{d\underline{\grave{x}}, \underline{\grave{y}}\}$, i.e., we have relations (R5) and (R6). As to (R7) we have that by (v) –and the fact that $dx_j\grave{}$ are commuting variables– that

$$[d\underline{\grave{x}}, d\underline{y}] = \sum dx_j\grave{}\,e_j\grave{}\,dy_k\grave{}\,e_k\grave{} - \sum dy_k\grave{}\,e_k\grave{}\,dx_j\grave{}\,e_j\grave{}$$
$$= -\sum dx_j\grave{}\,dy_k\grave{}(e_j\grave{}\,e_k\grave{} - e_k\grave{}\,e_j\grave{}) = -\sum dx_j\grave{}\,dy_k\grave{}\,f_{jk}$$

so that, in view of (v) and (i), $[d\underline{\grave{x}}, d\underline{y}]$ commutes with $\underline{\grave{z}}$.

It remains to be checked that the rules (i) up to (vi) –together with the fact that $x_j\grave{}\,y_k\grave{} = -y_k\grave{}\,x_j\grave{}$ and $dx_j\grave{}\,dy_k\grave{} = dy_k\grave{}\,dx_j\grave{}$– are consistent, knowing that (i)–(iii) lead to a consistent definition of differential calculus in anti-commuting variables. Now it is clear that if we would assume as extra conditions, $e_j\grave{}\,x_k\grave{} = x_k\grave{}\,e_j\grave{}$, $e_j\grave{}\,dx_k\grave{} = dx_k\grave{}\,e_j\grave{}$ and $d(e_j\grave{}\,G) = e_j\grave{}\,dG$, we have a consistent calculus which is just the tensor product of the scalar calculus with the symplectic Clifford algebra. By introducing two more standard Clifford numbers E_1, E_2 we can replace the elements $e_j\grave{}$ by $E_1\,e_j\grave{}$ (which does not alter the defining relations for the $e_j\grave{}$'s) and the operator d by $E_2\,d$ (which also leaves (i)–(iii) invariant). If we now denote these new calculus elements still by $e_j\grave{}, d$, we clearly obtain conditions (iv), (v) and (vi), showing that what we consider as "nearly the same" is, in fact, a tensor product.

Next one may investigate the "super-vector representation"

$$x = \underline{\grave{x}} + \underline{x} = \sum x_j\grave{}\,e_j\grave{} + \sum x_j\,e_j$$

and assume the already available rules for differential calculus for the commuting variables x_j as well as for the anti-commuting variables $x_j\grave{}$, together with the obvious relations $x_j\, x_k\grave{} = x_k\grave{}\, x_j,\, x_j\, dx_k\grave{} = dx_k\grave{}\, x_j$ and

(vii) $dx_j\, x_k\grave{} = -x_k\grave{}\, dx_j, \quad dx_j\, dx_k\grave{} = dx_k\grave{}\, dx_j.$

Then we have to extend these to rules for the Clifford basis elements $e_j, e_j\grave{}$ together with $x_j, x_j\grave{}, dx_j, dx_j\grave{}$ in order for (R1)–(R7) to be satisfied and this can only be done in one way. We leave the formulation of correct rules for the calculus generated by $x_j, x_j\grave{}, dx_j, dx_j\grave{}, e_j, e_j\grave{}$ as an exercise to the reader.

Remark 4. *The fact that the differentials $dx_j\grave{}$ do not commute with the values $e_j\grave{}$ is correct although somewhat awkward to work with. This situation may be altered by replacing (i), (ii), (iii) by an equivalent system of axioms, namely*

(i) $x_j\grave{}\, dx_k\grave{} = -dx_k\grave{}\, x_j\grave{}$

(ii) $d(x_j\grave{}\, G) = dx_j\grave{}\, G + x_j\grave{}\, dG$

(iii) $d(dx_j\grave{}\, G) = -dx_j\grave{}\, dG,$

leading to the more natural relations $e_j\grave{}\, dx_k\grave{} = dx_k\grave{}\, e_j\grave{}$ and $d(e_k\grave{}\, G) = e_k\grave{}\, dG$.

We finish this section with the definition of the abstract contraction operator $\partial_x|$ which will be based on what we already know about the standard contraction operator $\partial_{\underline{x}}| = \sum e_j\, \partial_{x_j}|$ (see also [8]). It will be defined as an endomorphism on $R(S \cup dS)$ like the vector derivative $\partial_{\underline{x}}$ which is defined on $R(S)$ and extends naturally to an endomorphism on $R(S \cup dS)$ by using the fact that every $F \in R(S \cup dS)$ which is a pure product may be written into the form

$$F = \sum f G + x \sum f G$$

whereby the factors $f \in S\,\mathrm{Alg}(S \cup dS)$ and the variable x does not appear in the elements G.

It hence suffices to extend the rules for ∂_x to

$$(V1) \quad \partial_x[g\, G] = \partial_x[g]\, G,$$
$$(V2) \quad \partial_x[x\, g\, G] = \partial_x[g]\, x\, G + \partial_x[x]\, g\, G$$

to reduce the problem to the evaluation of ∂_x on $S\,\mathrm{Alg}(S \cup dS)$ and to do something similar for the action of ∂_x from the right.

Moreover, the evaluation of ∂_x on $S\,\mathrm{Alg}(S \cup dS)$ is reduced to

$$(V3) \quad \partial_x[f\, g] = \partial_x[f]\, g + f\, \partial_x[g], \quad \partial_x[f] = [f]\partial_x,$$
$$(V4) \quad \partial_x\{x, y\} = 2y, \quad x \neq y, \quad \partial_x x^2 = 2x,$$
$$\partial_x\{x, dy\} = 2dy, \quad \partial_x[dy, dz] = 0.$$

The contractor $\partial_x|$ will be defined using the same policy, i.e., we first consider products of elements from $S \cup dS$ and try to bring the factors dx appearing in F in front of the expression, using the fact that $\{y, z\}, \{dy, z\}, [dy, dz]$ generate $S\,\text{Alg}(S \cup dS)$ and may also be brought in front by using (R1)–(R7). This leads to a canonical rewriting of products F of elements in $S \cup dS$ as

$$F = \sum f\,G + dx \sum f\,G + \ldots + dx^l \sum f\,G$$

whereby again dx does not appear in G and the factors $f \in S\,\text{Alg}(S \cup dS)$.

Hence, we only have to formulate the rules

$$(V1) \quad \partial_x|[dx^l f\,G] = \partial_x|[dx^l]f\,G + \partial_x|[f']dx^l F,$$

whereby $f' \in S\,\text{Alg}(S \cup dS)$ is determined by $dx^l f = f'dx^l$.

This brings us to the evaluation of $\partial_x|[dx^l]$ and $\partial_x|[f]$, $f \in \text{Alg}(S \cup dS)$. Now $\partial_x|[f]$ is determined by the rules

$$(V2) \quad \partial_x|[f\,g] = \partial_x|[f]g + (-1)^{\deg f}f\partial_x|[g]$$

whereby f is a pure product of elements $\{y, z\}, \{y, dz\}, [dy, dz]$ and $\deg f$ refers to the number of elements $\{y, dz\}$.

$$(V3) \quad \partial_x|\{y, z\} = 0, \quad y, z \in S,$$
$$\partial_x|\{dy, z\} = 0 \text{ for } y \neq x, \quad \partial_x|[dy, dz] = 0 \text{ for } y, z \neq x,$$
$$(V4) \quad \partial_x|\{dx, y\} = 2y, \quad \partial_x|[dx, dy] = 2dy.$$

Hence, we still have to evaluate $\partial_x|[dx^l]$ which, using the Clifford vector model $d\underline{x} = \sum e_j dx_j$, $\partial_x \to -\partial_{\underline{x}}| = -\sum e_j \partial_{x_j}|$, is given by

$$(V5) \quad \partial_x|[dx^l] = c(l, M)dx^{l-1}$$

whereby the constant $c(l, M)$ is determined by the consistency of the calculus together with the law

$$(V6) \quad \partial_x|[dx] = \partial_x[x] = M.$$

The right action of $\partial_x|$ may be defined similarly.

By using the fermionic resp. super-representation $x \to \underline{x}$, $x \to \underline{x} + \underline{x}$ there are canonically defined contraction operators which correspond to the abstractly defined operators $\partial_x|$ the calculation of which is left as an exercise to the reader.

REFERENCES

[1] F. A. Berezin, *Introduction to Algebra and Analysis with Anti-Commuting Variables*, Moscow University, Moscow, 1983; English transl. *Introduction to Superanalysis*, Reidel, Dordrecht, 1987.

[2] A. Crumeyrolle, Orthogonal and Symplectic Clifford Algebras, *Mathematics and Its Applications* **57**, Kluwer Acad. Publ., Dordrecht, 1990.

[3] W. Chan, G. C. Rota, and J. A. Stein, The power of positive thinking, *Invariant Methods in Discrete and Computational Geometry (Curaçao, 1994)*, Kluwer Acad. Publ., Dordrecht, 1995, 1–36.

[4] R. Delanghe, F. Sommen, and V. Souček, Clifford algebra and spinor valued functions: a function theory for the Dirac operator, *Math. and Its Appl.* **53**, Kluwer Acad. Publ., Dordrecht, 1992.

[5] K. Habermann, The Dirac operator on symplectic spinors, *Annals of Global Analysis and Geometry* **13** (1995), 155–168.

[6] D. Hestenes and G. Sobczyk, *Clifford Algebra to Geometric Calculus*, Reidel, Dordrecht, 1985.

[7] V. P. Palamodov, Cogitations over Berezin's integral, preprint series No. 14, Univ. of Oslo, 1994.

[8] F. Sommen, Monogenic differential calculus, *Trans. Amer. Math. Soc.* **326** (1991), 613–632.

[9] F. Sommen, An algebra of abstract vector variables, *Portugaliae Math.* **54** Fasc. 3, (1997), 287–310.

[10] F. Sommen and V. Souček, Monogenic differential forms, *Complex Variables: Theory Appl.* **19** (1992), 81–90.

[11] F. Sommen and P. Van Lancker, A product for special classes of monogenic functions and tensors, *Z.A.A.* **16**, No. 4, (1997), 1013–1026.

[12] V. Souček, Monogenic forms on manifolds, in *Spinors, Twistors, Clifford Algebras, and Quantum Deformations*, Z. Oziewicz, et al., Kluwer Acad. Publ., 1993, 159–166.

[13] V. S. Vladimirov and I. V. Volovich, *Superanalysis*, translated from *Teoreticheskaya i Matematicheskaya Fizika*, **59**, No. 1 (1984), 3–27.

Frank Sommen
Department of Mathematical Analysis
University of Ghent
Galglaan 2, B-9000 Ghent, Belgium
E-mail: fs@cage.rug.ac.be

Received: October 14, 1999; Revised: November 30, 1999

The Geometry of Generalized Dirac Operators and the Standard Model of Particle Physics

Jürgen Tolksdorf

ABSTRACT In this paper I will present a review on a mathematical frame, based on the geometric algebra of Clifford, which permits us to describe the Standard Model of particle physics in a unified way. In this frame the fundamental objects are generalized Dirac operators, and the geometrical setup is that of a Clifford module bundle over an even dimensional closed Riemannian manifold.
Keywords: Clifford modules, generalized Dirac operators, Wodzicki residue, standard model.

1 The Standard Model. Part I

To get started, let me briefly remind you of some aspects of the standard model of elementary particles. From the point-of-view of perturbation theory, this model can be considered as consisting of a certain input and a set of "rules," called quantization. Whereas the rules are not well understood mathematically – but, nevertheless, very efficient in correctly describing the phenomenology of elementary particles – the input of the model can be made very precise. This input basically consists of three assumptions:

1. A **Lie group** G:

$$G = SU(3) \times SU(2) \times U(1), \qquad (1.1)$$

with three *Yang-Mills* coupling constants $(g_{(3)}, g_{(2)}, g_{(1)})$, which parameterize the most general Killing form of the Lie algebra of G;

2. A **unitary representation** of this group for the *fermions*:

$$V_F \equiv V_L \oplus V_R \qquad (1.2)$$

AMS Subject Classification: 15A66, 15A69, 81T13.

$$V_L := \bigoplus_1^3 \left[(1,2,-\tfrac{1}{2}) \oplus (3,2,\tfrac{1}{6}) \right], \tag{1.3}$$

$$V_R := \bigoplus_1^3 \left[(1,1,-1) \oplus (3,1,-1/3) \oplus (3,1,\tfrac{2}{3}) \right], \tag{1.4}$$

and the *Higgs field*

$$V_H := (1,2,\tfrac{1}{2}), \tag{1.5}$$

where (n_3, n_2, n_1) denote the tensor product, respectively, of an n_3 dimensional representation of SU(3), an n_2 dimensional representation of SU(2) and a one dimensional representation of U(1);

3. An **action functional**:

$$\begin{aligned}
\mathcal{I} &\equiv \mathcal{I}_{\text{Fermion}} + \mathcal{I}_{\text{Boson}} \\
&:= \mathcal{I}_{\text{Dirac}} + \mathcal{I}_{\text{Yukawa}} + \mathcal{I}_{\text{Yang-Mills}} + \mathcal{I}_{\text{Higgs}} \\
&:= \int_{\mathcal{M}} * \left(\psi^{(l)*} i D_{(l)} \psi^{(l)} + \psi^{(q)*} i D_{(q)} \psi^{(q)} \right) \\
&\quad - \int_{\mathcal{M}} * \Big(\sum_{i,j=1}^{N} (\mathbf{g}_y^{(l)})_{ij}\, \bar{\psi}_{L_i}^{(l)} (\gamma_5 \varphi)\, \psi_{R_j}^{(l)} + (\mathbf{g}_y^{(q)})_{ij}\, \bar{\psi}_{L_i}^{(u)} (\gamma_5 \varphi)\, \psi_{R_j}^{(u)} + \\
&\qquad + (\mathbf{g'}_y^{(q)})_{ij}\, \bar{\psi}_{L_i}^{(d')} (\gamma_5 \epsilon\bar{\varphi})\, \psi_{R_j}^{(d')} \Big) + \text{comp. conj.} \\
&\quad + \tfrac{1}{g_{(3)}^2} \int_{\mathcal{M}} \text{tr}(\mathbf{C} \wedge *\mathbf{C}) + \tfrac{1}{g_{(2)}^2} \int_{\mathcal{M}} \text{tr}(\mathbf{W} \wedge *\mathbf{W}) + \\
&\qquad + \tfrac{1}{2g_{(1)}^2} \int_{\mathcal{M}} B \wedge *B + \int_{\mathcal{M}} (\nabla\varphi)^* \wedge *(\nabla\varphi) + \int_{\mathcal{M}} *V, \tag{1.6}
\end{aligned}$$

with the genuine *Higgs potential*

$$V := \lambda(\varphi\varphi^*)^2 - \mu^2 \varphi\varphi^*, \quad \lambda, \mu > 0. \tag{1.7}$$

There are 27 *Yukawa* coupling constants, each g_y for every one dimensional invariant subspace in the decomposition of the representation

$$(V_L^* \otimes V_R \otimes V_H) \oplus (V_L^* \otimes V_R \otimes V_H^*). \tag{1.8}$$

However, these are not all independent. In fact, the standard model can be parameterized by 18 free constants, c.f. [13]. Note that the traces in the definition of the Yang-Mills action are taken with respect to the corresponding fundamental representations of SU(3) and SU(2). Also note that $\mathcal{M}$ denotes a Riemannian manifold, which explains the occurrence of the apparently wrong relative sign in front of the Higgs potential and the occurence of γ_5 in the Yukawa coupling term in (1.6).

Let me stress that it is by no means a trivial matter to get out of this input physically meaningful statements. This is exactly where the "rules," called quantization, come in. Indeed, the action functional $\mathcal{I}$ only defines an approximation of what is called the *effective action* Γ_{eff}. The latter describes the dynamics of the full quantized theory and might be thought of, mathematically, as a formal asymptotic power series in Planck's constant $\hbar$ with leading coefficient $\mathcal{I}$, i.e.,

$$\Gamma_{\text{eff}} \sim \mathcal{I} + \mathcal{O}(\hbar). \tag{1.9}$$

In order to calculate the higher coefficients one uses certain quantization rules, which turns out to be highly sophisticated, mathematically. In fact, there is not yet a rigorous mathematical setup of these rules (although there have been several attempts). Despite mathematical peculiarity, the model based on both the rules and the above given input works extremely accurately, phenomenologically.

The functional $\mathcal{I}$ is called the *semi-classical* approximation. It describes the dynamics and the particle content of the standard model up to order $\hbar^0$ of the full effective action Γ_{eff}. Indeed, the asymptotic expansion (1.9) may be thought of as an analogue to the stationary phase expansion of an appropriate integral. This point of view may give another idea of the importance of the functional $\mathcal{I}$. It is this functional we are going to understand, geometrically, in what follows. In fact, there are two kinds of geometrical setups underlying this functional. First, the geometrical frame of the Yang-Mills part is that of a G-*principle bundle*, with G denoting the above given semi-simple Lie group. The Higgs field can be considered as a section in an associated vector bundle with typical fibre V_H. On the other hand, the fermionic part of the action $\mathcal{I}$ may be understood, geometrically, in terms of a *Clifford module* bundle, with the twisting part defined by the G-representation space V_F. Correspondingly, two different kinds of first order differential operators are involved in order to describe the dynamics of the various fields involved in the standard model. The dynamics of the bosons is described in terms of connections, whereas the dynamics of the fermions is given by a Dirac operator. Of course, both operators are related, somehow, and it is one aim of this paper to clarify the relation between both operators, which is well known not to be one-to-one. In fact, I will try to convince you that Dirac operators and Clifford modules are more profound in some respects than connections and principle bundles.

2   Generalized Dirac operators

In this part, I will summarize some basic facts about Clifford modules and generalized Dirac operators, which are necessary in understanding the geometrical setup introduced afterwards. This setup will give us an understanding of the action functional of the standard model as a specific example of a certain class of gauge theories described within the setting of Clifford algebras. The mathematical details can be found, e.g., in [2] and [14], where both references might serve as a kind of "general reference", in what follows.

To get started, let us denote by $(\mathcal{M}, g)$ a smooth (compact) *Riemannian manifold of even dimension* $2n$ and without boundary. In particular, we are interested in the case of dim $\mathcal{M} = 4$. Note that naturally associated with $(\mathcal{M}, g)$ is a non-trivial algebra bundle, the **Clifford bundle** $\mathcal{C}(\mathcal{M}) \xrightarrow{\tau} \mathcal{M}$, with typical fibre $\mathcal{C}_{0,2n}(\mathbb{R})$. Also, let us denote by

$$\mathcal{E} = \mathcal{E}_+ \oplus \mathcal{E}_- \xrightarrow{\pi} \mathcal{M} \tag{2.1}$$

a smooth $\mathbb{Z}_2$-graded vector bundle. A **generalized Dirac operator** on this setting is defined as any odd first order differential operator

$$\Gamma(\mathcal{E}^{\pm}) \xrightarrow{D} \Gamma(\mathcal{E}^{\mp}), \tag{2.2}$$

satisfying

$$D^2 = \Delta^{\hat{\nabla}^{\mathcal{E}}} + \mathcal{V}. \tag{2.3}$$

The operator $\Delta^{\hat{\nabla}^{\mathcal{E}}} := -ev_g(\hat{\nabla}^{T^*\mathcal{M} \otimes \mathcal{E}} \hat{\nabla}^{\mathcal{E}})$ denotes the **Bochner Laplacian** associated with the connection

$$\Gamma(\mathcal{E}) \xrightarrow{\hat{\nabla}^{\mathcal{E}}} \Gamma(T^*\mathcal{M} \otimes \mathcal{E});$$

the zero order operator $\mathcal{V} \in \Omega^0(\mathcal{M}, \mathrm{End}(\mathcal{E}))$ denotes a **potential** and "ev_g" means the evaluation map with respect to the metric g. Note that both the Bochner-Laplacian, as well as the potential, are uniquely determined by the operator D, c.f. [2]. In other words: a generalized Dirac operator is the "square root" of a Hamiltonian, where this root is again a differential operator, now of order one.

One may ask, when does such an operator exist? It can be shown that the existence of such an operator is equivalent to the existence of a *Clifford map*

$$T^*\mathcal{M} \xrightarrow{\gamma} \mathrm{End}(\mathcal{E}), \tag{2.4}$$

c.f. [2]. The operator D exists, iff there is a Clifford map γ, so that

$$(\mathcal{E}, \gamma) \xrightarrow{\pi} (\mathcal{M}, g) \tag{2.5}$$

denotes a **Clifford module bundle**. As a consequence, the corresponding endomorphism bundle, $\mathrm{End}(\mathcal{E})$, globally decomposes

$$\mathrm{End}(\mathcal{E}) \simeq \mathcal{C}(\mathcal{M}) \otimes \mathrm{End}_{\mathcal{C}\ell}(\mathcal{E}), \tag{2.6}$$

where $\text{End}_{C\ell}(\mathcal{E})$ denotes the algebra bundle of bundle endomorphisms of $\mathcal{E}$, super commuting with the action of $C(\mathcal{M})$.

$$\text{End}_{C\ell}(\mathcal{E}) := \{\sigma \in \text{End}(\mathcal{E}) \mid [c(a), \sigma]_{pm} = 0, \ \forall\, a \in C(\mathcal{M})\}. \tag{2.7}$$

In order to have a definite example in mind, let us assume in addition that $(\mathcal{M}, g)$ has a spin-structure. In this case, every Clifford module bundle is a *twisted spinor bundle*.

$$\mathcal{E} = S \otimes E \overset{\pi}{\longrightarrow} \mathcal{M}, \tag{2.8}$$

where S denotes the total space of the corresponding spinor module and $E \overset{\pi}{\longrightarrow} \mathcal{M}$ is an appropriate vector bundle. In this particular case there is an isomorphism between the (complexified) Clifford bundle and $\text{End}(S)$. Moreover, the Clifford map γ is given by left multiplication and

$$\text{End}_{C\ell}(\mathcal{E}) \simeq \text{End}(E).$$

Let us return to the more general case and denote by

$$\mathcal{D}(\mathcal{E}) := \{\tilde{D} \mid [\tilde{D}, f] = \gamma(df), \ \forall\, f \in C^{\infty}(\mathcal{M})\} \tag{2.9}$$

the set of all Dirac operators compatible with the Clifford action (induced by) γ. This set is an affine space, with vector space $\Omega^0(\mathcal{M}, \text{End}^-(\mathcal{E}))$. Also, let us denote by

$$\mathcal{A}(\mathcal{E}) := \{\nabla^{\mathcal{E}} \mid \Gamma(\mathcal{E}) \overset{\nabla^{\mathcal{E}}}{\longrightarrow} \Gamma(T^*\mathcal{M} \otimes \mathcal{E})\} \simeq \Omega^1(\mathcal{M}, \text{End}^+(\mathcal{E})) \tag{2.10}$$

the affine space of all linear connections on $\mathcal{E}$.

Because of the decomposition (2.6), on any Clifford module bundle their exists a distinguished affine subset within the affine set of all connections on $\mathcal{E}$

$$\mathcal{A}_{C\ell}(\mathcal{E}) := \{\nabla^{\mathcal{E}} \in \mathcal{A}(\mathcal{E}) \mid [\nabla^{\mathcal{E}}, c(a)] = c(\nabla^{C\ell}a), \ \forall\, a \in \Gamma(\mathcal{C}(\mathcal{M}))\}$$

$$\simeq \Omega^1(\mathcal{M}, \text{End}_{C\ell}^+(\mathcal{E})). \tag{2.11}$$

Here, $\nabla^{C\ell}$ denotes the induced Levi-Civita connection on $C(\mathcal{M})$. Since an element of this subset is a connection, which is compatible with the Clifford action, it is called a **Clifford connection**. Note that, in the case of a twisted spinor bundle, any Clifford connection takes the form of a *tensor product connection* (c.f. [2]).

We have now two first order differential operators on a Clifford module bundle and one may ask how they are related. This is answered with the help of the following lemma.

Lemma 1. *Let* $(\mathcal{E}, \gamma) \overset{\pi}{\longrightarrow} (\mathcal{M}, g)$ *be a Clifford module bundle. Then, the set of all Dirac operators, which are compatible with the Clifford action, is*

*related to the set of all connections on the above Clifford module bundle as
follows:*

$$\mathcal{D}(\mathcal{E}) \simeq \mathcal{A}(\mathcal{E}) / \ker(\gamma). \tag{2.12}$$

For the proof of this Lemma see [14]. Thus, to every Dirac operator corresponds a whole class of connections on the Clifford module bundle in question.

$$\mathrm{D} \leftrightarrow [\nabla^{\mathcal{E}}]. \tag{2.13}$$

Note that each connection class contains at most one Clifford connection. The corresponding Dirac operators are called **Standard Dirac Operators** (SDO), c.f. loc site. Also note that one can algebraically generalize the notion of connections and show that there is a one-to-one correspondence between arbitrary Dirac operators and *Clifford super-connections*, c.f. [2] and [14].

Again, because of the decomposition (2.6) of the algebra of endomorphisms associated with the Clifford module $(\mathcal{E}, \gamma)$, there exists a **canonical one-form** $\xi \in \Omega^1(\mathcal{M}, \mathrm{End}^-(\mathcal{E}))$, satisfying the crucial relations

$$\nabla^{T^*\mathcal{M} \otimes \mathrm{End}(\mathcal{E})} \xi \equiv 0, \qquad \text{for all} \quad \nabla^{\mathcal{E}} \in \mathcal{A}_{C\ell}(\mathcal{E}),$$

$$\gamma(\xi) = \mathrm{Id}_{\mathcal{E}}. \tag{2.14}$$

Locally, it reads (here, we adopt Einstein's convention for summation!)

$$\xi := -\frac{1}{2n} g(e_a, e_b) e^a \otimes \gamma(e^b) \otimes \mathrm{Id}_{\mathrm{End}_{C\ell}(\mathcal{E})}, \tag{2.15}$$

where $\{e^a\}_{1 \leq a \leq 2n}$ is a local basis in $T^*\mathcal{M}$ and $\{e_a\}_{1 \leq a \leq 2n}$ its dual.

Note that the above lemma is a direct consequence of the second relation, (2.14). Moreover, the canonical one-form ξ induces an (even) mapping

$$\delta_\xi : \Omega^p(\mathcal{M}, \mathrm{End}^{\mp}(\mathcal{E})) \to \Omega^{p+1}(\mathcal{M}, \mathrm{End}^{\pm}(\mathcal{E}))$$

$$\alpha \mapsto \xi \wedge \alpha. \tag{2.16}$$

Using this mapping, we may associate with a given Dirac operator $\tilde{\mathrm{D}}$ a representative of its corresponding connection class as follows

$$\tilde{\nabla}^{\mathcal{E}} := \nabla^{\mathcal{E}} + \delta_\xi \left(\tilde{\mathrm{D}} - \mathrm{D} \right), \tag{2.17}$$

where $\nabla^{\mathcal{E}}$ denotes an arbitrary Clifford connection and D the corresponding SDO. This will be used next while discussing the **Bochner-Lichnerowicz-Weitzenböck (BLW-) decomposition** of the square of a generalized Dirac operator, which plays an important role in the proposed Clifford frame to which I will turn afterwards (c.f. 3rd section).

By definition, each Dirac operator $\tilde{D}$ is uniquely associated with a connection $\hat{\nabla}^{\mathcal{E}}$ and an endomorphism $\mathcal{V}$ on the Clifford module $(\mathcal{E}, \gamma)$, such that $\tilde{D}^2 = \Delta^{\hat{\nabla}^{\mathcal{E}}} + \mathcal{V}$. Since we start with a Dirac operator, how do we get this connection and, in particular, the potential $\mathcal{V}$? Let, therefore, $\nabla^{\mathcal{E}}$ be an arbitrary Clifford connection and D the corresponding SDO. Using the above formula, (2.17), defining an appropriate connection of the given Dirac operator, it can be shown that (cf. [1])

$$\hat{\nabla}^{\mathcal{E}} := \tilde{\nabla}^{\mathcal{E}} + \omega_{\tilde{\nabla}\mathcal{E}}, \tag{2.18}$$

$$\mathcal{V} := \gamma\left(F^{\tilde{\nabla}^{\mathcal{E}}}\right) + \mathrm{ev}_g\left(\tilde{\nabla}^{T^*M \otimes \mathrm{End}(\mathcal{E})} \omega_{\tilde{\nabla}\mathcal{E}} + \omega_{\tilde{\nabla}\mathcal{E}}^2\right), \tag{2.19}$$

where the one-form $\omega_{\tilde{\nabla}\mathcal{E}} \in \Omega^1(\mathcal{M}, \mathrm{End}^+(\mathcal{E}))$ is locally given by

$$\omega_{\tilde{\nabla}\mathcal{E}} := -\frac{1}{2} g(e_\mu, e_\nu) e^\mu \otimes \gamma(e^\lambda)\left([\tilde{\nabla}^{\mathcal{E}}_\lambda, \gamma(e^\nu)] + \Gamma^\nu_{\sigma\lambda} \gamma(e^\sigma)\right). \tag{2.20}$$

The $F^{\tilde{\nabla}^{\mathcal{E}}}$ denotes the curvature on $\mathcal{E}$ with respect to the connection (2.17) and the Γ's are the Christoffel symbols defined by the metric g.

This, we call the generalized BLW-decomposition of $\tilde{D}^2$. Let me stress once more that both the connection and the potential only depend on the Dirac operator at hand and not on the specific connection used to represent this Dirac operator, c.f. [14]. Also, let me stress that two representatives of a given Dirac operator are not gauge equivalent, in general! Therefore, the potential is not only gauge invariant but only depends on the class of connections defining a certain Dirac operator.

The given formula for the potential $\mathcal{V}$ can be made much more explicit, using an appropriate (chiral) representation of the (complexified) Clifford algebra. It, thereby, can be calculated, using an algebraic computer program. In fact, locally it reads

$$\mathcal{V} = \frac{1}{4} r_{\mathcal{M}} 1_{\tilde{\mathcal{E}}} + \frac{1}{2} \gamma^{\mu\nu} \otimes F_{\mu\nu}$$
$$+ \frac{1}{2}[\gamma^\mu[\omega_\mu, \gamma^\nu], \omega_\nu] + \gamma^{\mu\nu}('\nabla_\mu \omega_\nu) - \frac{1}{2}\gamma^\mu[('\nabla_\nu \omega_\mu), \gamma^\nu]$$
$$+ \frac{1}{2}\gamma^{\mu\nu}[\omega_\mu, \omega_\nu] + \frac{1}{4} g_{\mu\nu}\gamma^\alpha[\omega_\alpha, \gamma^\mu]\gamma^\beta[\omega_\beta, \gamma^\nu], \tag{2.21}$$

where we have used the shorthand notation $\gamma^\mu \equiv \gamma(e^\mu)$ (Dirac matrices in the chiral representation) and $\gamma^{\mu\nu} := \frac{1}{2}[\gamma^\mu, \gamma^\nu]$. Also, $'\nabla_\mu$ means the covariant derivative associated with the connection $\nabla^{T^*M \otimes \mathrm{End}(\mathcal{E})}$ on the tensor bundle $T^*M \otimes \mathrm{End}(\mathcal{E}) \to \mathcal{M}$.

Before we proceed, let me note that on the space $\mathcal{D}(\mathcal{E})$ there is a "natural" one-parameter family of functionals

$$\mathcal{D}(\mathcal{E}) \longrightarrow C^\infty(\mathbb{R}_+, \mathbb{C})$$
$$D \mapsto \mathrm{Tr}(\exp(-t\,D^2)). \tag{2.22}$$

It is well known that this functional has an asymptotic expansion

$$\mathrm{Tr}(\exp(-t\,\mathrm{D}^2)) \sim_{t\to 0} \sum_{k\geq 0} a_k(\mathrm{D})\, t^{\frac{k-2n}{2}}, \tag{2.23}$$

$$a_k(\mathrm{D}) \equiv \int_{\mathcal{M}} *\mathrm{tr}_{\mathcal{E}}\, a_k(\mathrm{D}, x), \tag{2.24}$$

where the coefficients $a_k(\mathrm{D})$ contain geometric information via $a_0(\mathrm{D}) \sim \mathrm{vol}(\mathcal{M})$. In particular, the *sub-leading* term reads

$$a_2(\mathrm{D}) = \int_{\mathcal{M}} *\mathrm{tr}_{\mathcal{E}} \left(\frac{1}{6} r_{\mathcal{M}} - \mathcal{V} \right) \tag{2.25}$$

$$\sim \mathcal{W}res\left(\mathrm{D}^{-2n+2} \right). \tag{2.26}$$

Here, $\mathcal{W}res$ means the **Wodzicki residue** and is defined by

$$\mathcal{W}res : \Psi\mathrm{DO}(\mathcal{E}) \longrightarrow \mathbb{C}$$
$$A \mapsto \mathrm{ord}(P)\,\mathrm{res}|_{s=0}\mathrm{Tr}(A\,P^{-s}), \tag{2.27}$$

with $P \in \Psi\mathrm{DO}(\mathcal{E})$ arbitrary but positive and elliptic. Here, $\Psi\mathrm{DO}(\mathcal{E})$ denotes the associative algebra of (classical) *pseudo-differential operators*, acting on sections into $\mathcal{E} \xrightarrow{\pi} \mathcal{M}$, and Tr denotes the operator trace in the Hilbert space $L_2(\mathcal{M}, \mathcal{E})$. It has been shown by Wodzicki that this linear functional is a *trace* functional. Actually, it is the trace (up to multiplication by constants) in this algebra, c.f. [15] and [16]; see also [8]! Note that, in the case at hand, it is simply related to the ζ-function associated with D^2:

$$\mathcal{W}res(\mathrm{D}^{-2}) = -2\,\mathrm{res}|_{s=-1}\,\zeta_{\mathrm{D}^2}(s) \tag{2.28}$$

$$\zeta_{\mathrm{D}^2}(s) := \frac{1}{\Gamma(s+2)} \int_0^\infty t^{s-1}\,\mathrm{Tr}(\exp(-t\,\mathrm{D}^2))\,dt. \tag{2.29}$$

This means that the Wodzicki residue of the propagator of a Dirac operator is the (usual) residue of the Mellin transform of the heat trace associated with the (square of the) Dirac operator, c.f. [7]. We, therefore, consider

$$\mathcal{D}(\mathcal{E}) \longrightarrow \mathbb{C}$$
$$\mathrm{D} \mapsto \mathcal{W}res\left(\mathrm{D}^{-2} \right) \tag{2.30}$$

$$\mathrm{D}^{-2} : \quad \text{(Square of the) } \textbf{Fermion propagator} \tag{2.31}$$

as a kind of "universal functional."

Before we introduce our universal setup, let me answer the obvious question:

Why do we need NSDO?

In fact, there are at least two good reasons for dealing with non-standard Dirac operators (NSDO). The first one comes from mathematics. In considering SDO's only we obtain for $D \in \mathcal{D}(\mathcal{E})$,

$$Wres\left(D^{-2}\right) \sim \int_{\mathcal{M}} *r_{\mathcal{M}} \sim \mathcal{I}_{EH}. \tag{2.32}$$

This has been mentioned by Connes, c.f. [3] and later been proved, independently, by Kastler, c.f. [11] and Kalau/Walze, c.f. [12]. Here, $\mathcal{I}_{EH}$ means the Einstein-Hilbert action functional. Therefore, the fermion propagator, with respect to an SDO, only gives rise to the Einstein-Hilbert functional and, thus, the information about the Yang-Mills part is lost. This holds true for any SDO because of the trace in the definition of a_2, (2.25), and the BLW decomposition of an SDO, c.f. [1] and [14]! The second reason comes from physics. It is well known that the Yukawa coupling, together with an SDO, defines a new Dirac operator: the **Dirac-Yukawa operator**, which is an NSDO. Therefore, we have to consider NSDO anyway!

3   Standard Model. Part II

Next, let us introduce our **universal setup** and then summarize its application to the standard model. This setup consists of

$$(G, \rho, D), \tag{3.1}$$

where G denotes a semi-simple (compact) **Lie group**, representing the gauge degrees of freedom; A **unitary representation** ρ of this group, representing the fermionic degrees of freedom and a (generalized) **Dirac operator** D, defining the dynamics of the fermions up to order $\hbar^0$. With respect to this input, we consider the following universal functional

$$\mathcal{I}_D : \Gamma(\mathcal{E}) \times \mathcal{A}(\mathcal{E}) \to \mathbb{C}$$
$$(\psi, \nabla^{\mathcal{E}}) \mapsto \langle \psi, D\psi \rangle + Wres\left(D^{-2}\right), \tag{3.2}$$

where $\gamma(\nabla^{\mathcal{E}}) = D$ and

$$(\mathcal{E}, \gamma) \xrightarrow{\pi} (\mathcal{M}, g) \tag{3.3}$$

denotes a hermitian **Clifford module bundle** over a (closed, compact) four dimensional Riemannian manifold.

Using this Clifford frame one can prove that there exists a "natural" generalization of the Dirac-Yukawa operator, so that

$$\mathcal{I}_D \sim \mathcal{I}_{Fermion} + \mathcal{I}_{EHYMH}, \tag{3.4}$$

where $\mathcal{I}_{\mathrm{EHYMH}}$ denotes the semi-classical approximation of the full *bosonic action functional* of the standard model (with the Einstein-Hilbert functional included), c.f. [14].

The main features may be summarized as follows (c.f. [14]):

- The Yang-Mills functional comes together with the Einstein-Hilbert functional;

- The *Higgs* field becomes a part of a connection on the Clifford module bundle; consequently, the *Higgs potential* becomes a part of the curvature on $\mathcal{E}$;

- One obtains some *relations* between the various parameters involved in the model and which are not known in the usual description;

- Finally, the *length scale* is defined by the mass of the Higgs field

$$l \sim m_{\mathrm{H}}^{-1}. \tag{3.5}$$

Obviously, some of these results parallel the corresponding results within Connes' frame of non-commutative geometry, c.f. [3] and [4]. Concerning the mass relations see, e.g., [9]. In particular, see [6] and [10] for how the Einstein-Hilbert action is incorporated in Connes' non-commutative geometry.

Since there is only one representation involved in this model, the fermionic representation, all the other fields also have to fit into this representation. This restricts the class of possible gauge theories described within our Clifford frame.

4   Summary

Starting with an even-dimensional Riemannian manifold, I have shown how one may formulate a gauge theory using the geometrical setup of a Clifford module bundle. The fundamental objects in this frame are generalized Dirac operators, defining the dynamics of the fermions to some approximation. However, not only do these operators define the fermionic functional, but their corresponding propagators also define the dynamics of the bosons. In the case of the standard model this Clifford frame permits a geometrical understanding of the semi-classical approximation of the appropriate effective action. In particular, it yields a unified geometrical description of the Einstein-Hilbert-Yang-Mills functional. If the gauge symmetry is spontaneously broken, the Higgs sector is also geometrically incorporated. Needless to say that there are a lot of questions still left open. For example, how "natural" is the Dirac-Yukawa operator and its generalization used to describe the standard model within the Clifford frame? Or, how can the gauge group and its highly reducible fermionic representation be understood within Clifford algebras? Hopefully, we will know more about this by the year 2002 when the sixth conference on Clifford algebras and their applications in mathematical physics will take place.

Acknowledgments

I would like to express my gratitude to the organizer for the beautiful conference and, in particular, to thank them for inviting me.

REFERENCES

[1] Th. Ackermann and J. Tolksdorf, The generalized Lichnerowicz formula and analysis of Dirac operators, *J. Reine Angew. Math.* **471** (1996), 23–42.

[2] N. Berline, E. Getzler, and M. Vergne, *Heat Kernels and Dirac operators*, Springer, 1992.

[3] A. Connes and J. Lott, The metric aspect of non-commutative geometry, in *The Proceedings of the 1991 Cargese Summer Conference*, J. Fröhlich, ed., Plenum Press, 1992.

[4] A. Connes, *Non-Commutative Geometry*, Academic Press, 1994.

[5] A. Connes, Noncommutative geometry and physics, in *Proceeding of the 1992 Les Houches Summer School*, B. Julia and J. Zinn-Justin, eds., North Holland, 1995.

[6] A. H. Chamseddine and A. Connes, The spectral action principle, hep-th/9606001, *Commun. Math. Phys.* **186** (1997), 731–750.

[7] P. B. Gilkey, *Invariance Theory, the Heat Equation and the Atiyah-Singer Theorem, Second Edition*, CRC Press, 1995.

[8] V. Guillemin, A new proof of Weyl's formula on the asymptotic distribution of eigenvalues, *Adv. Math.* **55** (1985), 131–160.

[9] B. Iochum, D. Kastler, and Th. Schücker, Fuzzy mass relations in the standard model, CPT-95/P.3235, hep-th/9507150, *J. Math. Phys.* **36** (1995), 6232–6254.

[10] B. Iochum, D. Kastler, and Th. Schücker, On the universal Chamseddine-Connes action; 1. details of the action computation, *J. Math. Phys.* **38** (1997), 4929–4950, hep-th/9607158.

[11] D. Kastler, The Dirac operator and gravitation, *Commun. Math. Phys.* **166** (1995), 633–643.

[12] W. Kalau and M. Walze, Gravity, non-commutative geometry and the Wodzicki residue, *J. Geom. Phys.* **16** (1995), 327–344.

[13] O. Nachtmann, *Elementarteilchen Physik, Phänomene und Konzepte*, Vieweg, 1986.

[14] J. Tolksdorf, The Einstein-Hilbert-Yang-Mills-Higgs-Action and the Dirac-Yukawa operator, *J. Math. Phys.* **39** (1998), 2213–2241.

[15] M. Wodzicki, Local invariants of spectral asymmetry, *Inv. Math.* **75** (1984), 143–178 .

[16] M. Wodzicki, Non-commutative residue 1, *Lect. Notes Math.* **1289** (1987), 320–399.

Jürgen Tolksdorf
Department of Mathematics
University of Mannheim, D7-27
D-68131 Mannheim, Germany
E-mail: tolkdorf@riemann.math.uni-mannheim.de

Received: September 28, 1999; Revised: February 18, 2000

4.

MÖBIUS TRANSFORMATIONS AND MONOGENIC FUNCTIONS

The Schwarzian and Möbius Transformations in Higher Dimensions

Masaaki Wada and Osamu Kobayashi

ABSTRACT The relationship between the Schwarzian and Möbius transformations is well-known in dimensions 1 and 2. However, in dimension $n \geq 3$, even the "proper" definition of the Schwarzian is not clear. In this paper, we introduce a "natural" generalization of the Schwarzian using the Clifford algebra and show that it vanishes exactly for Möbius transformations. The situation is simplest for non-singular transformations of the Euclidean space although the framework can be applied, with as light modification, to maps as general as immersions between any Riemannian manifolds.
Keywords: Clifford algebra, Schwarzian transformations, Möbius transformations.

1 Introduction

The Schwarzian derivative of a holomorphic function f of the complex plane is defined by the formula

$$S_f(z) = \frac{f'''(z)}{f'(z)} - \frac{3}{2}\left(\frac{f''(z)}{f'(z)}\right)^2 .$$

Of the many interesting properties of the Schwarzian derivative, the most characteristic is the following well-known fact:

Proposition 1. *The following are equivalent.*

(i) *The Schwarzian derivative $S_f(z)$ is constantly zero.*

(ii) *The function f is a Möbius transformation.*

(iii) *The function f is a linear fractional transformation, $f(z) \equiv \frac{az+b}{cz+d}$, for some complex numbers a, b, c, d with $ad - bc = 1$.*

AMS Subject Classification: 15A66, 53A55, 53A04, 53A30.

Several authors ([1], [2], [5], [6], [8], [11], [12]) have invented methods of expressing higher dimensional Möbius transformations as linear fractional transformations using Clifford numbers. Although they worked somehow independently, all the implementations are basically the same. For a historical remark on the subject, the reader is referred to [1].

On the other hand, there seems to be a variety of generalizations of the notion of "Schwarzian". Ahlfors [3] and Ryan [10] define Schwarzian derivatives of transformations of $\mathbb{R}^n$ using directional derivatives. The difficulty of their implementations is that the generalized Schwarzian derivatives vanish not only for Möbius transformations but also for arbitrary linear transformations.

In [13], the first author of this paper introduced the Schwarzian of (germs of) curves in $\mathbb{R}^n$ and used it to define the Schwarzian of transformations of $\mathbb{R}^n$. We show in Section 4 that the generalized Schwarzian of [13] vanishes exactly for Möbius transformations.

There are also generalizations of the Schwarzian from the viewpoint of differential geometry ([4], [9]). In the last section we show how to modify the framework of [13] to include these generalizations.

2   Möbius transformations and linear fractional transformations

The group of Möbius transformations of $\widehat{\mathbb{R}}^n = \mathbb{R}^n \cup \{\infty\}$ is generated by the Euclidean isometries and the inversions in the spheres. The Möbius transformations are also characterized as the transformations of $\widehat{\mathbb{R}}^n$ mapping spheres (including hyperplanes) to themselves, or alternatively as those mapping circles (including straight lines) to themselves.

It is well-known that the orientation preserving Möbius transformations of $\widehat{\mathbb{R}} = \mathbb{R} \cup \{\infty\}$ and $\widehat{\mathbb{C}} = \mathbb{C} \cup \{\infty\}$ are represented by $\mathrm{PSL}(2,\mathbb{R})$ and $\mathrm{PSL}(2,\mathbb{C})$ respectively. Namely, every Möbius transformation f of $\widehat{\mathbb{R}}$ (resp. $\widehat{\mathbb{C}}$) is expressed as

$$f(z) \equiv \frac{az+b}{cz+d},$$

for some real (resp. complex) numbers a, b, c, d with $ad - bc = 1$.

Here we follow Ahlfors for a generalization of linear fractional transformations to higher dimensions. But instead of adopting the implementation of [1] and [2] directly, we introduce the modification by the first author [12] which can deal with not only orientation preserving Möbius transformations but also orientation reversing ones.

Let us fix some notation first. We denote by $C\ell_n$ the Clifford algebra of the Euclidean n-space $\mathbb{R}^n$ with the minus squared norm as the associated quadratic form, i.e., $v^2 = -|v|^2$ for vectors $v \in \mathbb{R}^n$. An element of $C\ell_n$ is called a Clifford number. We denote by $C\ell_n^{(k)}$ the subspace of $C\ell_n$ spanned by the k-vectors; we identify $C\ell_n^{(0)}$ with the real numbers and $C\ell_n^{(1)}$ with the Euclidean space $\mathbb{R}^n$ itself. Thus, we have a direct sum decomposition

$$C\ell_n = C\ell_n^{(0)} \oplus C\ell_n^{(1)} \oplus C\ell_n^{(2)} \oplus \cdots \oplus C\ell_n^{(n)}.$$

Accordingly, every Clifford number $a \in C\ell_n$ is written as

$$a = a_{(0)} + a_{(1)} + a_{(2)} + \cdots + a_{(n)}.$$

The reversion $\tilde{\ }: C\ell_n \to C\ell_n$ is the anti-automorphism of $C\ell_n$ characterized by

$$(v_1 v_2 \cdots v_k)^{\tilde{\ }} = v_k \cdots v_2 v_1 \quad \text{for vectors} \quad v_1, v_2, \ldots, v_k \in \mathbb{R}^n.$$

Definition 1. *A matrix*

$$g = \begin{pmatrix} a & b \\ c & d \end{pmatrix} \tag{2.1}$$

is called a Clifford matrix if

(i) $a, b, c, d \in \Gamma^n \cup \{0\}$,

(ii) $a\tilde{b}, c\tilde{d}, \tilde{a}c, \tilde{b}d \in \mathbb{R}^n$, *and*

(iii) $a\tilde{d} - b\tilde{c} = \tilde{d}a - \tilde{b}c = 1$,

where Γ^n denotes the Clifford group, i.e., the multiplicative group of $C\ell_n$ generated by the non-zero vectors.

Then we have:

Proposition 2. *Every Möbius transformation of $\widehat{\mathbb{R}}^n$ is given by some Clifford matrix g as*

$$g(x) = (ax + b)(cx + d)^{-1}.$$

Two Clifford matrices g and g' give the same Möbius transformation if and only if $g = \pm g'$.

3   The Schwarzian derivative of curves in $\mathbb{R}^n$

Let I be an interval of $\mathbb{R}$, and $x : I \longrightarrow \mathbb{R}^n$ a regular curve, i.e., a smooth curve with $\dot{x} \neq 0$.

Definition 2. *The Schwarzian derivative of the curve x is defined by*

$$s^3 x = \dddot{x} - \frac{3}{2}\ddot{x}\dot{x}^{-1}\ddot{x},$$

$$s^2 x = (s^3 x)\dot{x}^{-1} = \dddot{x}\dot{x}^{-1} - \frac{3}{2}(\ddot{x}\dot{x}^{-1})^2.$$

Thus, for $t \in I$, we have $s^3x(t) \in C\ell_n^{(1)} = \mathbb{R}^n$ and $s^2x(t) \in C\ell_n^{(0)} \oplus C\ell_n^{(2)}$. The Schwarzian derivative of curves is closely related to circles. A direct calculation using the Frenet-Serret formula shows the following.

Proposition 3. *Let κ and τ denote, respectively, the geodesic curvature and the torsion vector of x. Then, $s^2x_{(2)} = 0$ if and only if $\dot{\kappa}(t) = \tau(t) = 0$.*

Corollary 1. *The 2-part of s^2x is constantly 0 if and only if the image of x is a circle.*

Definition 3. *We call a curve x satisfying $s^3x \equiv s^2x \equiv 0$ a Möbius circle.*

Therefore, the image of a Möbius circle is a circle. Conversely, every circle is a Möbius circle if appropriately parameterized. The 2-part of s^2x may be considered to control the shape of the curve. As for the 0-part of s^2x, we have the following:

Theorem 1. *If $s^2x_{(0)} \leq 0$, then x is injective.*

This interesting theorem, which relates a local property of the curve x to the global injectivity, is not needed for our purpose here. But this becomes the key when we apply the Schwarzian to injectivity problems. For the proof of this theorem, the reader is referred to [7].

Theorem 2. *Let g be a Möbius transformation given by a Clifford matrix (2.1). Let x be a regular curve in $\mathbb{R}^n$, and $y = g \circ x$ its image under g. Then,*

(i) $\dot{y} = \widetilde{(cx+d)}^{-1}\dot{x}(cx+d)^{-1}$,

(ii) $s^3y = \widetilde{(cx+d)}^{-1}s^3x(cx+d)^{-1}$.

Therefore, s^3x is mapped onto s^3y by the tangential map of g.

Proof. Differentiating both sides of

$$y(cx+d) = ax + b$$

and multiplying by $\widetilde{cx+d}$ from the left, one obtains

$$(\widetilde{cx+d})\dot{y}(cx+d) = (x\tilde{c} + \tilde{d})a\dot{x} - (ax + \tilde{b})c\dot{x}$$
$$= (x(\tilde{c}a - \tilde{a}c) + (\tilde{d}a - \tilde{b}c))\dot{x}.$$

For $\tilde{c}a = \widetilde{(\tilde{a}c)} = \tilde{a}c$ and $\tilde{d}a - \tilde{b}c = 1$, we have (i).

Since showing (ii) is easy for the case $c = 0$, let us assume $c \neq 0$. Now denote the vector $c^{-1}d \in \mathbb{R}^n$ by v, and the formula (i) may be written as

$$\dot{y} = \tilde{c}^{-1}(x+v)^{-1}\dot{x}(x+v)^{-1}c^{-1}.$$

Differentiating both sides of this using the rule $d(x^{-1})/dt = -x^{-1}\dot{x}x^{-1}$, one obtains

$$\ddot{y} = -2\bar{c}^{-1}(x+v)^{-1}\dot{x}(x+v)^{-1}\dot{x}(x+v)^{-1}c^{-1} + \\ \bar{c}^{-1}(x+v)^{-1}\ddot{x}(x+v)^{-1}c^{-1},$$

and

$$\dddot{y} = 6\bar{c}^{-1}(x+v)^{-1}\dot{x}(x+v)^{-1}\dot{x}(x+v)^{-1}\dot{x}(x+v)^{-1}c^{-1} - \\ 3\bar{c}^{-1}(x+v)^{-1}\ddot{x}(x+v)^{-1}\dot{x}(x+v)^{-1}c^{-1} - \\ 3\bar{c}^{-1}(x+v)^{-1}\dot{x}(x+v)^{-1}\ddot{x}(x+v)^{-1}c^{-1} + \\ \bar{c}^{-1}(x+v)^{-1}\dddot{x}(x+v)^{-1}c^{-1}.$$

A straightforward computation then gives (ii). ☐

Corollary 2. *Every Möbius transformation maps Möbius circles to Möbius circles.*

A straight line of constant speed is a Möbius circle, so are its images under Möbius transformations. One can easily see that there are as "many" Möbius transformations as the solutions to the equation $s^3x = 0$. Hence we have:

Corollary 3. *Every Möbius circle is the image of a straight line of constant speed under some Möbius transformation of $\widehat{\mathbb{R}}^n$.*

4   The Schwarzian of transformations of $\mathbb{R}^n$

Let $f : \mathbb{R}^n \to \mathbb{R}^n$ be a diffeomorphism. In fact, since we are interested in local properties here, f needs only be defined on some open subset of $\mathbb{R}^n$. Now let $x : I \to \mathbb{R}^n$ be any regular curve, and $y = f \circ x$ its image under f.

Definition 4. *We put*

$$S^3f = s^3y - f_*(s^3x), \quad S^2f = (S^3f)\dot{y}^{-1},$$

where f_* is the tangential map of f, and call S^2f the Schwarzian of f.

For each $t \in I$, we have $S^3f(t) \in C\ell_n^{(1)}$ and $S^2f(t) \in C\ell_n^{(0)} \oplus C\ell_n^{(2)}$. Although S^3f and S^2f are defined in terms of a regular curve x, their values actually depend only on the derivatives $X = \dot{x}$ and $Y = \ddot{x}$. Thus it is reasonable to use the notation $S^3f(X,Y)$ and $S^2f(X,Y)$ for arbitrary tangent vectors X, Y with $X \neq 0$.

To see that the Schwarzian defined above is a generalization of the classical Schwarzian, identify $\mathbb{C}$ with $\mathbb{R}^2$ and consider a holomorphic function

$f : \mathbb{R}^2 \to \mathbb{R}^2$. The Schwarzian $S^2 f$ takes values in $C\ell_2^{(ev)} = C\ell_2^{(0)} \oplus C\ell_2^{(2)}$. Let us identify the subalgebra $C\ell_2^{(ev)}$ with $\mathbb{C}$ by the correspondence $1 \leftrightarrow 1, e_1 e_2 \leftrightarrow i$, where e_1, e_2 are the canonical basis for $\mathbb{R}^2$. Then the Schwarzian becomes $S^2 f = S_f(z)\dot{z}^2$. The computation is straightforward; refer to [13] for the detail.

Now we are ready for the main theorem:

Main Theorem. *The Schwarzian $S^2 f$ is constantly zero, i.e.,*

$$S^2 f(X, Y) = 0$$

for all tangent vectors X, Y at each point if and only if f is a Möbius transformation.

Proof. The sufficiency is a direct consequence of Theorem 2.

To prove the necessity, let us assume that $S^2 f \equiv S^3 f \equiv 0$. It is then clear from the definition of the Schwarzian that f maps Möbius circles to Möbius circles. Since every circle (minus a point) can be parameterized by a Möbius circle, this shows that f maps circles to circles. Thus f must be a Möbius transformation. $\qquad \square$

5   The Schwarzian of immersions between Riemannian manifolds

The framework developed so far for the transformations of the Euclidean space can be modified so that it applies to as general maps as immersions between Riemannian manifolds.

Let (M, g) be a Riemannian manifold and $x : I \to M$ a regular curve.

Definition 5. *We define the Schwarzian of the curve x by*

$$s^3 x = \nabla_{\dot{x}} \nabla_{\dot{x}} \dot{x} - \frac{3}{2} \nabla_{\dot{x}} \dot{x} \dot{x}^{-1} \nabla_{\dot{x}} \dot{x} - \frac{R_g}{2n(n-1)} \dot{x}^3,$$
$$s^2 x = (s^3 x)\dot{x}^{-1},$$

where ∇ and R_g denote the Riemannian connection and the scalar curvature of g, respectively.

For $t \in I$, we have $s^3 x(t) \in T_{x(t)}M$ and $s^2 x(t) \in \mathbb{R} \oplus C\ell^{(2)}(T_{x(t)}M)$. Now, let N be another Riemannian manifold and $f : M \to N$ an immersion.

Definition 6. *Let $x : I \to M$ be any regular curve, and $y = f \circ x : I \to N$ its image under f. We put*

$$S^3 f = s^3 y - f_*(s^3 x),$$

where $f_* : T_{x(t)}M \to T_{y(t)}N$ is the tangential map of f. The Schwarzian of f is then defined to be

$$S^2 f = (S^3 f)\dot{y}^{-1}.$$

As before, the values of $S^3 f$ and $S^2 f$ depend only on $X = \dot{x}$ and $Y = \nabla_{\dot{x}}\dot{x}$, and we use the notation $S^3 f(X, Y)$ and $S^2 f(X, Y)$ for arbitrary tangent vectors $X, Y \in T_p M$ with $X \neq 0$. It can also be shown by computation that $S^3 f$ and $S^2 f$ do not depend on $Y = \nabla_{\dot{x}}\dot{x}$ if f is a conformal immersion.

Finally to give a flavor of what kind of results may be obtained as an application of the Schwarzian, let us quote an injectivity theorem from [7].

Theorem 3. *Let (M^n, g) be a Riemannian manifold and C some fixed real number. Suppose that every pair of points of M can be joined by a curve whose geodesic curvature κ and length ℓ satisfy*

$$\frac{1}{2}\left(\kappa^2 - \frac{4\pi^2}{\ell^2}\right) \leq -C.$$

If a conformal immersion

$$f : M \longrightarrow \mathbb{R}^n \ (or \ S^n)$$

of class C^3 *satisfies*

$$\frac{S^2 f^{(0)}(X)}{g(X, X)} \leq C - \frac{R_g}{2n(n-1)}$$

for all tangent vectors X, f is injective.

REFERENCES

[1] L. V. Ahlfors, Möbius transformations and Clifford numbers, *Differential Geometry and Complex Analysis*, dedicated to H. E. Rauch, Springer-Verlag, 1985, pp. 65–73.

[2] L. V. Ahlfors, Old and new in Möbius groups, *Ann. Acad. Sci. Fenn.* **9** (1984), 93–105.

[3] L. V. Ahlfors, Cross-ratios and Schwarzian derivatives in $\mathbb{R}^n$, *Complex Analysis* J. Hersch and A. Huber, eds., articles dedicated to Albert Pfluger on the occasion of his 80th birthday, Birkhäuser, 1988, 1–15.

[4] K. Carne, The Schwarzian derivative for conformal maps, *J. Reine Angew. Math.* **408** (1990), 10–33.

[5] R. Fueter, Sur les groupes improprement discontinus, *C. R. Acad. Sci. Paris* **182** (1926), 432–434.

[6] R. Fueter, Über automorphe Funktionen in bezug auf Gruppen, die in der Ebene uneigentlich diskontinuierlich sind, *J. Reine Angew. Math.* **157** (1927), 66–78.

[7] O. Kobayashi, and M. Wada, Circular geometry and the Schwarzian, preprint.

[8] H. Maass, Automorphe Funktionen von mehreren Veränderlichen und Dirichletsche Reihen, *Abh. Math. Sem. Univ. Hamburg* **16** (1949), 53–104.

[9] B. Osgood, and D. Stowe, The Schwarzian derivative and conformal mapping of Riemannian manifolds, *Duke Math. J.* **67** (1992), 57–99.

[10] J. Ryan, Generalized Schwarzian derivatives for generalized fractional linear transformations, *Ann. Polon. Math.* **LVII** (1992), 29–44.

[11] K. T. Vahlen, Über Bewegungen und complexe Zahlen, *Math. Ann.* **55** (1902), 585–593.

[12] M. Wada, Conjugacy invariants of Möbius transformations, *Complex Variables* **15** (1990), 125–133.

[13] M. Wada, A generalization of the Schwarzian via Clifford numbers, *Ann. Acad. Sci. Fenn.* **23** (1998), 453–460.

Masaaki Wada
Department of Information and Computer Sciences
Nara Women's University
Kita-Uoya Nishimachi
Nara 630-8506, Japan
E-mail: wada@ics.nara-wu.ac.jp

Osamu Kobayashi
Department of Computational Science
Kanazawa University
Kakuma, Kanazawa 920-1192, Japan
E-mail: o.kobayashi@kappa.s.kanazawa-u.ac.jp

Received: October 1, 1999; Revised: January 6, 2000

The Structure of Monogenic Functions

Yakov Krasnov

ABSTRACT We study the structure of monogenic functions using symmetries of the Dirac operator.
Keywords: Monogenic functions, Clifford algebra, symmetries.

1 Preliminaries

Since all the functions considered in this paper are supposed to take their values in a Clifford algebra, in this section we present some notation and definitions concerning the real Clifford algebra $\mathcal{A}_m$ and infinitesimal analysis on this algebra.

The real Clifford algebra $\mathcal{A}_m$ is a real vector space with 2^m basis elements $\mathbf{e}_0, \mathbf{e}_1, \ldots, \mathbf{e}_{2^m-1}$, defined by

$$\mathbf{e}_0 \equiv e_0 = 1, \ \mathbf{e}_1 = e_1, \ldots, \mathbf{e}_m = e_m,$$

$$\mathbf{e}_{12} = e_1 e_2, \ \mathbf{e}_{13} = e_1 e_3, \ \ldots, \mathbf{e}_{m-1,m} = e_{m-1} e_m, \ldots, \mathbf{e}_{12\ldots m} = e_1 e_2 \ldots e_m,$$

where $e_1, \ldots, e_m$ are standard hypercomplex numbers satisfying

$$e_i e_j + e_j e_i = -2 e_0 \delta_{ij},$$
$$(e_i e_j) e_k = e_i (e_j e_k), \quad i, j, k = 1, \ldots, m. \tag{1.1}$$

As usual, we identify the canonical basis $\{e_i\}_{i=1}^m$ in $\mathbb{R}^m$ with m imaginary units generating $\mathcal{A}_m$ and such that every element in $\mathcal{A}_m$ can be (uniquely) represented in the form

$$u = \sum_{k=0}^m \sum_{|I|=k} u_I \mathbf{e}_I, \quad u_I \in \mathbb{R}. \tag{1.2}$$

Hereafter, the summation index in the inner sum runs over all strictly increasing ordered k-tuples of elements from the set $\omega = \{1, 2, \ldots, m\}$. In particular, $I = (i_1, i_2, \ldots i_k)$ in (1.2) is defined in such a way that $1 \leq i_1 < i_2 < \ldots < i_k \leq m$.

AMS Subject Classification: 30G35, 31B05, 41A10.

In a standard way we can define a Clifford algebra valued function f : $\mathbb{R}^{m+1} \to \mathcal{A}_m$ by the formula

$$f(x) = \sum_{k=0}^{m} \sum_{|I|=k} \varphi_I(x) e_I, \qquad (1.3)$$

where $\varphi_I(x)$ is a real valued function for each multi index I.

Denote by $\mathbf{C}^n(\Omega; \mathcal{A}_m)$ the space of all Clifford algebra valued functions (1.3) which are n times differentiable in some open connected set $\Omega \subseteq \mathbb{R}^{m+1}$. We shall call $f(x) \in \mathbf{C}^2(\Omega; \mathcal{A}_m)$ harmonic if every component in (1.3) is a harmonic function.

We define the conjugate of $f(x)$ to be the function $\bar{f}(x)$ given by the formula

$$\bar{f}(x) = \sum_{k=0}^{m} (-1)^k \sum_{|I|=k} \varphi_I(x) e_I. \qquad (1.4)$$

In this notation, the function conjugate to $f(x) \equiv \mathbf{x}$, where $\mathbf{x} = x_0 + x_1 e_1 + \ldots + x_m e_m$ is $\bar{\mathbf{x}} = x_0 - x_1 e_1 - \ldots - x_m e_m$.

In Clifford analysis the generalization of the Cauchy-Riemann operator is the Dirac operator D

$$D := \partial_{x_0} + e_1 \partial_{x_1} + \ldots + e_m \partial_{x_m} \qquad (1.5)$$

(see [1], [2], [3]).

Definition 1 (Left monogenic function). *The Clifford algebra valued function $f(x) \in \mathbf{C}^1(\Omega; \mathcal{A}_m)$ is called monogenic in Ω if and only if $Df(x) = 0$ in Ω.*

Hereafter, by the operator action we mean a left operator action on the Clifford value function $f(x)$ and we leave the word "left" in "left monogenic function". The space of all monogenic functions f in Ω is denoted by $\mathcal{M}_m(\Omega)$. By the same token, $\mathcal{M}_m(\Omega)$ is the null space of the Dirac operator D and, obviously, every monogenic function is harmonic.

Given linear operator $Q : \mathbf{U} \to \mathbf{V}$ and let $\ker(Q) \subset \mathbf{U}$ be a set of the solutions of the homogenious equation $Qu = 0$.

Definition 2 (Symmetry Operator). *$L_Q : \mathbf{U} \to \mathbf{U}$ is a symmetry operator of the operator Q, (see [4], [5]), if L_Q maps the space $\ker(Q) = \{u \in \mathbf{U} : Qu = 0\}$ of the solutions of the homogenious equation $Qu = 0$ into itself.*

In other words L is symmetry of Q if $Lu \in \ker(Q)$ whenever $u \in \ker(Q)$. According to the Definition 2 we will define the symmetries of the Dirac operator.

Definition 3. *L is a symmetry operator of the Dirac operator D if Lu is a monogenic function for every monogenic function u.*

We denote by $S(D)$ the space of all symmetries of the Dirac operator. In ([6]-[9], [11]) it is shown that all symmetry operators of the Dirac operator D actually form an infinite dimensional finitely generated real Lie algebra with standard addition and composition for multiplication.

Definition 4. *A symmetry operator L of the Dirac operator D is said to be a 1-symmetry operator if it is a first order partial differential operator with Clifford algebra valued coefficients.*

The space of 1-symmetry operators for the Dirac operator D, denoted by $S_1(D)$, is a subalgebra of $S(D)$. Since the coefficients of the first order partial differential operators in $S_1(D)$ are not necessary real, the latter fact is not evident. The Dirac operator D and the identity operator I are trivial 1-symmetries. Except for these trivial symmetries, the following four classes of first order partial differential operators with Clifford algebra valued coefficients (1-symmetries) constitute the basis of the set of all symmetry operators for the Dirac operator D (compare [6]-[9]):

- the generators of translation group in $\mathbb{R}^{m+1}$

$$D_k = \partial_{x_k}, \quad k = 0, 1, \ldots, m; \tag{1.6}$$

- the dilatations

$$R_0 = x_0 \partial_{x_0} + x_1 \partial_{x_1} + \ldots + x_m \partial_{x_m} + \frac{1}{2} m; \tag{1.7}$$

- the generators of the rotation group

$$J_{ij} = -J_{ji} = x_j \partial_{x_i} - x_i \partial_{x_j} + \frac{1}{2} e_{ij}, \quad i, j = 1, 2, \ldots, m, \quad i \neq j$$

$$J_{i0} = -J_{0i} = x_0 \partial_{x_i} - x_i \partial_{x_0} + \frac{1}{2} e_i, \quad i = 1, 2, \ldots, m, \tag{1.8}$$

- and the generators of the "conformal group"

$$K_i = \sum_{s=0}^{m} 2 x_i x_s \partial_{x_s} - \mathbf{x} \bar{\mathbf{x}} \partial_{x_i} + (m+1) x_i - \bar{\mathbf{x}} e_i, \tag{1.9}$$

for $i = 0, 1, \ldots, m$. Here $\mathbf{x} = x_0 + x_1 e_1 + \ldots + x_m e_m$ and $\bar{\mathbf{x}}$ are conjugate in the above sense of Clifford algebra valued functions.

The real Lie algebra $S_1(D)$ with the natural basis (1.6)-(1.9) is isomorphic to the Lie algebra $so(m+2, 1)$ of all real valued $(m+3) \times (m+3)$ matrices A such that $AG^{m+2,1} + G^{m+2,1} A^t = 0$, where

$$G^{m+2,1} = \sum_{k=1}^{m+2} \mathcal{E}_{kk} - \mathcal{E}_{m+3,m+3}.$$

Here $\mathcal{E}_{kl}$ is the $(m+3) \times (m+3)$ matrix with the unit in the k-th row, l-th column, and zeros everywhere else.

As is well-known, the following elements

$$\Gamma_{kl} = -\Gamma_{lk} = \mathcal{E}_{kl} - \mathcal{E}_{lk}, \quad 1 \le k, l \le m+2,$$
$$\Gamma_{k,m+3} = \Gamma_{m+3,k} = \mathcal{E}_{k,m+3} + \mathcal{E}_{m+3,k}, \quad 1 \le k \le m+2,$$

constitute a basis for $so(m+2,1)$. In addition, the formulae

$$[\Gamma_{kl}, \Gamma_{rs}] = \delta_{lr}\Gamma_{ks} + \delta_{ks}\Gamma_{lr} + \delta_{rk}\Gamma_{sl} + \delta_{sl}\Gamma_{rk},$$
$$[\Gamma_{k.m+3}, \Gamma_{rs}] = -\delta_{ks}\Gamma_{r,m+3} + \delta_{kr}\Gamma_{s,m+3}, \quad [\Gamma_{k,m+3}, \Gamma_{l,m+3}] = \Gamma_{kl},$$

present the commutation relations for $so(m+2,1)$ where (δ_{kl} is the Kronecker delta function). Set

$$D_0 = \Gamma_{12} + \Gamma_{2,m+3}, \qquad\qquad K_0 = \Gamma_{12} - \Gamma_{2,m+3},$$
$$D_1 = \Gamma_{13} + \Gamma_{3,m+3}, \qquad\qquad K_1 = \Gamma_{13} - \Gamma_{3,m+3},$$
$$\vdots \qquad\qquad\qquad\qquad \vdots$$
$$D_m = \Gamma_{1m+2} + \Gamma_{m+2,m+3}, \qquad K_m = \Gamma_{1,m+2} - \Gamma_{m+2,m+3},$$
$$J_{01} = \Gamma_{23}, \qquad\qquad\qquad J_{12} = \Gamma_{34},$$
$$J_{02} = \Gamma_{24}, \qquad\qquad\qquad J_{13} = \Gamma_{35},$$
$$\vdots \qquad\qquad\qquad\qquad \vdots$$
$$J_{0,m} = \Gamma_{2,m+2}, \qquad\qquad J_{ij} = \Gamma_{i+2,j+2},$$

and $R_0 = \Gamma_{1,m+3}$. One can verify that the above relations determine the Lie algebra isomorphism of $S_1(D)$ and $so(m+2,1)$.

Remark 1. *Note that all symmetries of the Dirac operator are the symmetries of the Laplace operator Δ in the space of Clifford algebra valued functions. This means that $S(D) \subset S(\Delta)$. Therefore, the natural symmetry operators bases for the Dirac operator D considered in the space of Clifford algebra valued functions and for the Laplace operator Δ considered in the space of real valued function are algebraically isomorphic. However, only the infinitesimal generators of translation and dilatations coincide identically for both of them. At the same time, the generators of rotations and the conformal group differ for the Laplace operator in the space of real valued functions and have the following form.*

The generators of the rotation group are

$$J'_{ij} = -J'_{ji} = x_j \partial_{x_i} - x_i \partial_{x_j}, \quad i, j = 1, 2, \ldots, m, \quad i \neq j,$$
$$J'_{i0} = -J'_{0i} = x_0 \partial_{x_i} - x_i \partial_{x_0}, \quad i = 1, 2, \ldots, m. \tag{1.10}$$

The generators of the "conformal group" in the space of real harmonic functions are

$$K_i' = \sum_{s=0}^{m} 2x_i x_s \partial_{x_s} - \mathbf{x}\bar{\mathbf{x}}\partial_{x_i} + (m-1)x_i, \quad i = 0, 1, \dots, m. \qquad (1.11)$$

Hence, the rotational symmetries of the Dirac operators differ only from the corresponding symmetries of the Laplace operator by standard hypercomplex numbers multiplication

$$\mathbf{e}_i = 2J_{i0} - 2J_{i0}', \quad \mathbf{e}_{ij} = 2J_{ij} - 2J_{ij}', \quad i, j = 1, 2, \dots, m, \quad i \neq j.$$

The generators of the "conformal group" differ only in the variable $\mathbf{x}$ by a hypercomplex number left multiplication

$$\mathbf{x} = K_0 - K_0', \quad \mathbf{e}_i\mathbf{x} = K_i' - K_i, \quad i = 1, \dots, m.$$

It is well-known that the standard algebraic operations (multiplications by a Clifford algebra valued constant or by a Clifford algebra valued variable) are inferred from the space of monogenic functions.

The above observations allow us to redefine the multiplication by a constant $\mathbf{a} \in \mathcal{A}_m$ and by a variable $\mathbf{x} \in \mathcal{A}_m$ in a way that preserves the monogenic structure of the resultant. All we need in order to correct multiplication is to make use of the additional harmonic amendments.

Proposition 1. *For every Clifford constants* $\mathbf{a}, \mathbf{b}, \mathbf{c}$ *there exists a 1-symmetry operator* $S \in S_1(\Delta)$ *for the Laplace operator* Δ *such that*

$$(\mathbf{a}(\mathbf{x} + \mathbf{b}) + \mathbf{c})f(x) + Sf(x)$$

is monogenic for every monogenic function $f(x)$.

Proof. It follows from straightforward computation that

$$S = a_0 K_0' - \sum_{i=0}^{m} a_i K_i' + 2\sum_{i,j=1}^{m} a_i b_j J_{ij} + 2\sum_{i=1}^{m}(a_0 b_i + a_i b_0 + c_i)J_{0i}' + (a_0 b_0 + c_0)I.$$

$\square$

Example 1. *Let* $f(x)$ *be monogenic. To multiply* $f(x)$ *by* $\mathbf{e}_i\mathbf{x}$ *one can use either*

$$F_1(x) \equiv (\mathbf{e}_i\mathbf{x} - K_i')f(x) := -K_i[f(x)]$$

or

$$F_2(x) \equiv (\mathbf{e}_i + 2J_{i0}')(\mathbf{x} + K_0')f(x) := 2J_{i0}K_0[f(x)].$$

$F_1(x)$ *and* $F_2(x)$ *are obviously monogenic if* $f(x)$ *is monogenic. However even for* $f(x) \equiv 1$, *the result* $F_1(x)$ *of multiplying* $f(x)$ *by* $\mathbf{e}_i\mathbf{x}$ *is different from* $F_2(x)$. *Hence we need a new multiplication that preserves the structure of monogenic functions and is independent of "harmonic amendments".*

2 Basic commutation relations

We focus here on commutation relations between 1-symmetry operators defined in the previous section. We start with a classification of the commutative subalgebra of $S_1(D)$. This approach is the key tool in describing the structure of monogenic functions as a theory of functions in commutative variables.

Theorem 1. L *is a 1-symmetry operator,* $L \in S_1(D)$ *if and only if there exists the first order partial differential operator L' with Clifford algebra valued coefficients (not nessesarily 1-symmetry) and such that*

$$DL = L'D. \tag{2.1}$$

Proof. If (2.1) holds, then for every $f \in \mathcal{M}_m(\Omega)$, obviously, $Lf \in \mathcal{M}_m(\Omega)$. On the other hand, without loss of generality, we can choose L in the form

$$L = a_0(x)D + \sum_{k=1}^{m} a_k(x)\partial_{x_k} + b(x).$$

Here $a_k(x), k = 0, \ldots, m, \; b(x) \in \mathbf{C}^1(\Omega, \mathcal{A}_m)$.

Now let L^* be chosen such that the operator $R_L = DL - L^*D$ does not depend on ∂_{x_0}, for example, in the following form

$$L^* = Da_0(x) + \sum_{k=1}^{m} a_k(x)\partial_{x_k} + b(x).$$

Then, obviously,

$$R_L f(x) = D(Lf(x)) - L^*(Df(x))$$
$$= \sum_{k,l=1}^{m} a_{kl}(x)\partial_{x_k}\partial_{x_l} f(x) + \sum_{k=1}^{m} b_k(x)\partial_{x_k} f(x) + c(x)f(x). \tag{2.2}$$

For (2.2) to be valid when applied to an arbitrary monogenic function $f(x)$, it is nessesary and sufficient that all the coefficients of $R_L f$ be zero. This means that for $f(x) \equiv 1$ (every constant is a trivial monogenic function), $c(x) \equiv 0$. Consequently, taking $f(x) = z_k$, for

$$z_k = x_k - x_0\mathbf{e}_k, \quad k = 1, 2, \ldots, m,$$

in (2.2), we obtain that all Clifford algebra valued coefficients $b_k(x) \equiv 0, \; k = 1, 2, \ldots, m$. Now if $f(x)$ is chosen equal to one of the following functions

$$z_k^2 := x_0^2 - x_k^2 + 2x_0x_k\mathbf{e}_k, \quad k = 1, 2, \ldots, m,$$

(actually hypercomplex second order powers, that are, obviously, monogenic), we obtain $a_{kk}(x) \equiv 0$, $k = 1, 2, \ldots, m$. In order to complete the proof, we can substitute the following monogenic functions $f(x)$ in (2.2):

$$z_{12}^2 := x_1^2 - x_2^2 - 2x_1x_2\mathbf{e}_{12},$$
$$z_{13} := x_1^2 - x_3^2 - 2x_1x_3\mathbf{e}_{13}, \ldots,$$
$$z_{m-1,m} := x_{m-1}^2 - x_m^2 - 2x_{m-1}x_m\mathbf{e}_{m-1,m}.$$

Now all the remaining coefficients of R_L are equal to 0. The theorem is proven for $L' = L^*$. $\qquad\square$

A straightforward calculation of the commutators of 1-symmetry basis operators (1.6)-(1.9) with the Dirac operator D leads to

$$[D, D_i] = 0, \quad [D, J_{0i}] = \mathbf{e}_i D, \quad [D, R_0] = D, \quad [D, K_i] = 2(x_i + x_0\mathbf{e}_i)D$$

for all indices $i = 0, 1, \ldots, m$, and

$$[D, J_{i,j}] = 0$$

for all $i, j = 1, 2, \ldots, m$ with $j \neq i$.

The commutation relations between the basis operators are

$$[K_i, K_j] = 0, \quad [D_i, D_j] = 0, \quad [R_0, J_{ij}] = 0, \qquad (2.3)$$

$$[D_i, R_0] = D_i, \quad [R_0, K_i] = K_i, \quad [D_i, K_i] = 2R_0, \quad [D_i, K_j] = 2J_{ij} \qquad (2.4)$$

for $i, j = 0, \ldots, m$, $i < j$, and $J_{ij} = -J_{ji}$. The other commutation relations are

$$[J_{kl}, J_{rs}] = \delta_{lr}J_{ks} + \delta_{ks}J_{lr} + \delta_{rk}J_{sl} + \delta_{sl}J_{rk},$$

$$[K_i, J_{jk}] = \delta_{ik}K_j - \delta_{ij}K_k, \quad [D_i, J_{jk}] = \delta_{ik}D_j - \delta_{ij}D_k$$

for all multi indices $i, j, k, l, r, s \in \omega$.

It follows from (2.3) that the subalgebra of all conformal group generators (denoted by $S_{1,K}(D)$) in $S_1(D)$ with the basis $K_0, K_1, \ldots, K_m$ is commutative. Actually, any translation of the algebra preserves the above commutativity as well.

Denote by $S_{1,K-A}(D)$ a subalgebra of $S_1(D)$ as a span of a translations of basis conformal group generators

$$K_0 - A_0, K_1 - A_1, \ldots, K_m - A_m,$$

where

$$K_i - A_i = \exp(-a_0 D_0 - a_1 D_1 - \ldots - a_m D_m)K_i. \qquad (2.5)$$

By construction, the subalgebra $S_{1,K-A}(D)$ is commutative for every **a**. Direct calculations, using (2.5), give an explicit form of the symmetry operators A_i, $i = 0, 1, \ldots, m$:

$$A_i = 2 \sum_{j \neq i, j=0}^{m} a_j J_{ji} - 2a_i \sum_{j=0}^{m} a_j D_j + 2a_i R_0 + \bar{\mathbf{a}} \mathbf{a} D_i. \tag{2.6}$$

Obviously, $A_i \in S_1(D)$. Here the multiplication on a Clifford algebra valued constant is thought of as the operator action. The operator A_i depends on $\mathbf{a} = a_0 + a_1 \mathbf{e}_1 + \ldots + a_m \mathbf{e}_m$. If $\mathbf{a} = 0$, then $A_i = 0$ for all indices $i = 0, 1, \ldots, m$. We can see that A is a symmetry operator that preserves the monogenic structure of a function. Moreover, to some extent it can be used to multiply the functions by a constant.

Example 2. *The following polynomial operator*

$$P_n(x) = \sum_{|\alpha| < n} (K - A)^{\alpha} c_{\alpha} \tag{2.7}$$

with Clifford algebra valued coefficients c_{α} is correctly defined since all of the operators $K_i - A_i$ are pairwise commutative.

3   Representation of monogenic functions

It is well-known that an entire monogenic function can be represented as a convergent series of monogenic homogeneous polynomial functions defined in $\mathcal{M}_m(\mathbb{R}^{m+1})$. It is homogeneous polynomial of degree n (see [1],[2],[7])

$$Y^n(x_0, x_1, \ldots, x_m) = \sum_{|\alpha| = n} c_{\alpha} x^{\alpha}, \quad n = 0, 1, \ldots,$$

where α is a multi index and $|\alpha| = \alpha_0 + \alpha_1 + \ldots + \alpha_m$, $x^{\alpha} = x_0^{\alpha_0} x_1^{\alpha_1} \ldots x_m^{\alpha_m}$, c_{α} are the Clifford algebra valued constants such that

$$\sum_{|\alpha| = n} \left(\frac{\alpha_0}{x_0} + \frac{\alpha_1}{x_1} \mathbf{e}_1 + \frac{\alpha_2}{x_2} \mathbf{e}_2 + \ldots + \frac{\alpha_m}{x_m} \mathbf{e}_m \right) c_{\alpha} x^{\alpha} \equiv 0.$$

All homogeneous monogenic polynomials of degree n constitute a space of monogenic functions $\mathcal{M}_m^n(\mathbb{R}^{m+1}) \subset \mathcal{M}_m(\mathbb{R}^{m+1})$.

The following theorem shows how one can construct a Clifford algebra valued harmonic $Y^n(x) \in \mathcal{M}_m^n(\mathbb{R}^{m+1})$ when its gradient is known.

Theorem 2. *Every Clifford algebra valued, homogeneous of order n, harmonic $Y^n(x)$ may be represented in the form*

$$(2n + m - 1) n Y^n(x) = \sum_{i=0}^{m} K_i(\partial_{x_i} Y^n(x)), \tag{3.1}$$

or, equivalently,

$$(2n + m - 1)nY^n(x) = \sum_{i=1}^{m} Z_i(\partial_{x_i} Y^n(x)), \qquad (3.2)$$

where $K_0, K_1, \ldots, K_m$ *are symmetry operators defined in (1.9) and* $Z_1, Z_2,$ $\ldots, Z_m$ *are (not necessarily symmetry operators) defined by*

$$Z_i = K_i - K_0 e_i$$

for all $i = 1, 2, \ldots, m$.

Proof. Let R_0 be, as above, the generator (1.7) of in $\mathbb{R}^{m+1}$. The following relation is satisfied in the algebra $S_1(D)$:

$$K_0 D_0 + K_1 D_1 + \ldots + K_m D_m = 2R_0^2 - (m+1)R_0 + \frac{1}{2}mI - x\bar{x}\Delta - \bar{x}D.$$

Every homogeneous function $Y_n(x)$ is an eigenfunction for the operator R_0 with the eigenvalue $n + \frac{1}{2}m$, i.e., $R_0 Y^n(x) = (n + \frac{1}{2}m)Y^n(x)$, for $n = 0, 1, 2, \ldots$.

Using these relations, we obtain

$$\sum_{s=0}^{m} K_s(\partial_{x_s} Y^n(x))$$

$$= [2(n + \frac{m}{2})^2 - (m+1)(n + \frac{m}{2}) + \frac{m}{2}]Y^n(x) - x\bar{x}\Delta Y^n(x) - \bar{x}DY^n(x).$$

Here Δ is the Laplace operator, $\Delta Y^n(x) = 0$, and $DY^n(x) = 0$ for every monogenic function $Y^n(x)$; and (3.1) follows. Moreover, taking into account the monogenic structure of $Y^n(x)$, we obtain $\partial_{x_0} Y^n(x) = -\sum_{s=1}^{m} \partial_{x_s} Y^n(x)$. Substituting $K_0 D_0 \equiv -K_1 e_1 D_1 - K_2 e_2 D_2 - \ldots - K_m e_m D_m$ in (3.1) and using the notation $Z_i = K_i - K_0 e_i$, we obtain (3.2) explicitly. $\qquad \square$

Theorem 3. *Any Clifford harmonic* $Y^n(x)$ *can be represented as follows:*

$$Y^n(x) = \sum_{s=0}^{m} \sum_{|I|=s} P_I^n(K_0, K_1, \ldots, K_m)[e_I] \qquad (3.3)$$

where $P_I^n(x_0, x_1, \ldots, x_m)$ *are homogeneous polynomials of degree* n *with real coefficients for every* $n \in \mathbb{N}$ *and* $I \subset \omega$.

Proof. The polynomials in (3.3) are defined correctly because they depend on the commutative variables $K_0, K_1, \ldots, K_m$. They act on Clifford constants $\{e_0, e_1, \ldots, e_\omega\}$.

Let (3.3) be fulfilled for all k-homogeneous monogenic functions ($k < n$). The gradient of an n-homogeneous monogenic function $Y^n(x)$ is a homogeneous monogenic function of order $n-1$, i.e., it can be represented in the form (3.3). Using (3.1), we obtain that (3.3) is satisfied for $Y^n(x)$ as well. The main theorem is proved by induction because every linear monogenic function $Y^1(x)$ may be represented by the formulae

$$Y^1(x) = \sum_{s=0}^{m} a_s x_s,$$

where Clifford algebra valued constants $a_0, a_1, \ldots, a_m$ are parameters satisfying the relation $a_0 + e_1 a_1 + \ldots e_m a_m = 0$. On the other hand,

$$\sum_{s=0}^{m} K_s[a_s] = (m+1) \sum_{s=0}^{m} a_s x_s - \bar{x}(a_0 + e_1 a_1 + \ldots + e_m a_m).$$

Thus all linear Clifford harmonics

$$Y^1(x) = \frac{1}{m+1}(K_0[a_0] + K_1[a_1] + \ldots + K_m[a_m])$$

may be represented in the form (3.3).   □

The same is true for operators K "translated" by A.

Example 3. *The following representation of a monogenic function $Y^n(x)$ (cf. 2.6, 3.3) holds*

$$Y^n(x) = \sum_{s=0}^{m} \sum_{|I|=s} Q_I^n(K_0 - A_0, K_1 - A_1, \ldots, K_m - A_m)[e_I].$$

Here Q_I^n are polynomials of degree less than or equal to n for every set of multi indices $I \subset \omega$. So one can use these two representations of the same monogenic function $Y^n(x)$ as a rule for the coefficients of the corresponding polynomial $P(x_0, x_1, \ldots, x_m)$ recalculation.

The next result shows how one can define every monogenic function when their partial derivatives in a given point are known.

Theorem 4. *Let $f(x)$ be a monogenic function analytic in neighborhood of a given point $a = (a_0, a_1 \ldots, a_m)$. The following "Taylor like" expansion holds:*

$$f(x) = f(a) + \sum_{n=1}^{\infty} \frac{(m-1)!!}{(2n+m-1)!!} \sum_{|\alpha|=n} \frac{1}{\alpha!}(K-A)^\alpha[\partial_\alpha f(a)], \qquad (3.4)$$

where $(K-A)^\alpha = (K_0 - A_0)^{\alpha_0}(K_1 - A_1)^{\alpha_1} \ldots (K_m - A_m)^{\alpha_m}$, A_i is defined in (2.6) and $\partial_\alpha = \partial_{x_0}^{\alpha_0} \partial_{x_1}^{\alpha_1} \ldots \partial_{x_m}^{\alpha_m}$, $\alpha! = \alpha_0! \alpha_1! \ldots \alpha_m!$.

Proof. This is provided by Théorem 3. $\qquad\square$

We can say that (3.4) gives rise to the one-to-one correspondences between $\mathcal{M}_m(\Omega)$ and the vector space of 2^m *real analytic vector functions in some open neighborhood of* $\mathbf{a} \in \mathbb{R}^{m+1}$.

Moreover, using the basic condition on partial derivatives of $f(x)$, we can exclude, for example, the partial derivatives D_0 from (3.4). This means one can use (3.4) to solve the Dirichlet boundary value problem for the Dirac equation.

Example 4. *Let* $\varphi(x_1, x_2, \ldots, x_m)$ *be a Clifford algebra valued and real analytic in the* $\mathbb{R}^m$ *function. Consider the Dirichlet boundary value problem in the upper half space* $\mathbb{R}^{m+1}_+ = \{x_0 > 0\}$

$$Df(x_0, x_1, \ldots, x_m) = 0, \quad x_0 > 0,$$
$$f(0, x_1, x_2, \ldots, x_m) = \varphi(x_1, x_2 \ldots, x_m).$$

The solution of the Dirichlet problem is given by the formula

$$f(x) = \varphi(0) + \sum_{n=1}^{\infty} \frac{(m-1)!!}{(2n+m-1)!!} \sum_{|\alpha|=n} \frac{1}{\alpha!} K^{\alpha}[\partial_{\alpha}\varphi(0)]. \tag{3.5}$$

Example 5. *The formula*

$$f(x) = f(0) + \sum_{n=1}^{\infty} K_0^n[c_n] \tag{3.6}$$

gives the representation of all axial symmetric Clifford algebra valued functions (cf. [8])

$$f(x) = \sum_I [\varphi_I(x_0, r) + \psi_I(x_0, r)\mathbf{v}]\mathbf{e}_I. \tag{3.7}$$

Here $\mathbf{v}$ *is defined by*

$$\mathbf{v} = \frac{1}{2}(\mathbf{x} - \bar{\mathbf{x}}) = \sum_{s=1}^{m} x_s \mathbf{e}_s,$$

and $r^2 = -\mathbf{v}^2 = \mathbf{x}\bar{\mathbf{x}} - x_0^2 = x_1^2 + \ldots + x_m^2$. *Since*

$$K_0\varphi(x_0, r) = (x_0^2 - r^2)\partial_{x_0}\varphi + 2x_0 r \partial_r \varphi + \varphi\mathbf{v},$$
$$K_0\mathbf{v} = -r^2 + (m+2)x_0\mathbf{v},$$

we obtain

$$K_0[\varphi(x_0, r) + \psi(x_0, r)\mathbf{v}] = \Phi(x_0, r) + \Psi(x_0, r)\mathbf{v},$$

where

$$\Phi(x_0, r) = (x_0^2 - r^2)\partial_{x_0}\varphi + 2x_0 r\partial_r\varphi - 2r^2\psi,$$

$$\Psi(x_0, r) = \varphi + (x_0^2 - r^2)\partial_{x_0}\psi + 2x_0 r\partial_r\psi + (m+2)x_0\psi.$$

This means that the operator action $K_0 f(x)$ on every axial symmetric (3.7) function $f(x)$ preserves its structure. In addition we have to use the fact that $f(x)$ is monogenic. In other words, from $Df = 0$, it follows that $\varphi_I(x_0, r), \psi_I(x_0, r)$ must be a solution of the following two-dimensional real system of linear first order partial differential equations

$$\partial_{x_0}\varphi_I(x_0, r) = -r\partial_r\psi_I(x_0, r) + m\psi_I(x_0, r),$$

$$\partial_r\varphi_I(x_0, r) + r\partial_{x_0}\psi_I(x_0, r) = 0.$$

Using the complex variable $\zeta = x_0 + ir$, we obtain a nonlinear complex equation

$$\frac{\partial w(\zeta)}{\partial\zeta} = \frac{2}{\zeta - \bar{\zeta}}(w - \bar{w})$$

in the sense of a generalized complex analytic function [10], where $w(\zeta) = \varphi_I(x_0, r) + \frac{1}{r}\psi_I(x_0, r)$ by definition.

Using results given in [10], we can see that there exists a one-to-one correspondence between complex analytic functions and the radial symmetric monogenic functions represented by (3.6) with operator valued variables.

Example 6. *For $m = 1$ and $\mathbf{a} = 0$ from (3.4) follows the exact Taylor expansion of functions $f(z)$. Here $z = x_0 + e_1 x_1$. Since $K_0 D_0 + K_1 D_1 = 2z\partial_z z\partial_z$ in (3.2), only one operator $Z_1 = 2z\partial_z z$ is present. Thus, (3.4) may be rewritten in the form*

$$f(z) = f(0) + \sum_{n=1}^{\infty} \frac{1}{2^n (n!)^2} Z_1^n[f^{(n)}(0)]$$

where $Z_1^n[1] = 2^n n! z^n$. Hence for every complex analytic function $f(z)$ defined in a neighborhood of the origin, the formula (3.4) is equivalent to the exact Taylor formula

$$f(z) = f(0) + \sum_{n=1}^{\infty} \frac{1}{n!} z^n[f^{(n)}(0)].$$

4   Multiplication

This section defines a multiplication in the space of monogenic functions $\mathcal{M}_m(\Omega)$.

We have seen earlier that (3.4) constitute the one-to-one correspondence between $\mathcal{M}_m(\Omega)$ and the vector space of 2^m *real analytic vector functions*. Define a subclass of monogenic functions $\mathcal{M}_m^0(\mathbb{R}^{m+1})$ represented by the formula

$$F(x_0, x_1, \ldots, x_m) = \text{ by definition } = F(K_0, K_1, \ldots, K_m)[1]. \qquad (4.1)$$

Here $F(x_0, x_1, \ldots, x_m)$ is a real analytic function.

Example 7. *Set x_0^2 to be equal to $K_0^2[1]$,*

$$x_0^2 := (m+1)[(m-1)x_0^2 + \mathbf{x}^2].$$

Set also $x_0 x_1$ to be equal to $K_0 K_1[1]$,

$$x_0 x_1 := (m+1)x_0[(m+1)x_1 - \bar{\mathbf{x}}\mathbf{e}_1] + \mathbf{x}[(m+1)x_1 - 2\bar{\mathbf{x}}\mathbf{e}_1].$$

Definition 5. *For any two functions $F(x), G(x) \in \mathcal{M}_m^0(\mathbb{R}^{m+1})$ their product $F * G$ is defined as a composition*

$$F(K)[G(x)] \equiv G(K)[F(x)] \equiv F(K)G(K)[1].$$

One can verify that the product $F * G$ belongs to $\mathcal{M}_m^0(\mathbb{R}^{m+1})$. The following list describes the properties of the multiplication in $\mathcal{M}_m^0(\mathbb{R}^{m+1})$:

1. Multiplication is associative.

2. Multiplication is distributive with respect to addition.

3. Multiplication is commutative.

4. There is a unit element for multiplication; it is $1 + 0\mathbf{e}_1 + \ldots + 0\mathbf{e}_m$.

Acknowledgments

The author gratefully acknowledges fruitful conversations with Prof. C. Berenstein, University of Maryland. The author is grateful to Prof. J. Ryan, University of Arkansas, for useful discussion concerning the results of this article.

REFERENCES

[1] F. Brackx, R. Delanghe, and F. Sommen, *Clifford Analysis*, Pitman Research Notes in Math **76**, 1982.

[2] A. McIntosh, Review of Clifford algebra and spinor-valued functions, a function theory for the Dirac operator, by R. Delanghe, F. Sommen, and V. Soucek, *Bull. of Amer. Math. Soc.*, Vol. **32** (1995), 344–348.

[3] J. Ryan, ed., *Clifford Algebras in Analysis and Related Topics*, CRC Press, Boca Raton, 1996.

[4] P. J. Olver, *Equivalence, Invariants, and Symmetry*, Cambridge University Press, 1995.

[5] W. Miller, Symmetry and separation of variables, *Encyclopedia of Mathematics and its Applications* **4** Addison-Wesley, 1977.

[6] P. Lounesto, Spinor valued regular functions in hypercomplex analysis, Thesis, Helsinki, 1979.

[7] H. Leutwiler, Modified Clifford analysis, *Complex Variables Theory Appl.* **17** No. 3-4 (1992), 153-171.

[8] P. Lounesto and P. Bergh, Axially symmetric vector fields and their complex potentials, *Complex Variables Theory Appl.* **2** (1983), 139-150.

[9] F. Sommen and N. Van Acker, Monogenic differential operators, *Results in Math.*, Vol. **22**, 1992, 781-798.

[10] I. Vekua, *Generalized Analytic Functions*, London, Pergamon, 1962.

[11] Y. Krasnov, Symmetries of Cauchy-Riemann-Fueter equation, *Complex Variables*, 2000, to appear.

Yakov Krasnov
Department of Mathematics and Computer Science
Bar-Ilan University
52900 Ramat Gan, Israel
E-mail: krasnov@macs.biu.ac.il

Received: October 1, 1999; Revised: February 2, 2000

On the Radial Part of the Cauchy-Riemann Operator

Thomas Hempfling

ABSTRACT Using the Clifford algebra C_n with generators $e_1, \ldots, e_n$, $e_i^2 = -1$, the Cauchy-Riemann operator in $\mathbb{R}^{n+1}$ is defined as $\overline{\partial} = \frac{\partial}{\partial x_0} + \sum_{i=1}^{n} \frac{\partial}{\partial x_i} e_i$, where $x = x_0 + \sum_{i=1}^{n} x_i e_i$ is a paravector. We consider Fueter-type paravector functions $f = u(x_0, \rho) + I(x) v(x_0, \rho)$, where $\rho^2 = \sum_{i=1}^{n} x_i^2$, $I(x) := \frac{1}{\rho} \sum_{i=1}^{n} x_i e_i$, and u, v are real-valued. The equation $\overline{\partial} f = 0$ splits into two parts. One of them depends only on x_0, ρ. This leads to a system of partial differential equations which coincides with the system defining hypermonogenic functions. These functions arise for example as solutions of the Dirac equation in the upper half space $\mathbb{R}_+^{n+1}$ endowed with the Poincaré metric.
Keywords: Cauchy-Riemann operator, Dirac operator, hypermonogenic functions, axially symmetric functions, Clifford analysis.

1 Introduction

The Cauchy-Riemann operator in complex analysis is defined by

$$\overline{\partial} := \tfrac{1}{2}\left(\tfrac{\partial}{\partial x} + \tfrac{\partial}{\partial y} i\right),$$

making sense for functions of the complex variable $z = x + yi$, say $f = u + vi$, $u = u(x, y)$, $v = v(x, y)$ real-valued, which are defined in an open subset U of $\mathbb{R}^2 \cong \mathbb{C}$. If u, v are differentiable functions in U, $\overline{\partial}$ has the well-known property

$$\overline{\partial} f = 0 \text{ in } U \;\Leftrightarrow\; f \text{ holomorphic in } U$$
$$\Leftrightarrow\; \tfrac{\partial u}{\partial x} = \tfrac{\partial v}{\partial y}, \quad \tfrac{\partial u}{\partial y} = -\tfrac{\partial v}{\partial x} \quad \text{in } U.$$

This work has been supported by the Swiss National Fonds and the Deutsche Forschungsgemeinschaft.
AMS Subject Classification: 30G35, 53C20, 58A14, 58J05.

We recall the definition of the higher dimensional analog of this operator using a Clifford algebra framework in Section 2. Solutions of the equation $\overline{\partial} f = 0$ in this generalized context are called monogenic functions. In this paper we are mainly dealing with special classes of axially symmetric functions (Section 5). Taking spherical coordinates with respect to this axis, the operator $\overline{\partial}$ splits into a radial and a spherical part, $\overline{\partial} = \overline{\partial}_{\mathrm{rad}} + \overline{\partial}_{\mathrm{sph}}$ (Section 4). The solutions of $\overline{\partial}_{\mathrm{rad}} f = 0$, f being axially symmetric, coincide with so-called hypermonogenic functions (Section 3) if the regarded domain lies inside upper half space (Section 5). An interpretation with Dirac operators is given in Section 6.

2   The Cauchy-Riemann operator in $\mathbb{R}^{n+1}$

Firstly, we define the standard Clifford algebra C_n : Let $e_1, \ldots, e_n$ be the canonical basis of $\mathbb{R}^n$. Then a vector space basis for C_n is given by the unit element $e_0 := 1 \in \mathbb{R}$ and the products $e_{i_1} \cdots e_{i_k}$, $1 \leq i_1 < \ldots < i_k \leq n$, $k = 1, \ldots, n$. C_n is the associative real algebra with unit e_0 due to the relations $e_i e_j + e_j e_i = -2\delta_{ij}$ $(i, j = 1, \ldots, n)$. Elements of C_n having the form $x = \sum_{i=0}^{n} x_i e_i =: x_0 + \sum_{i=1}^{n} x_i e_i$, $x_i \in \mathbb{R}$, are called paravectors; x_0 is the real part, $\sum_{i=1}^{n} x_i e_i$ the vector part of a paravector. They will be identified with elements of $\mathbb{R}^{n+1}$. Analogously, we regard functions $f : U \to \mathbb{R}^{n+1}$ defined in an open subset U of $\mathbb{R}^{n+1}$ with values in $\mathbb{R}^{n+1}$ as paravector functions, i.e., $f = f_0 + \sum_{i=1}^{n} f_i e_i$ where $f_i : U \to \mathbb{R}$ is real-valued $(i = 0, \ldots, n)$.

The generalized Cauchy-Riemann operator in $\mathbb{R}^{n+1}$ is now defined by $\overline{\partial} := \frac{\partial}{\partial x_0} + \sum_{i=1}^{n} e_i \frac{\partial}{\partial x_i}$. Differentiable paravector functions which are solutions of $\overline{\partial} f = 0$ are called monogenic functions (differentiable means here at least partially differentiable with respect to $x_0, \ldots, x_n$). For paravector functions there is no difference between the solutions of $\overline{\partial} f = 0$ and $\overline{\partial}_r f = 0$ with $\overline{\partial}_r := \frac{\partial}{\partial x_0} + \sum_{i=1}^{n} \frac{\partial}{\partial x_i} e_i$. This changes of course if looking at functions with values in a bigger part of the Clifford algebra (actually, $\overline{\partial} f = 0$ defines left monogenic functions). Simple calculations show that for differentiable $f = f_0 + \sum_{i=1}^{n} f_i e_i : U \to \mathbb{R}^{n+1}$

$$\overline{\partial} f = 0 \Leftrightarrow \begin{cases} \dfrac{\partial f_0}{\partial x_0} - \sum_{i=1}^{n} \dfrac{\partial f_i}{\partial x_i} = 0 \\[2mm] \dfrac{\partial f_0}{\partial x_i} = -\dfrac{\partial f_i}{\partial x_0}, \quad \dfrac{\partial f_i}{\partial x_j} = \dfrac{\partial f_j}{\partial x_i} \quad (i, j = 1, \ldots, n). \end{cases}$$

This system is often referred to as the Riesz system (originally using $-f_0$ instead of f_0) (cf. [18]) or, generalized, Cauchy-Riemann system.

3   Hypermonogenic functions

In [15] Leutwiler introduced the concept of so-called hypermonogenic functions; these are solutions of the *modified* Cauchy-Riemann system given by

$$x_n \overline{\partial} f + (n-1) f_n = 0,$$

or, equivalently,

$$x_n \left(\frac{\partial f_0}{\partial x_0} - \sum_{i=1}^{n} \frac{\partial f_i}{\partial x_i} \right) + (n-1) f_n = 0$$

$$\frac{\partial f_0}{\partial x_i} = -\frac{\partial f_i}{\partial x_0}, \quad \frac{\partial f_i}{\partial x_j} = \frac{\partial f_j}{\partial x_i} \quad (i, j = 1, \dots, n).$$

If defined in an open subset of upper half space of paravectors, $\mathbb{R}_+^{n+1} := \{ x = x_0 + \sum_{i=1}^{n} x_i e_i \in \mathbb{R}^{n+1} \mid x_n > 0 \}$, hypermonogenic functions can be interpreted as harmonic 1-forms

$$f \equiv f_0 dx_0 - \sum_{i=1}^{n} f_i dx_i \tag{3.1}$$

with respect to the hyperbolic (Poincarè) metric given by the line element

$$ds^2 = \frac{dx_0^2 + \sum_{i=1}^{n} dx_i^2}{x_n^2}.$$

Harmonicity means

$$df = 0 = d^* f,$$

$d^* = (-1)^\alpha * d*$, here $\alpha = 2(n+1)$, $*$ being the Hodge star operator mapping p-forms to $(n+1-p)$-forms (see, e.g., [22]). The 1-form (3.1) can also be detected as the real part of the formal Clifford product $f \cdot dx$:

$$f \equiv \mathrm{Re}(f \cdot dx) = \mathrm{Re}((f_0 + \sum_{i=1}^{n} f_i e_i)(dx_0 + \sum_{i=1}^{n} dx_i e_i)).$$

Locally, hypermonogenic functions are given by the existence of a *hyperbolically harmonic* function h, i.e., $\Delta_{\mathrm{LB}} h = 0$ and $\frac{\partial h}{\partial x_0} = f_0$, $\frac{\partial h}{\partial x_i} = -f_i$ $(i = 1, \dots, n)$, where

$$\Delta_{\mathrm{LB}} := x_n^2 \Delta - (n-1) x_n \frac{\partial}{\partial x_n}$$

is the Laplace-Beltrami operator associated to the Poincarè metric. The co-ordinate functions of a hypermonogenic function are eigenfunctions of Δ_{LB}.

In contrast to the solutions of the Riesz system, elementary functions in terms of paravector powers are among the hypermonogenic ones: x^n, for all $n \in \mathbb{Z}$, is hypermonogenic; this makes it possible to define hypermonogenic power series with real coefficients such as $\exp(x) := \sum\limits_{i=0}^{\infty} \frac{x^i}{i!}$. As partial derivatives of hypermonogenic functions are also hypermonogenic (if not operating with respect to x_n), the class of *elementary* polynomials of type

$$L_m^k(x) := \frac{1}{k!} \frac{\partial^{|k|}}{\partial x_1^{k_1} \cdots \partial x_{m-1}^{k_{m-1}}} (x^{m+|k|}),$$

$$k = (k_1, \ldots, k_{n-1}), \quad |k| = \sum_{i=1}^{n-1} k_i, \quad k! = k_1! \cdots k_{n-1}!$$

is hypermonogenic. For even n these build a *basis* for all hypermonogenic homogeneous polynomials of degree m (cf. [4], [12]).

More information about this subject can be found in [5], [9]–[17].

4 The 'radial part' of $\overline{\partial}$

Fix the real part x_0 of the paravector variable $x = x_0 + \sum\limits_{i=1}^{n} x_i e_i$ and take spherical coordinates with respect to the vector part $\sum\limits_{i=1}^{n} x_i e_i$, setting

$$\rho := \sqrt{\sum_{i=1}^{n} x_i^2} \quad \text{and} \quad I := \frac{1}{\rho} \sum_{i=1}^{n} x_i e_i. \qquad (4.1)$$

Then $\overline{\partial}$ admits the representation

$$\overline{\partial} = \overline{\partial}_{\mathrm{rad}} - \frac{I}{\rho} \partial_{\mathrm{sph}} \quad (\rho \neq 0),$$

where

- $\partial_{\mathrm{sph}} := \sum\limits_{1 \leq i < j \leq n} e_i e_j (x_i \frac{\partial}{\partial x_j} - x_j \frac{\partial}{\partial x_i})$ is the *spherical* Cauchy-Riemann operator (as introduced in [1])

- $\overline{\partial}_{\mathrm{rad}} := \frac{\partial}{\partial x_0} + \frac{I}{\rho} \sum\limits_{i=1}^{n} x_i \frac{\partial}{\partial x_i}$ will be called the *radial* Cauchy-Riemann operator.

Proof.

$$\frac{I}{\rho}\left(\sum_{i=1}^{n} x_i \frac{\partial}{\partial x_i} - \partial_{\mathrm{sph}}\right)$$

$$= \frac{1}{\rho^2} \sum_{\ell=1}^{n} \left[e_\ell x_\ell \left(\sum_{j=1}^{n} x_j \frac{\partial}{\partial x_j}\right) + \sum_{i=1}^{\ell-1} e_\ell \left(x_i^2 \frac{\partial}{\partial x_\ell} - x_\ell x_i \frac{\partial}{\partial x_i}\right) - \right.$$

$$\left. \sum_{i=\ell+1}^{n} e_\ell \left(x_i x_\ell \frac{\partial}{\partial x_i} - x_i^2 \frac{\partial}{\partial x_\ell}\right) \right] + R$$

$$= \sum_{\ell=1}^{n} e_\ell \frac{\partial}{\partial x_\ell} + R = \partial - \frac{\partial}{\partial x_0} + R$$

where $R = 0$ because $\rho^2 R = \sum_{i=1}^{n} \sum_{\substack{j<k, \\ i\neq j, i\neq k}} e_i e_j e_k \left(x_i x_j \frac{\partial}{\partial x_k} - x_i x_k \frac{\partial}{\partial x_j}\right) = 0$. □

An analogous construction is possible by taking the radial part with respect to the $(n+1-k)$-dimensional subspace $\mathrm{span}(e_k, \ldots, e_n)$ of $\mathbb{R}^{n+1}$, fixing $x_0, \ldots, x_{k-1}$. Therefore exchange ρ and I by

$$\rho_k := \sqrt{\sum_{i=k}^{n} x_i^2} \quad \text{and} \quad I_k := \frac{1}{\rho_k} \sum_{i=k}^{n} x_i e_i. \tag{4.2}$$

Then $\bar{\partial} = \bar{\partial}_{\mathrm{rad},k} - \frac{I_k}{\rho_k} \partial_{\mathrm{sph},k}$ where

- $\partial_{\mathrm{sph},k} := \sum_{k \leq i < j \leq n} e_i e_j \left(x_i \frac{\partial}{\partial x_j} - x_j \frac{\partial}{\partial x_i}\right)$

- $\bar{\partial}_{\mathrm{rad},k} := \frac{\partial}{\partial x_0} + \sum_{i=1}^{k-1} e_i \frac{\partial}{\partial x_i} + \frac{I_k}{\rho_k} \sum_{i=k}^{n} x_i \frac{\partial}{\partial x_i}$.

Obviously, $\bar{\partial}_{\mathrm{rad},1} = \bar{\partial}_{\mathrm{rad}}$.

5 Fueter-type functions

Let $\tilde{U} \subseteq \mathbb{C}$ be an open set and $\tilde{f} \colon \tilde{U} \to \mathbb{C}$ be holomorphic in $\tilde{U}$, $\tilde{f} = u_0 + i u_1$, $u_0, u_1 \colon \tilde{U} \to \mathbb{R}$. Associate to $\tilde{f}$ the axially symmetric function

$$f \colon U \to \mathbb{R}^{n+1}, \quad f(x) = u_0(x_0, \rho) + I\, u_1(x_0, \rho),$$

where $U := \{x \in \mathbb{R}^{n+1} \mid (x_0, \rho) \in \tilde{U}\}$. Fueter proved in [6] that Δf is a monogenic function if $n = 3$, $\Delta := \sum_{i=0}^{3} \frac{\partial^2}{\partial x_i^2}$ being the Laplace operator in $\mathbb{R}^4$. The same function f can be used to construct hypermonogenic functions out of holomorphic ones if restricting to upper half space $\mathbb{R}^{n+1}_+$ and $\mathbb{C}_+$, respectively (cf. Leutwiler [15]):

$$\tilde{f} \text{ holomorphic in } \tilde{U} \subset \mathbb{C}_+ \;\Leftrightarrow\; f \text{ hypermonogenic in } U \subset \mathbb{R}^{n+1}_+.$$

We call these axially symmetric functions f to be of Fueter-type. For example, the complex powers $\tilde{f}(z) = z^n$ are transformed to paravector powers $f(x) = x^n$. In this case, the domain of definition can be extended to the whole of $\mathbb{C}$ and $\mathbb{R}^{n+1}$, respectively.

The connection to the Cauchy-Riemann operator is given in

Theorem 1. *For a differentiable Fueter-type function $f = u_0 + Iu_1$, defined in a domain in $\mathbb{R}^{n+1}$ where $\rho > 0$, the following are equivalent:*

(i) $\overline{\partial}_{\mathrm{rad}} f = 0$

(ii) $\frac{\partial u_0}{\partial x_0} = \frac{\partial u_1}{\partial \rho}$, $\quad \frac{\partial u_0}{\partial \rho} = -\frac{\partial u_1}{\partial x_0}$, $\quad$ *i.e., $\tilde{f} = u_0 + iu_1$ is holomorphic*

(iii) $\rho \overline{\partial} f + I \partial_{\mathrm{sph}} f = 0$

(iv) $\rho \overline{\partial} f + (n-1) u_1 = 0$;

if the domain is contained in the set $\{x \in \mathbb{R}^{n+1} \mid x_\ell \neq 0\}$ for some $\ell \in \{1, \ldots, n\}$, then additionally equivalent is

(v) $x_\ell (\frac{\partial f_0}{\partial x_0} - \sum_{i=1}^{n} \frac{\partial f_i}{\partial x_i}) + (n-1) f_\ell = 0$, $\quad \frac{\partial f_0}{\partial x_i} = -\frac{\partial f_i}{\partial x_0}$, $\quad \frac{\partial f_i}{\partial x_j} = \frac{\partial f_j}{\partial x_i}$,

where f is written out in coordinates: $f = f_0 + \sum_{i=1}^{n} f_i e_i$, i.e. $f_0 = u_0$, $f_i = \frac{x_i}{\rho} u_1$ $(i = 1, \ldots, n)$, according to the definition of I in (4.1).

Proof. (i) $\Leftrightarrow$ (ii):

We have $\frac{1}{\rho} \sum_{i=1}^{n} x_i \frac{\partial f}{\partial x_i} = \frac{\partial u_0}{\partial \rho} + I \frac{\partial u_1}{\partial \rho}$ from $\frac{\partial I}{\partial x_i} = \frac{e_i}{\rho} - \frac{x_i I}{\rho^2}$ $(i = 1, \ldots, n)$.

Thus,

$$\overline{\partial}_{\mathrm{rad}} f = 0 \;\Leftrightarrow\; \frac{\partial u_0}{\partial x_0} + I \frac{\partial u_1}{\partial x_0} + I(\frac{\partial u_0}{\partial \rho} + I \frac{\partial u_1}{\partial \rho}) = 0$$

$$\overset{I^2 = -1}{\Longleftrightarrow} \frac{\partial u_0}{\partial x_0} = \frac{\partial u_1}{\partial \rho}, \quad \frac{\partial u_0}{\partial \rho} = -\frac{\partial u_1}{\partial x_0}.$$

(i) $\Leftrightarrow$ (iii): It is just the definition of $\overline{\partial}_{\mathrm{rad}}$ and ∂_{sph}.

(iii) $\Leftrightarrow$ (iv): follows from

$$\partial_{\mathrm{sph}} I = \sum_{i<j} e_i e_j (x_i e_j - x_j e_i) = -\frac{1}{\rho} \sum_{i<j} (x_i e_i + x_j e_j) = -(n-1) I$$

and $\partial_{\mathrm{sph}} u_0 = 0 = \partial_{\mathrm{sph}} u_1$;

(iv) $\Leftrightarrow$ (v): It is also straightforward using the definition of f_i, $(i = 0, \ldots, n)$. $\qquad \square$

Condition (v) coincides with the definition of hypermonogenic functions for the case $\ell = n$ and a domain in $\mathbb{R}^{n+1}_+$.

Remark 1. *For a twice differentiable Fueter-function f fulfilling Theorem 1, we have*

(a) $\rho^2 \Delta f = (n-1) \sum_{i=1}^{n} x_i \frac{\partial f}{\partial x_i} - (n-1) I u_1,$

(b) $\rho \Delta f_0 = (n-1) \frac{\partial f_0}{\partial \rho},$

(c) $x_\ell \Delta f_i = (n-1) \frac{\partial f_i}{\partial x_\ell}$ *if* $i \neq \ell,$

(d) $x_\ell^2 \Delta f_\ell = (n-1) x_\ell \frac{\partial f_\ell}{\partial x_\ell} - (n-1) f_\ell,$

i.e., the coordinate functions of f are eigenfunctions of a Laplace-Beltrami equation (as indicated in [20]).

More generally, we consider paravector functions of the form

$$f = u_0 + \sum_{i=1}^{k-1} u_i e_i + I_k u_k,$$

where $u_i = u_i(x_0, x_1, \ldots, x_{k-1}, \rho_k)$, $(i = 0, 1, \ldots, k)$ and ρ_k, I_k are defined by (4.2). Write $\tilde{f} = u_0 + \sum_{i=1}^{k} u_i e_i$, $u_i = u_i(x_0, \ldots, x_k)$; then

$\tilde{f}$ hypermonogenic in $\mathbb{R}^{k+1}$ $\Leftrightarrow$ f hypermonogenic in $\mathbb{R}^{n+1}$.

The corresponding result is

Theorem 2. *Let $k \in \{1, \ldots, n\}$. For a differentiable Fueter-type function*

$$f = u_0 + \sum_{i=1}^{k-1} u_i e_i + I_k u_k =: f_0 + \sum_{i=1}^{n} f_i e_i,$$

defined in a domain of $\mathbb{R}^{n+1}$ where $\rho_k \neq 0$, the following are equivalent:

(i) $\overline{\partial}_{\mathrm{rad},k} f = 0$

(ii) $\begin{cases} \frac{\partial u_0}{\partial x_0} - \sum_{i=1}^{k-1} \frac{\partial u_i}{\partial x_i} - \frac{\partial u_k}{\partial \rho_k} = 0 \\[2mm] \frac{\partial u_0}{\partial x_i} = -\frac{\partial u_i}{\partial x_0} \quad (i = 1, \ldots, k-1), \quad \frac{\partial u_0}{\partial \rho_k} = -\frac{\partial u_k}{\partial x_0} \\[2mm] \frac{\partial u_i}{\partial x_j} = \frac{\partial u_j}{\partial x_i}, \quad \frac{\partial u_i}{\partial \rho_k} = \frac{\partial u_k}{\partial x_i} \quad (i, j = 1, \ldots, k-1) \end{cases}$

(iii) $\rho_k \overline{\partial} f + I_k \partial_{\mathrm{sph},k} f = 0$

(iv) $\rho_k \overline{\partial} f + (n-k) u_k = 0;$

if the domain is contained in $\{x_\ell \neq 0\}$ for some $\ell \in \{k, \ldots, n\}$, then additionally equivalent is

(v) $x_\ell\left(\frac{\partial f_0}{\partial x_0} - \sum_{i=1}^{n} \frac{\partial f_i}{\partial x_i}\right) + (n-k)f_\ell = 0 \quad \frac{\partial f_0}{\partial x_i} = -\frac{\partial f_i}{\partial x_0}, \quad \frac{\partial f_i}{\partial x_j} = \frac{\partial f_j}{\partial x_i}.$ □

6   Interpretation with Dirac operators

We start with an $(n+1)$-dimensional Riemannian manifold (M, g_{ij}) and a local basis $(\frac{\partial}{\partial x_0}, \ldots, \frac{\partial}{\partial x_n})$ of the tangential space. Orthonormalized we get the basis $e^0, \ldots, e^n$ given by $e^i := \sum_{j=0}^{n} \sqrt{g^{ij}} \frac{\partial}{\partial x_j}$ $(i = 0, \ldots, n)$. Then the Dirac operator is defined by

$$D := \sum_{i=0}^{n} e^i \nabla_{e^i},$$

where ∇_{e^i} is the covariant derivative in direction e^i, explicitly (for $f = f_0 \frac{\partial}{\partial x_0} + \sum_{i=1}^{n} f_i \frac{\partial}{\partial x_i}$)

$$\nabla_{\frac{\partial}{\partial x_k}} f = \sum_{i=0}^{n} \left(\frac{\partial f_i}{\partial x_k} + \sum_{j=0}^{n} \Gamma^i_{kj} f_j\right) \frac{\partial}{\partial x_i},$$

and the multiplication is done in the (standard) Clifford algebra C_{n+1} generated by $e^0, \ldots, e^n$. Our interest lies especially in the (diagonal) metric defined via

$$g_{ij} = \frac{\delta_{ij}}{\alpha}, \quad \alpha = \alpha(x) > 0, \text{ e.g.,}$$

- $\alpha = 1$: Euclidean $(M = \mathbb{R}^{n+1})$

- $\alpha = x_n^2$: Poincarè $(M = \mathbb{R}_+^{n+1})$

- $\alpha = \rho^2$: radial (e.g. $M = \mathbb{R}_+^{n+1}$).

D operates on vector fields; having hypermonogenic functions in mind we want to calculate Df for the vector field corresponding to the 1-form (3.1). As the isomorphism between tangential and cotangential space is given by

$$\sqrt{\alpha}\frac{\partial}{\partial x_i} \mapsto \frac{1}{\sqrt{\alpha}}dx_i \quad (i = 0, \ldots, n),$$

we look at the vector field

$$f \equiv \alpha f_0 \frac{\partial}{\partial x_0} - \alpha \sum_{i=1}^{n} f_i \frac{\partial}{\partial x_i} = \sqrt{\alpha}f_0 e^0 - \sqrt{\alpha} \sum_{i=1}^{n} f_i e^i.$$

Then

$$
\begin{aligned}
Df &= \sum_{i=0}^{n} e^i \nabla_{e^i} f = \sum_{i=0}^{n} \sqrt{\alpha} \tfrac{\partial}{\partial x_i} \sqrt{\alpha} \nabla_{\tfrac{\partial}{\partial x_i}} f \\
&= -\alpha \tfrac{\partial f_0}{\partial x_0} - \tfrac{\partial \alpha}{\partial x_0} f_0 + \sum_{i=1}^{n} (\alpha \tfrac{\partial f_i}{\partial x_i} + \tfrac{\partial \alpha}{\partial x_i} f_i) \\
&\quad + \tfrac{n+1}{2} (-\tfrac{\partial \alpha}{\partial x_0} f_0 + \sum_{i=1}^{n} \tfrac{\partial \alpha}{\partial x_i} f_i) + \alpha \sum_{i=1}^{n} e^0 e^i (\tfrac{\partial f_0}{\partial x_i} + \tfrac{\partial f_i}{\partial x_0}) \\
&\quad + \alpha \sum_{1 \le i < j \le n} e^i e^j (\tfrac{\partial f_j}{\partial x_i} - \tfrac{\partial f_j}{\partial x_i});
\end{aligned}
$$

hence, $Df = 0$ if and only if

$$
\alpha(\tfrac{\partial f_0}{\partial x_0} - \sum_{i=1}^{n} \tfrac{\partial f_i}{\partial x_i}) + \tfrac{n-1}{2}(-\tfrac{\partial \alpha}{\partial x_0} f_0 + \sum_{i=1}^{n} \tfrac{\partial \alpha}{\partial x_i} f_i) = 0
$$
$$
\tfrac{\partial f_0}{\partial x_i} = -\tfrac{\partial f_i}{\partial x_0}, \quad \tfrac{\partial f_i}{\partial x_j} = \tfrac{\partial f_j}{\partial x_i} \quad (i, j = 1, \dots, n)
$$

Regarding our special cases we get:

- if $\alpha = 1$: f monogenic
- if $\alpha = x_n^2$: f hypermonogenic
- if $\alpha = \rho^2$: first equation becomes

$$
\rho^2 (\tfrac{\partial f_0}{\partial x_0} - \sum_{i=1}^{n} \tfrac{\partial f_i}{\partial x_i}) + (n-1) \sum_{i=1}^{n} x_i f_i = 0;
$$

taking $f = u_0 + I u_1$, $u_i = u_i(x_0, \rho)$ results in

$$
\sum_{i=1}^{n} x_i f_i = \sum_{i=1}^{n} x_i \cdot \tfrac{x_i}{\rho} u_1 = \rho u_1;
$$

thus $\rho(\tfrac{\partial f_0}{\partial x_0} - \sum_{i=1}^{n} \tfrac{\partial f_i}{\partial x_i}) + (n-1)u_1 = 0$; hence, Fueter-type functions (for $k = 1$) are simultaneous solutions of the Dirac equations for $\alpha = \rho^2$ and $\alpha = x_n^2$.

- $\alpha = \rho_k^2$: $\rho_k^2(\tfrac{\partial f_0}{\partial x_0} - \sum_{i=1}^{n} \tfrac{\partial f_i}{\partial x_i}) + (n-1)\sum_{i=k}^{n} x_i f_i = 0$; taking $f = u_0 + \sum_{i=1}^{k-1} u_i e_i + I_k u_k$ yields $\sum_{i=k}^{n} x_i f_i = \rho_k u_k$, implying

$$
\rho_k (\frac{\partial f_0}{\partial x_0} - \sum_{i=1}^{n} \frac{\partial f_i}{\partial x_i}) + (n-1)u_k = 0,
$$

which says that these Fueter-type functions are simultaneous solutions of the Dirac equations for $\alpha = \rho_k^2$ and $\alpha = x_n^2$.

Of course these properties mimic the coordinate change $x_n \longleftrightarrow \rho$.

Another rather trivial example: The simultaneous solutions of the Dirac equations for $\alpha = 1$ and $\alpha = x_n^2$ are exactly the paravector functions $f = f_0 + \sum_{i=1}^{n} f_i e_i$ with vanishing f_n.

REFERENCES

[1] F. Brackx, R. Delanghe, and F. Sommen, *Clifford Analysis*, Pitman, London, 1983.

[2] P. Cerejeiras, Decomposition of analytic hyperbolically harmonic functions, *Proc. of the 4th Conf. on Clifford Algebras and Their Appl. in Math. Phys.*, Aachen, Germany, Kluwer, May 1996, 45–52.

[3] J. Cnops, Hurwitz pairs and applications of Möbius transformations, Habilitation Thesis, Univ. Gent., 1994.

[4] S. -L. Ericsson-Bisque, Comparison of quaternionic analysis and modified Clifford analysis, *Dirac Operators in Analysis*, J. Ryan and D. Struppa, eds., Pitman Research Notes in Math. **394** (1998), 109–121.

[5] S. -L. Ericsson-Bique and H. Leutwiler, On modified quaternionic analysis in $\mathbb{R}^3$, *Arch. Math.* **70** No. 3, (1998), 228–234.

[6] R. Fueter, Die funktionentheorie der differentialgleichungen $\Delta u = 0$ und $\Delta\Delta u = 0$ mit vier reellen Variablen, *Comment. Math. Helv.* No. 7, (1934/35), 307–330.

[7] J. E. Gilbert and M. A. M. Murray, *Clifford Algebras and Dirac Operators in Harmonic Analysis*, Cambridge University Press, 1991.

[8] K. Gürlebeck and W. Sprössig, *Quaternionic and Clifford Calculus for Physicists and Engineers*, J. Wiley, 1996.

[9] Th. Hempfling, Quaternionale analysis in $\mathbb{R}^4$, Diplomarbeit, Univ. Erlangen-Nürnberg, February 1993.

[10] Th. Hempfling, Aspects of modified Clifford analysis, symposium *Analytical and Numerical Methods in Quaternionic and Clifford Analysis*, K. Gürlebeck and W. Sprössig, eds., Seiffen, June 1996, 49–59.

[11] Th. Hempfling, Multinomials in modified Clifford analysis, *C. R. Math. Rep. Acad. Sci.* **(2,3)** No. 18, Canada, 1996, 99–102.

[12] Th. Hempfling, Beiträge zur modifizierten Clifford analysis, Ph.D. Thesis, Univ. Erlangen-Nürnberg, August 1997.

[13] Th. Hempfling, The Dirac operator in $\mathbb{R}^{d+1}_+$ with hyperbolic metric and modified Clifford analysis, in *Dirac Operators in Analysis*, J. Ryan and D. Struppa, eds., Pitman Research Notes in Math. 394, 1998, 95–108.

[14] Th. Hempfling and H. Leutwiler, Modified quaternionic analysis in $\mathbb{R}^4$, *Proc. of the 4th Conf. on Clifford Algebras and Their Appl. in Math. Phys.*, Aachen, Germany, Kluwer, May 1996, 227–238.

[15] H. Leutwiler, Modified Clifford analysis, *Complex Variables Theory Appl.* No. 17 (1992), 153–171.

[16] H. Leutwiler, Modified quaternionic analysis in $\mathbb{R}^3$, *Complex Variables Theory Appl.* No. 20 (1992), 19–51.

[17] H. Leutwiler, Rudiments of a function theory in $\mathbb{R}^3$, *Exposition. Math.* No. 14 (1996), 97–123.

[18] M. Riesz, Sur les fonctions conjugées, *Math. Z.* **27** (1927), 218–244.

[19] M. Riesz, Clifford numbers and spinors, *Lecture Series* No. 38, Institute for Physical Science and Technology, Maryland, 1958.

[20] W. Sprössig, On operators and elementary functions in Clifford analysis, *ZAA,* **18** No. 2 (1999), 349-360.

[21] E. M. Stein and G. Weiss, Generalization of the Cauchy-Riemann equations and representations of the rotation group, *Amer. J. Math.* **90** (1968), 163–196.

[22] F. W. Warner, *Foundations of Differentiable Manifolds and Lie Groups*, Scott, Foresman and Co., Glenview, Illinois, London, 1971.

Thomas Hempfling
Institute for Applied Mathematics I
Freiberg University of Mining and Technology
Bernhard-von-Cotta Str. 2
D-09596 Freiberg
Germany
E-mail: hempfl@math.tu-freiberg.de

Received: September 28, 1999; Revised: February 14, 2000

Hypercomplex Derivability – The Characterization of Monogenic Functions in $\mathbb{R}^{n+1}$ by Their Derivative

Helmuth R. Malonek

ABSTRACT From the concept of hypercomplex differentiability (introduced at the end of the 1980's ([6], [7]) by using local linear approximation properties of monogenic functions) the existence of a monogenic derivative does not directly follow. We show that if some relation between higher order differential forms are introduced then, (as in the complex case) the conjugated Cauchy-Riemann operator again gives the monogenic derivative of a monogenic function in $\mathbb{R}^{n+1}$.
Keywords: Monogenic functions, hypercomplex derivative, hypercomplex differential forms.

1 Introduction

We are interested in defining the derivative of a monogenic function in $\mathbb{R}^{n+1}$ in an appropriate way. The definition and main properties of monogenic functions in terms of null-solutions of a generalized Cauchy-Riemann operator can be found in [1]. These main properties are analogous to that of holomorphic functions in $\mathbb{C} \cong \mathbb{R}^2$. Our main tool is the application of special hypercomplex differential forms. For that purpose we analyze relations between the derivative and the differential of a complex-valued function f of one complex variable. In particular, we consider the form of the differential df under the influence of special types of complexifications of $\mathbb{R}^2$ which are different from the usual ones (and in such a form almost never used). But, as it sometimes happens, old and approved tools can still hide some lesser known but, nevertheless, very useful meaning. We will show that this is also the case with the representation of the differential df by Wirtinger's partial complex derivatives. In our case we had to have in view its capability of adaptation to higher dimensional extensions by using Clifford algebras. In our opinion, the explanation of basic relations in $\mathbb{C}$ from

AMS Subject Classification: 30G35.

different points of view seems to be a good introduction to the problem that we are going to attack.

First, let Ω be a domain in the complex plane $\mathbb{C}$. With the help of Wirtinger's formal complex partial derivatives

$$f_z := \frac{\partial f}{\partial z} = \frac{1}{2}\left(\frac{\partial f}{\partial x_0} - i\frac{\partial f}{\partial x_1}\right) \text{ and } f_{\bar{z}} := \frac{\partial f}{\partial \bar{z}} = \frac{1}{2}\left(\frac{\partial f}{\partial x_0} + i\frac{\partial f}{\partial x_1}\right), \quad z = x_0 + x_1,$$

the differential

$$df = \frac{\partial f}{\partial x_0}dx_0 + \frac{\partial f}{\partial x_1}dx_1 \tag{1.1}$$

of a real differentiable function $f : \Omega \mapsto \mathbb{C}$ in a neighborhood of $z \in \Omega$ can be rearranged in the form

$$df = f_z dz + f_{\bar{z}} d\bar{z}, \tag{1.2}$$

which corresponds to choosing the complex variables z and $\bar{z}$. In the case of a holomorphic function $f = f(z)$ the Cauchy-Riemann equations can be expressed in complex form by $f_{\bar{z}} = 0$. In this case the formal partial complex derivative f_z coincides with the ordinary complex derivative, which is defined by the limit of the corresponding difference quotient, i. e.,

$$f_z = \frac{df}{dz} = f'(z) = \lim_{\Delta z \to 0} \frac{f(z + \Delta z) - f(z)}{\Delta z}.$$

Obviously, in these terms the differential of a holomorphic function is simply given by

$$df = f'(z)dz. \tag{1.3}$$

Along the same line of reasoning the differential of a holomorphic function of several complex variables $z_1, \ldots, z_n$, $n > 1$, is obtained by

$$df = f_{z_1}dz_1 + \ldots + f_{z_n}dz_n. \tag{1.4}$$

Taking into account that now in formula (1.4) the partial derivatives are defined in the ordinary way as partial differential quotients, this formula looks like the most natural extension of (1.3) to higher dimensions.

We are concerned with other approaches to higher dimensional generalizations of differentiability and derivability. They are different from the use of several complex variables and avoid, for instance, the limitation to an even (real) dimension as happens in the $\mathbb{C}^n$ case. The first non-usual approach to the representation of the differential of a holomorphic function can be described in the following way. Using the Cauchy-Riemann differential operator D defined by

$$D := 2\frac{\partial}{\partial \bar{z}} = \frac{\partial}{\partial x_0} + i\frac{\partial}{\partial x_1}$$

and the variables

$$z_0 := x_0 \text{ and } z_1 := -iz = x_1 - ix_0$$

instead of z and $\bar{z}$, it is possible to rearrange the differential (1.1) of f simply as

$$\begin{aligned} df &= f_{x_0} dx_0 + f_{x_1} dx_1 \\ &= (f_{x_0} + if_{x_1})dx_0 + f_{x_1}(dx_1 - idx_0) \\ &= Df dz_0 + f_{x_1} dz_1. \end{aligned} \tag{1.5}$$

In the holomorphic case, i.e., if $Dw = 2f_{\bar{z}} = 0$, formula (1.5) reduces to

$$df = f_{x_1} dz_1. \tag{1.6}$$

The fact that for a holomorphic function it holds that $f_{x_1} = i\frac{df}{dz}$ again leads, together with $idz_1 = dz$, to formula (1.3). In [9] it is shown that formula (1.5) allows an extension to higher dimensions in the framework of hypercomplex differentiability. This is done with the help of a calculus of alternating differential forms based on the approach developed in [7].

It is worth noticing that, though hypercomplex differentiability and hypercomplex derivability in $\mathbb{R}^{n+1}$ define the same class of functions, the corresponding properties have their expression in relations of different kind. Whereas differentiability (with its connection to linearization) relates to a differential form of first degree, the property of derivability (as possessing a well-defined derived monogenic function) relates to a certain differential form of degree n (Section 3). If the values of coincide, i.e., if $n = 1$ (the complex case), then we should expect that concepts relate to the same form of degree 1 and this should be df. Nevertheless, we will see that from the viewpoint of codimension 1 in $\mathbb{R}^2$, it makes sense to consider still a second and slightly modified approach. In fact, the representation of the differential df of a holomorphic function can be also derived from some combination of (1.2) and (1.5). More exactly, it is a modified form of df written with respect to $z_1 = -iz$ and $\overline{z_1} = i\bar{z}$. The easiest way to obtain the desired form is by multiplying equation (1.2) by $-i$. We obtain

$$\begin{aligned} d(-if) &= f_z d(-iz) + f_{\bar{z}} d(-i\bar{z}), \\ &= f_z dz_1 - f_{\bar{z}} d\overline{z_1} \\ &= (\frac{1}{2}\overline{D}f)dz_1 - (\frac{1}{2}Df)d\overline{z_1}. \end{aligned} \tag{1.7}$$

For the sake of symmetry, we have used the notation

$$\overline{D} := 2\frac{\partial}{\partial z} = \frac{\partial}{\partial x_0} - i\frac{\partial}{\partial x_1}$$

for the conjugate Cauchy-Riemann operator. Now, in the case of holomorphy, we get

$$d(-if) = (\frac{1}{2}\overline{D}f)dz_1. \qquad (1.8)$$

The presence of the factor $-i$ will become clear in Section 3 when we arrive at the general case of $n > 1$. From the complex point of view it means nothing more than a change to new dependent and independent variables. They are consequences of rotations in the original and in the image plane by $-\frac{\pi}{2}$. It is obvious, that such a situation does not change any relative geometrical relation, which is expressed, for instance, in the conformal mapping property of holomorphic functions.

Resuming the different viewpoints with respect to the differential df of a real differentiable function $f = f(z)$ expressed by (1.2) and (1.7) and comparing with the corresponding formulae (1.3) resp. (1.8) (with the additional assumption of holomorphy $Dw = 2f_{\overline{z}} = 0$), we notice the following. The derivative $f'(z) = \frac{1}{2}\overline{D}f$ appears as a *differential coefficient* (a notion which was used also in [2] for a similar situation) between two forms of degree 1, i.e.,

$$(i) \quad \text{between} \qquad df \quad \text{and} \quad dz,$$
$$(ii) \quad \text{between} \quad d(-if) \quad \text{and} \quad dz_1.$$

In Section 3 we will see that the second relation (1) has exactly the desired counterpart in higher dimensions. It permits us to define monogenic functions by the existence of the hypercomplex derivative in an appropriate way.

On the contrary, the formal generalization of the first relation leads only to the small class of affine linear functions as regular functions (as is shown by [10] and [16] for the quaternionic case). The first indication of such a possibility of defining the derivative as a *differential coefficient* between two forms of higher degree than 1 was given by [16] in the case of quaternionic analysis (corresponding to $n = 3$). Unfortunately this approach to hypercomplex differential forms seems to work only for quaternions.[1] Due to the role of differential forms of degree $(n - 1)$ and n, our explanation makes use of the Hodge star operator. This operator has been already applied to hypercomplex differential forms by Habetha [5].

---

[1] After submitting this paper the author has been informed by a referee that in [12] Ryan describes a generalization of Sudbery's method of non-alternating differential forms to arbitrary dimensions. Therefore, less elementary tensor calculus methods in combination with the Hodge star operator can be employed. The author is very grateful for this reference. However he would like to notice that since [12] mainly deals with generalized elliptic functions the paper neither mentions the conjugate Cauchy-Riemann operator nor its interpretation referring to a monogenic derivative.

2   Notation and basic relations

As usual let $\{e_1,\dots,e_n\}$ be an orthonormal base of the Euclidean vector space $\mathbb{R}^n$ and $C\ell_{0,n}$ the 2^n-dimensional universal Clifford algebra over $\mathbb{R}$ according to the multiplication rules

$$e_k e_l + e_l e_k = -2\delta_{kl}e_0, \quad k,l = 1,\dots,n,$$

with δ_{kl} the Kronecker symbol. The set $\{e_A : A \subseteq \{1,\dots,n\}\}$, with $e_A = e_{h_1}e_{h_2}\dots e_{h_r}, 1 \le h_1 < \dots < h_r \le n, \ e_\emptyset = e_0 = 1$, is a basis of $C\ell_{0,n}$. Therefore, each element $\alpha \in C\ell_{0,n}$ can be represented in the form

$$\alpha = \sum_A \alpha_A e_A,$$

where α_A are real numbers. The conjugate element to α is defined by

$$\bar{\alpha} = \sum_A \alpha_A \bar{e}_A,$$

where $\bar{e}_A = \bar{e}_{h_r}\bar{e}_{h_{r-1}}\dots\bar{e}_{h_1}; \bar{e}_k = -e_k \ (k = 1,\dots,n), \ \bar{e}_0 = e_0 = 1$.

Identifying each element $x = (x_0, x_1,\dots,x_n)$ of $\mathbb{R}^{n+1}$ with

$$z = x_0 + x_1 e_1 + \dots + x_n e_n \in \mathcal{A} := \operatorname{span}_{\mathbb{R}}\{1, e_1,\dots,e_n\}$$

the conjugate to z is given by

$$\bar{z} = x_0 - \sum_{k=1}^{n} x_k e_k.$$

We distinguish between the scalar part

$$\operatorname{Sc} z := \frac{1}{2}(z + \bar{z})$$

and the vector part

$$\operatorname{Vec} z := \frac{1}{2}(z - \bar{z})$$

of z. The norm of $z \in \mathcal{A}$ is $|z| := \sqrt{z\bar{z}}$. It immediately follows that each non-zero $z \in \mathcal{A}$ is invertible and its inverse is

$$z^{-1} = \frac{\bar{z}}{|z|^2}.$$

The starting point of hypercomplex function theory is $C\ell_{0,n}$-valued functions of the form

$$f(z) = \sum_A f_A(z)e_A, \quad f_A(z) \in \mathbb{R},$$

considered as mappings of the form

$$f : \Omega \subset \mathbb{R}^{n+1} \cong \mathcal{A} \mapsto C\ell_{0,n}. \tag{2.1}$$

The class of regular functions generalizing the holomorphic functions of complex analysis is defined as the kernel of the generalized Cauchy-Riemann operator in $\mathbb{R}^{n+1}, n \geq 1$, given by

$$D = \frac{\partial}{\partial x_0} + e_1 \frac{\partial}{\partial x_1} + \ldots + e_n \frac{\partial}{\partial x_n}. \tag{2.2}$$

Functions $f = f(z)$ satisfying the equation $Df = 0$ are called left monogenic functions, and solutions of $fD = 0$ are called right monogenic functions, respectively. In the following only left monogenic functions are considered. The treatment of right monogenic functions is identical.

Remark 1. *It is easy to see that the aforementioned definition of monogenic functions relates to a generalized Cauchy-Riemann system of (real) differential equations and needs no hypercomplex structure on the pre-image set. Another situation occurs if we are considering the notion of hypercomplex derivability. Its use for an equivalent definition of monogenic functions by the existence of a hypercomplex derivative depends on the structure of the pre-image set. Whereas the hypercomplex structure described by the isomorphism between $\mathbb{R}^{n+1}$ and $\mathcal{A}$ is sufficient for that purpose, other equivalent definitions are based on another dual structure. For example, the property of hypercomplex differentiability depends essentially on the use of n different hypercomplex variables of the form $z_k := x_k - x_0 e_k$ $(k = 1, \ldots, n)$. The same happens in the case of generalized convergent power series (see, for example, [7], [8].) For more details about applications of the hypercomplex derivative in connection with the calculus of generalized power series see [3].*

For our purpose the generalized conjugate Cauchy-Riemann operator is of particular importance. Therefore, the same symbol as in the complex case corresponding to $n = 1$ is used, i.e.,

$$\overline{D} = \frac{\partial}{\partial x_0} - e_1 \frac{\partial}{\partial x_1} - \ldots - e_n \frac{\partial}{\partial x_n}. \tag{2.3}$$

3   Hypercomplex derivability

Following the ideas suggested by the examples for the complex case, we will look for hypercomplex differential forms which are connected with the generalized Cauchy-Riemann operator and its conjugate. Our aim is to show that it is possible to define the hypercomplex derivative of a (left) monogenic function $f = f(z)$ as the (left) differential coefficient between two differential forms of degree n. In this way it will be confirmed that the formulae (1.7) and (1.8) are those which have the capability of being generalized to the case of an arbitrary $n \geq 1$. This statement includes the fact that we get the expression for the (left) derivative in the same form, namely, with the help of the conjugate Cauchy-Riemann operator $\overline{D}$.

To be short we will concentrate only on those hypercomplex differential forms that are needed for our purpose. We will try to explain the essential relations by using the most economic and suggestive way. Applications will not be included, but we provide references where they can be found. For the general background about differential forms in Clifford analysis we refer to the extensive papers of Sommen, [13] and [14]. Particular questions are treated, for instance, in [5] and [9]. In [16] a quaternionic calculus of (non-alternating) differential forms is considered, where the exterior product is defined in terms of quaternion multiplication.

We consider real differentiable functions as maps $f : \mathcal{A} \mapsto C\ell_{0,n}$; their differential at a point $z \in \mathcal{A}$ is then a $\mathbb{R}$-linear map $df : \mathcal{A} \mapsto C\ell_{0,n}$. This means, that we regard the differential as a $C\ell_{0,n}$-valued 1-form

$$df = \frac{\partial f}{\partial x_0} dx_0 + \frac{\partial f}{\partial x_1} dx_1 \cdots + \frac{\partial f}{\partial x_n} dx_n. \tag{3.1}$$

Other special differential forms which we will need for our purpose are the elements of the real vector space of $\mathcal{A}$-valued p-forms $\bigwedge_{\mathcal{A}}^{p}$, with a basis of real p-forms consisting of exterior products of the real 1-forms dx_k, $k = 0, \ldots, n$. For their definition the differential of the identity function $f(z) = z$ given by

$$dz = dx_0 + e_1 dx_1 \cdots + e_n dx_n \tag{3.2}$$

is an essential tool.

Remark 2. *However, we should point out that in general the algebraic relations in $\mathcal{A}$ cause problems in the construction of a hypercomplex function theory by the same methods as in the plane case. For instance, it is easy to see that neither the identity function $f(z) = z$ nor any of its powers are monogenic if $n > 1$. Obviously, this means that in order to develop an analogue of the Weierstrass approach one has to look for other functions which could serve as a generalization of the holomorphic powers z^n (see, for example, [7], [8]).*

Another important tool that we will apply to $\mathcal{A}$-valued forms is the well-known Hodge star-operator which is defined as a linear transformation between the pairs of spaces $\bigwedge_{\mathcal{A}}^{p}$ and $\bigwedge_{\mathcal{A}}^{(n+1)-p}$, i.e.,

$$* : \bigwedge_{\mathcal{A}}^{p} \longrightarrow \bigwedge_{\mathcal{A}}^{(n+1)-p} \qquad p = 0, \ldots, n+1,$$

in the following way. If

$$\omega_p = dx_{i_0} \wedge dx_{i_1} \wedge \ldots \wedge dx_{i_{p-1}}$$

is one of the basic p-forms, then

$$*(\omega_p) := \operatorname{sgn}(i_0, i_1, \dots, i_{n+1}) dx_{i_p} \wedge dx_{i_{p+1}} \wedge \dots \wedge dx_{i_{n+1}} \qquad (3.3)$$

is known as the adjoint (or dual) form of ω_p; for a general differential form $\omega \in \bigwedge_A^p$ the Hodge star operator is extended by linearity (cf. [5]).

The following overview of some significant differential forms is sufficient for the next steps towards the notion of hypercomplex derivability. We will also mention the corresponding complex forms ($n = 1$; and $e_1 \cong i$) to show how the general hypercomplex case mimics the relations used in the introductory explanation. Whenever possible, we also mention the formulae with respect to the pairs of all three types of variables introduced there, i.e., for $(z, \bar{z}), (z_0, z_1)$, and $(z_1, \overline{z_1})$.

1. Let $f \equiv 1 \in \bigwedge_A^0$ be the constant function. Duality yields the volume element in $\mathbb{R}^{n+1}$, i.e., the $(n+1)$-form

$$*1 = dV|_{\mathbb{R}^{n+1}} := dx_0 \wedge dx_1 \wedge \dots \wedge dx_n$$

For $n = 1$:

$$*1 = dx_0 \wedge dx_1 = \frac{1}{2} i dz \wedge d\bar{z} = dz_0 \wedge dz_1 = \frac{1}{2} i dz_1 \wedge d\overline{z_1}.$$

2. Let $dz = dx_0 + e_1 dx_1 + \dots + e_n dx_n$ be the differential of the identity in (3.2). By application of the Hodge star operator we obtain the hypercomplex surface-element in the usual way as the n-form

$$*dz = d\sigma_{(n)} := d\hat{x}_0 - e_1 d\hat{x}_1 + \dots + (-1)^n e_n d\hat{x}_n$$

where, as usual, the notation $d\hat{x}_m$ $(m = 1, \dots, n)$ means that in the ordered outer product of the 1-forms dx_k $(k = 0, \dots, n)$ the factor dx_m is absent. For $n = 1$ we have

$$*dz = dx_1 - i dx_0 = -i dz = dz_1.$$

3. Consider now

$$\operatorname{Vec} dz = e_1 dx_1 + \dots + e_n dx_n.$$

This 1-form can be considered as related to a hyperplane $x_0 = c = \text{const}$. In this sense, it can be considered as a 1-form in $\bigwedge_{A|_{x_0=c}}^1 \cong \bigwedge_{\mathbb{R}^n}^1$. With respect to the effective variables $(x_1, \dots, x_n)$ the corresponding volume-form is given by

$$dV|_{\mathbb{R}^n} = dx_1 \wedge \dots \wedge dx_n$$

(for $n = 1$ it reduces to $dV = dx_1$), and the corresponding surface-element is

$$* (\text{Vec } dz) = d\sigma_{(n-1)} :=$$
$$- e_1 d\hat{x}_{0,1} + e_2 d\hat{x}_{0,2} + \cdots + (-1)^n e_n d\hat{x}_{0,n}. \quad (3.4)$$

Here the Hodge star operator has been applied to $\bigwedge_{\mathbb{R}^n}^1$ and the notation $d\hat{x}_{0,m}$ $(m = 1, \ldots, n)$ means that in the ordered outer product of the 1-forms dx_k $(k = 0, \ldots, n)$ the factors dx_0 and dx_m are absent. For $n = 1$ we obtain the constant 0-form $d\sigma_{(0)} = -e_1 \cong -i$.

We also need the definition of the outer derivative of a differential form.

Let $\nu = (\nu_0, \nu_1, \ldots, \nu_{p-1})$, $0 \le \nu_0 < \nu_1 < \cdots < \nu_{p-1} \le n$, be a multi-index. Every basic p-form $\omega_p \in \bigwedge_A^p$ can be written in a unique way as

$$\omega_p = dx_{\nu_0} \wedge dx_{\nu_1} \wedge \ldots \wedge dx_{\nu_{p-1}} =: dx_\nu.$$

Let $f_\nu = f_\nu(z)$ be $\binom{n+1}{p}$ given $C\ell_{0,n}$-valued continuous functions. Then $\omega_p = \sum_\nu dz_\nu f_\nu(z)$, resp. $\omega_p = \sum_\nu f_\nu(z) dz_\nu$, are called left, resp. right, $C\ell_{0,n}$-valued p-forms.

The definition of the (outer) derivative of a (left) hypercomplex differential form is the same as in the real case.

Definition 1. *Let*

$$\omega_p = \sum_\nu dz_\nu f_\nu(z)(z)$$

be a continuously real differentiable left p-form on $\Omega \in \mathcal{A}$. Then the derivative $d\omega_p$ is defined as the $(p+1)$-form

$$d\omega_p = \sum_\nu (-1)^p dz_\nu \wedge df_\nu(z)$$

where df_ν is the differential (3.1) of f_ν, i.e., the outer derivative of the 0-form f_ν.

The treatment of right p-forms or p-forms with coefficients on sides is straightforward.

With the help of Definition 1 we can easily show the significance of special differential forms $d\sigma_{(n)}$ and $dV|_{\mathbb{R}^{n+1}}$. The (left) n-form $\omega = d\sigma_{(n)} f$ is closed if f is monogenic (i.e., $Df = 0$) because

$$dw = (-1)^n d\sigma_{(n)} \wedge df$$
$$= (-1)^n (d\hat{x}_0 - e_1 d\hat{x}_1 + \cdots + (-1)^n e_n d\hat{x}_n) \wedge$$
$$(dx_0 \frac{\partial f}{\partial x_0} + dx_1 \frac{\partial f}{\partial x_1} \cdots \cdots + dx_n \frac{\partial f}{\partial x_n})$$
$$= dV(Df) = 0.$$

With the help of Stokes' theorem this gives origin to a generalized Cauchy integral theorem in hypercomplex analysis (cf. [1]).

This is the usual, but not unique, example of the relevance and usefulness of the study of monogenic functions via hypercomplex differential forms. In [7], the details can be found for the characterization of monogenic functions by the property of hypercomplex differentiability which is related to differential 1-forms (and their special structure).

For $n = 1$ this gives the complete analogue to the complex case with

$$z_1 = x_1 - e_1 x_0 \cong -iz = y - ix, \ \frac{\partial}{\partial x_1} \cong 2i\frac{\partial}{\partial z},$$

(cf. the explanation to formulae (1.5) and (1.6)). For $n > 1$ this concept leads to the existence of a hypercomplex gradient which plays the role of the derivative of the considered function, and so we have in fact a *vector of monogenic functions* but not a *single function* as equivalent to a derivative. Nevertheless, looking at the complex case $n = 1$ we expect that $\overline{D}f$ should play a special role and be (at least in some sense) the hypercomplex derivative of a monogenic function.

As we have seen, the remarks about the formulae (1.7) and (1.8) are directed to a somehow different way of reasoning. Indeed, the main potential of those unusual formulae lies in the fact that the involved terms (on the left-hand side the negative imaginary unit and on the right-hand side dz_1 and its conjugate) have an interpretation as differential forms of degree $0 = n - 1$ and $1 = (n+1) - 1 = n$ (cf. points 2 and 3 in the list above). We will now show that the intrinsic relations between hypercomplex differential forms of degree n (i.e., $d\sigma_{(n)}$) and $(n-1)$ (i.e., $d\sigma_{(n-1)}$) are responsible for the positive answer to the question about the notion of hypercomplex derivability.

Definition 2. *A function* $f : \mathcal{A} \mapsto C\ell_{0,n}$ *is left derivable (or left regular) at* $z \in \mathcal{A}$ *if it is real differentiable at* z *and there exists* $A_{f,L}(z) \in C\ell_{0,n}$ *such that*

$$d(d\sigma_{(n-1)}f) = d\sigma_{(n)}A_{f,L}(z). \tag{3.5}$$

It is right derivable (or right regular) if there exists $A_{f,R}(z) \in C\ell_{0,n}$ *such that*

$$d(fd\sigma_{(n-1)}) = A_{f,R}(z)d\sigma_{(n)}. \tag{3.6}$$

$A_{f,L}(z)$, *respectively* $A_{f,R}(z)$, *are called the left, resp. right, derivative of* f *at* z.

Remark 3. *As mentioned in the introduction, Sudbery [16] used a similar approach for generalizing the concept of regularity (derivability) for quaternionic functions. Unfortunately, the chosen treatment of the involved* *quaternionic differential forms does not give hints for an elementary treatment of the general case [2] of monogenic functions* (see, for example, [3] for a more detailed analysis).

The equivalence between the notions of derivability and monogenicity of a function $f : \mathcal{A} \mapsto C\ell_{0,n}$ is stated in the following theorem. We restrict ourselves to the case of left derivability. The right case can be treated in the same way.

Theorem 1. *A real differentiable function f is left derivable (regular) at $z \in \mathcal{A}$ if and only if f is left monogenic at z, i.e., if it holds at z that*

$$Df = \frac{\partial f}{\partial x_0} + e_1 \frac{\partial f}{\partial x_1} + \cdots + e_n \frac{\partial f}{\partial x_n} = 0. \tag{3.7}$$

Proof. Let f be left derivable at z. Then it follows by Definition 2 that there exists $A_{f,L}(z) \in C\ell_{0,n}$ such that

$$d(d\sigma_{(n-1)}f) = d\sigma_{(n)}A_{f,L}(z). \tag{3.8}$$

But

$$d(d\sigma_{(n-1)}f)$$
$$= (-1)^{n-1}d\sigma_{(n-1)} \wedge df$$
$$= (-1)^{n-1}(-e_1 d\hat{x}_{0,1} + e_2 d\hat{x}_{0,2} + \cdots + (-1)^n e_n d\hat{x}_{0,n}) \wedge$$
$$\left(dx_0 \frac{\partial f}{\partial x_0} + dx_1 \frac{\partial f}{\partial x_1} \cdots + dx_n \frac{\partial f}{\partial x_n}\right)$$
$$= (-e_1 d\hat{x}_1 + e_2 d\hat{x}_2 + \cdots + (-1)^n d\hat{x}_n)\left(\frac{\partial f}{\partial x_0}\right) +$$
$$dx_1 \wedge dx_2 \wedge \cdots \wedge dx_n\left(e_1 \frac{\partial f}{\partial x_1} + e_2 \frac{\partial f}{\partial x_2} + \cdots + e_n \frac{\partial f}{\partial x_n}\right)$$
$$= \frac{1}{4}(d\sigma_{(n)} - \overline{d\sigma_{(n)}})(Df + \overline{D}f) - \frac{1}{4}(d\sigma_{(n)} + \overline{d\sigma_{(n)}})(Df - \overline{D}f)$$
$$= \frac{1}{2}d\sigma_{(n)}\overline{D}f - \frac{1}{2}\overline{d\sigma_{(n)}}Df$$
$$= d\sigma_{(n)}A_{f,L}(z). \tag{3.9}$$

Since $\overline{d\sigma_{(n)}}$ and $d\sigma_{(n)}$ are $C\ell_{0,n}$-linearly independent, this implies $Df = 0$ and therefore

$$A_{f,L}(z) = \frac{1}{2}\overline{D}f.$$

The inverse relation (that $Df = 0$ is equivalent to left derivability) follows directly from (3.7) and Definition 2.   □

---

[2]See footnote 1.

Remark 4. *The theorem motivates the treatment of $\frac{1}{2}\overline{D}f$ as the derivative $f'_L(z)$ of the monogenic function f. Indeed, for $n = 1$ and setting again $e_1 \cong i$ we have the complex case (formula (1.8)). The monogenicity of $\frac{1}{2}\overline{D}f$ is a simple consequence of the fact that $D\overline{D}f = \overline{D}Df = 0$.*

Using Stokes' theorem it is also possible to express the relation (3.8) in the form of a limit of a quotient of two integrals (see, for example, [16], but also [9] where for the coordinate-free version of the gradient of hypercomplex differentiable functions vector-valued differential forms are used). Formula (3.9) also shows that this approach (and not the approach by hypercomplex differentiability developed in [7]) is precisely the approach which leads to the greatest formal similarity with the complex case expressed by (1.2). It would be recommended by formula (3.9) to change the use of D and $\overline{D}$ to get more consistent formulas. Maybe the further use and application of the hypercomplex derivative will be decisive for such a development. There are several examples, recent and not so recent, for its usefulness (see, for example, [3], [4] [11], [15] and others).

REFERENCES

[1] F. Brackx, R. Delanghe, and F. Sommen, *Clifford Analysis* **76**, Pitman, Boston-London-Melbourne, 1982.

[2] H. E. Edwards, *Advanced Calculus: A Differential Forms Approach, Third Edition*, Birkhäuser, Boston, Basel, Berlin, 1993.

[3] K. Gürlebeck and H. Malonek, A hypercomplex derivative of monogenic functions in $\mathbb{R}^{n+1}$ and its applications, *Complex Variables Theory Appl.*, Vol. **39** (1999), 199–228.

[4] K. Gürlebeck and W. Sprößig, *Quaternionic and Clifford Calculus for Engineers and Physicists*, John Wiley &. Sons, Chichester, 1997.

[5] K. Habetha, Eine bemerkung zur funktionentheory in algebren, in *Function Theoretic Methods for Partial Differential Equations*, V. E. Meister, N. Weck, and W. L. Wendland, eds., *Lecture Notes in Mathematics*, Vol. **561**, Springer, Berlin, Heidelberg, New York, 1976, 502–509.

[6] H. Malonek, Zum holomorphiebegriff in höheren dimensionen, Habilitation Thesis, PH Halle, 1987.

[7] H. Malonek, A new hypercomplex structure of the Euclidean space $\mathbb{R}^{m+1}$ and the concept of hypercomplex differentiability, *Complex Variables Theory Appl.*, Vol. **14** (1990), 25–33.

[8] H. Malonek, Power series representation for monogenic functions in $\mathbb{R}^{m+1}$ based on a permutational product, *Complex Variables Theory Appl.*, Vol. **15** (1990), 181–191.

[9] H. Malonek, The concept of hypercomplex differentiability and related differential forms, in *Studies in Complex Analysis and Its Applications to Partial Differential Equations 1*, R. Kühnau and W. Tutschke, eds., Pitman **256**, Longman, 1991, 193–202.

[10] A. S. Mejlihzon, Because of monogeneity of quaternions (in Russian), *Doklady Acad. Sc. USSR* **59** (1948), 431–434.

[11] I. M. Mitelman and M. V. Shapiro, Differentiation of the Martinelli-Bochner integrals and the notion of hyperderivability, *Math. Nachr.* **172** (1995), 211–238.

[12] J. Ryan, Clifford analysis with generalized elliptic and quasi-elliptic functions, *Applicable Analysis* **13** (1982), 151–171.

[13] F. Sommen, Monogenic differential forms and homology theory, *Proc. R. Irish Acad.*, Vol. **84A** No. 2 (1984), 87–109.

[14] F. Sommen, Monogenic differential calculus, *Trans. AMS*, Vol. **326** No. 2 (1991), 613–632.

[15] W. Sprößig, Über eine mehrdimensionale operatorrechnung über beschränkten gebieten des $\mathbb{R}^n$, Thesis, TH Karl-Marx-Stadt, 1979.

[16] A. Sudbery, Quaternionic analysis, *Math. Proc. Cambr. Phil. Soc.* **85** (1979), 199–225.

Helmuth R. Malonek
Universidade de Aveiro
Departamento de Matemática
P-3810-193 Aveiro, Portugal
E-mail: mop20116@mail.telepac.pt

Received: November 4, 1999; Revised: February 18, 2000

Hypermonogenic Functions

Sirkka-Liisa Eriksson-Bique
and Heinz Leutwiler

ABSTRACT Let Cl_n be the (universal) Clifford algebra generated by $e_1, \ldots, e_n$. The Cauchy-Riemann (or Dirac) operator in Cl_n is defined by $D = \sum_{i=0}^{n} e_i \frac{\partial}{\partial x_i}$. The functions $f : \Omega \to Cl_n$, $\Omega \subset \mathbb{R}^{n+1}$ open, satisfying $Df = 0$, called left monogenic functions, are widely studied in Clifford analysis. The power function x^m is not monogenic. The second author noticed that the power function is the solution of the modified Cauchy-Riemann system $x_n Df + (n-1) f_n = 0$ which has strong connections to the hyperbolic metric. We study solutions of the equation $x_n D_n f + (n-1) Q'_{n-1}(f) = 0$, where Qf is the function given by the decomposition $f(x) = Pf(x) + Qf(x) e_n$ for $Pf(x), Qf(x) \in Cl_{n-1}$. If $f = \sum_{i=0}^{n} f_i e_i$ for some real functions f_i, then f is the solution of the modified Cauchy-Riemann system stated above.
Keywords: Monogenic functions, hypermonogenic functions, hyperbolic metric, Laplace-Beltrami operator, Dirac operator.

1 Preliminaries

Let Cl_n be the universal Clifford algebra generated by the elements $e_1, \ldots, e_n$ satisfying the relation

$$e_i e_j + e_j e_i = -2\delta_{ij}, \tag{1.1}$$

where δ_{ij} is the usual Kronecker delta. The elements

$$x = x_0 + x_1 e_1 + \ldots + x_n e_n$$

for $x_0, \ldots, x_n \in \mathbb{R}$ are called (para) vectors. The set $\mathbb{R}^{n+1}$ is identified with the set of vectors. The vector space dimension of the Clifford algebra Cl_n is 2^n, and every element $a \in Cl_n$ may be presented as

$$a = \sum_{\nu} a_\nu e_\nu,$$

where $a_\nu \in \mathbb{R}$ and the sum ranges over all subsets $\nu = \{v_1, \ldots, v_k\}$ of $\{1, \ldots, n\}$ with $1 \leq \nu_1 < \ldots < \nu_k \leq n$, $e_\nu = e_{\nu_1} \ldots e_{\nu_k}$ and $e_\emptyset = 1$. A multi-index $\varepsilon_i = (\varepsilon_{i0}, \ldots, \varepsilon_{in})$ is introduced by $\varepsilon_{ij} = \delta_{ij}$.

This research is supported by the Academy of Finland.
AMS Subject Classification: 30G35.

The involution $' : C\ell_n \to C\ell_n$ is defined by

$$a' = \sum_\nu (-1)^{|\nu|} a_\nu e_\nu,$$

where usually $|\nu|$ is the number of elements in the set ν. Clearly, we have $e_0' = e_0 = 1$ and $e_i' = -e_i$ if $i = 1, \dots, n$. Moreover the mapping $'$ is an isomorphism since

$$(ab)' = a'b'.$$

It is also easy to calculate that for any $\nu \neq 0$

$$e_i e_\nu = \begin{cases} e_\nu' e_i, & \text{if } i \notin \nu, \\ -e_\nu' e_i, & \text{if } i \in \nu. \end{cases} \tag{1.2}$$

The following characterization of vectors is useful:

Lemma 1. *An element $x \in C\ell_n$ is a vector if and only if*

$$\sum_{i=0}^n e_i x e_i = (1 - n) x'. \tag{1.3}$$

Proof. It is easy to see that the equation (1.3) holds for all e_i with $i = 0, \dots, n$. Using the linearity we note that it holds for all vectors. Conversely, comparing the components of the left and right side of the equality (1.3) together with (1.2), we infer that the equality (1.3) implies that x has to be a vector. $\square$

The antiautomorphism $* : C\ell_n \to C\ell_n$, called *reversion*, is defined by

$$a^* = \sum_{\substack{\nu_i \in \{1, \dots, n\} \\ \nu_1 < \nu_2 \dots < \nu_k}} a_\nu e_{\nu_k} e_{\nu_{k-1}} \dots e_{\nu_1}.$$

Using (1.1) we note that

$$a^* = \sum_\nu (-1)^{\frac{1}{2}|\nu|(|\nu|-1)} a_\nu e_\nu.$$

Moreover, the mapping $*$ satisfies

$$(ab)^* = b^* a^*.$$

The conjugation $\bar{a}$ is defined by $\bar{a} = (a')^*$. Applying (1.1) we easily deduce that

$$\bar{a} = \sum_\nu (-1)^{\frac{1}{2}|\nu|(|\nu|+1)} a_\nu e_\nu.$$

If $x = x_0 + x_1 e_1 + \ldots + x_n e_n$ is a vector, then $x^* = x$ and

$$\bar{x} = x' = x_0 - x_1 e_1 - \ldots - x_n e_n.$$

Moreover, the equality

$$x\bar{x} = \bar{x}x = |x|^2$$

holds for any vector x, which implies that any non-zero vector is invertible and $x^{-1} = \bar{x}/|x|^2$.

Proposition 1. [7, Proposition 2.1.] *If x is a vector, then for any $m \in \mathbb{N}$*

$$x^m = \sum_{|\alpha|=m} \binom{m}{\alpha} c(\alpha)\, x^\alpha,$$

where the coefficients $c(\alpha)$ with $\alpha = (\alpha_0, \alpha_1, \ldots, \alpha_n)$ are given by

$$c(\alpha) = \begin{cases} \dfrac{\binom{\frac{m-\alpha_0}{2}}{\frac{\alpha-\alpha_0\varepsilon_0}{2}}}{\binom{m-\alpha_0}{\alpha-\alpha_0\varepsilon_0}} (-1)^{\frac{m-\alpha_0}{2}} & \text{if } \alpha - \alpha_0\varepsilon_0 \text{ is even;} \\[4ex] \dfrac{\binom{\frac{m-\alpha_0-1}{2}}{\frac{\alpha-\alpha_0\varepsilon_0-\varepsilon_i}{2}}}{\binom{m-\alpha_0}{\alpha-\alpha_0\varepsilon_0}} (-1)^{\frac{m-\alpha_0-1}{2}} e_i & \text{if } \alpha - \alpha_0\varepsilon_0 - \alpha_i e_i \text{ is even;} \\[4ex] 0, & \text{otherwise,} \end{cases}$$

with $\binom{r}{\alpha} = \dfrac{r!}{\alpha_0! \alpha_1! \ldots \alpha_n!}.$

Proof. Let $x = x_0 + x_1 e_1 + \ldots + x_n e_n$ be a vector. Since x_0 commutes with all e_i and

$$(x_1 e_1 + \ldots + x_n e_n)^2 = -\left(x_1^2 + \ldots + x_n^2\right),$$

we obtain the result. The details are explained in [7, Proposition 2.1]. □

Recall that any element $a \in C\ell_n$ may be uniquely decomposed as $a = b + ce_n$ for $b, c \in C\ell_{n-1}$. Using this decomposition we define the mappings $P_{n-1} : C\ell_n \to C\ell_{n-1}$ and $Q_{n-1} : C\ell_n \to C\ell_{n-1}$ by $P_{n-1}a = b$ and $Q_{n-1}a = c$. Note that if $a = \sum a_\nu e_\nu \in C\ell_n$, then

$$P_{n-1}a = \sum_{n \notin \nu} a_\nu e_\nu \qquad \text{and} \qquad Q_{n-1}a = \sum_{n \in \nu} a_\nu e_{\nu \setminus \{n\}}.$$

Clearly the mappings P_{n-1} and Q_{n-1} are linear and satisfy the properties $P_{n-1}^2 = P_{n-1}$, $P_{n-1}Q_{n-1} = Q_{n-1}$, $Q_{n-1}P_{n-1} = 0$, $Q_{n-1}^2 = 0$ and $I = P_{n-1} + Q_{n-1}e_n$. Since

$$P_{n-1}(a') + Q_{n-1}(a')e_n = a' = (P_{n-1}a + (Q_{n-1}a)e_n)'$$
$$= (P_{n-1}a)' - (Q_{n-1}a)'e_n,$$

we obtain the properties

$$P_{n-1}(a') = (P_{n-1}a)', \tag{1.4}$$
$$Q_{n-1}(a') = -(Q_{n-1}a)'. \tag{1.5}$$

When there is no confusion possible, we abbreviate P_{n-1} and Q_{n-1} by P and Q, respectively.

Lemma 2. *Let a and b belong to $C\ell_n$. Then*

$$P_{n-1}(ab) = (P_{n-1}a)P_{n-1}b + (Q_{n-1}a)Q_{n-1}(b'),$$
$$Q_{n-1}(ab) = (P_{n-1}a)Q_{n-1}b + (Q_{n-1}a)P_{n-1}(b').$$

Proof. Write $a = Pa + (Qa)e_n$ and $b = Pb + (Qb)e_n$. Applying (1.2) we obtain

$$ab = (Pa)(Pb) + (Qa)e_n(Qb)e_n + (Pa)(Qb)e_n + (Qa)e_n(Pb)$$
$$= (Pa)(Pb) - (Qa)(Qb)' + ((Pa)(Qb) + (Qa)(Pb)')e_n.$$

Recalling (1.4) and (1.5) we establish the desired equalities.   □

Let Ω be on open subset of $\mathbb{R}^{n+1}$. The Cauchy-Riemann operator in $C\ell_n$ is defined by modifying the Dirac operator as follows

$$D_n f = \sum_{i=0}^{n} e_i \frac{\partial f}{\partial x_i}$$

for a mapping $f : \Omega \to C\ell_n$, whose components are partially differentiable. Functions f satisfying $D_n f = 0$ are called (left) monogenic functions (see [2], [3]) or regular functions ([17]).

We use the notation $P_{n-1}f$ for the function from Ω into $C\ell_{n-1}$ defined by

$$(P_{n-1}f)(x) = P_{n-1}(f(x))$$

and similarly for $Q_{n-1}f$. It is clear that

$$\frac{\partial(P_{n-1}f)}{\partial x_i} = P_{n-1}\left(\frac{\partial f}{\partial x_i}\right),$$
$$\frac{\partial(Q_{n-1}f)}{\partial x_i} = Q_{n-1}\left(\frac{\partial f}{\partial x_i}\right).$$

The following lemma is useful.

Lemma 3. *Let Ω be an open subset of $\mathbb{R}^{n+1}$. If the components of a function $f : \Omega \to C\ell_n$ are partially differentiable, then*

$$P_{n-1}(D_n f) = D_{n-1}(P_{n-1}f) - \frac{\partial (Q_{n-1}f)'}{\partial x_n},$$

$$Q_{n-1}(D_n f) = D_{n-1}(Q_{n-1}f) + \frac{\partial (P_{n-1}f)'}{\partial x_n}.$$

Proof. Decomposing $f = P_{n-1}f + (Q_{n-1}f) e_n$, we obtain

$$D_n f = D_n (P_{n-1}f) + D_n ((Q_{n-1}f) e_n)$$

$$= D_{n-1}(P_{n-1}f) + e_n \frac{\partial (P_{n-1}f)}{\partial x_n} +$$

$$D_{n-1}(Q_{n-1}f) e_n + e_n \frac{\partial (Q_{n-1}f)}{\partial x_n} e_n.$$

Applying (1.2) we deduce

$$D_n f = D_{n-1}(P_{n-1}f) - \frac{\partial (Q_{n-1}f)'}{\partial x_n}$$

$$+ \left(D_{n-1}(Q_{n-1}f) + \frac{\partial (P_{n-1}f)'}{\partial x_n} \right) e_n,$$

completing the proof. $\square$

Besides D_n we also consider its conjugate operator $\overline{D}_n$ defined by

$$\overline{D}_n = \frac{\partial}{\partial x_0} - \sum_{k=1}^{n} e_j \frac{\partial}{\partial x_j}.$$

It has the property $\overline{D}_n D_n = D_n \overline{D}_n = \triangle_n$, where $\triangle_n = \sum_{i=0}^{n} \frac{\partial^2}{\partial x_i^2}$ is the usual Laplacian in $\mathbb{R}^{n+1}$. Note also that

$$D_n f + \overline{D}_n f = 2 \frac{\partial f}{\partial x_0}$$

and

$$\overline{D}_n f = \left(\frac{\partial f'}{\partial x_0} \right)' + \sum_{k=1}^{n} e'_j \left(\frac{\partial f'}{\partial x_j} \right)' = (D_n f')'. \tag{1.6}$$

Our main object is the modified Dirac operator M_n defined by

$$(M_n f)(x) = (D_n f)(x) + \frac{n-1}{x_n} Q'_{n-1}f(x),$$

where $Q'_{n-1}f = (Q_{n-1}f)'$. If f satisfies $M_n f = 0$ in some open subset $\Omega \setminus \{x \mid x_n = 0\}$ of $\mathbb{R}^{n+1}$, the function f is called *hypermonogenic*. If $M_n f = 0$ and $f = \sum_{i=0}^{n} f_i e_i$ for some real functions f_i the function f is called an H_n-*solution* introduced by the second author in [13] and [14]. When there is no confusion possible, we use the abbreviated notations $M_n = M$ and $H_n = H$.

We introduce, besides M_n, the operator

$$\overline{M}_n f (x) = (\overline{D}_n f)(x) - \frac{n-1}{x_n}(Q'_{n-1}f)(x).$$

Using the properties (1.5) and (1.6) we infer that

$$\overline{M}_n f = (M_n f')'.$$

The H-solutions are notably studied in [8], [9], [10], [11], [4], [13], [14], [15] and [16]. In the case $n = 2$, hypermonogenic functions, called hyper-holomorphic functions, are investigated by W. Hengartner and the second author in [12].

Lemma 4. Let $f : \Omega \to C\ell_n$ be twice continuously differentiable on an open subset Ω of $\mathbb{R}^{n+1}$. Then

$$\overline{M}Mf = M\overline{M}f = \triangle(P_{n-1}f) - \frac{n-1}{x_n}\frac{\partial P_{n-1}f}{\partial x_n}$$

$$+ \left(\triangle(Q_{n-1}f) - \frac{n-1}{x_n}\frac{\partial Q_{n-1}f}{\partial x_n} + \frac{n-1}{x_n^2}Q_{n-1}f\right)e_n,$$

where $\triangle$ is the Laplacian in $\mathbb{R}^{n+1}$.

Proof. Using Lemma 3 we obtain

$$QMf = QD_n f + \frac{n-1}{x_n}Q(Q'f) = D_{n-1}Qf + \frac{\partial P'f}{\partial x_n}.$$

Hence, we have

$$\overline{M}Mf = \overline{D}Df + (n-1)\overline{D}\left(\frac{Q'f}{x_n}\right) - \frac{n-1}{x_n}\left((D_{n-1}Qf)' + \frac{\partial Pf}{\partial x_n}\right).$$

Since

$$\overline{D}\left(\frac{Q'f}{x_n}\right) = \frac{1}{x_n}\frac{\partial Q'f}{\partial x_0} - \frac{1}{x_n}\sum_{i=1}^{n}e_i\frac{\partial Q'f}{\partial x_i} + \frac{e_n Q'f}{x_n^2}$$

$$= \frac{(D_{n-1}Qf)'}{x_n} + \frac{Qf}{x_n^2}e_n - \frac{1}{x_n}\frac{\partial Qf}{\partial x_n}e_n,$$

we have

$$\overline{M}Mf = \Delta Pf - \frac{n-1}{x_n}\frac{\partial Pf}{\partial x_n} + \left(\Delta Qf - \frac{n-1}{x_n}\frac{\partial Qf}{\partial x_n} + (n-1)\frac{Qf}{x_n^2}\right)e_n.$$

From the definitions we note that $Mf + \overline{M}f = 2\frac{\partial f}{\partial x_0}$. Hence, we obtain

$$M\overline{M}f + M^2 f = 2M\left(\frac{\partial f}{\partial x_0}\right) = 2\frac{\partial Mf}{\partial x_0} = (M+\overline{M})(Mf) = \overline{M}Mf + M^2 f,$$

which implies $M\overline{M}f = \overline{M}Mf$, completing the proof.   □

Corollary 1. *Let* $f : \Omega \to C\ell_n$ *be twice continuously differentiable on an open subset* Ω *of* $\mathbb{R}^{n+1}$. *If* $Mf = 0$ *in* $\Omega\backslash\{x\,|\,x_n = 0\}$, *then the components of* $P_{n-1}f$ *satisfy the equation*

$$x_n\Delta u - (n-1)\frac{\partial u}{\partial x_n} = 0, \tag{1.7}$$

and the components of $Q_{n-1}f$ *fulfill the equation*

$$x_n^2\Delta u - (n-1)x_n\frac{\partial u}{\partial x_n} + (n-1)u = 0. \tag{1.8}$$

Note also the following result.

Lemma 5. *Let* $v : \Omega \to \mathbb{R}$ *be twice continuously differentiable on an open subset* Ω *of* $\mathbb{R}^{n+1}$. *If* v *satisfies the equation (1.8), then the mapping* $g : \Omega \to \mathbb{R}$ *defined by*

$$g(x) = \begin{cases} \dfrac{v(x)}{x_n} & \text{if } x_n \neq 0, \\[2mm] \dfrac{\partial v}{\partial x_n}(x) & \text{if } x_n = 0 \end{cases}$$

is twice continuously differentiable and satisfies the equation

$$x_n\Delta g - (n-3)\frac{\partial g}{\partial x_n} = 0.$$

The preceding equations (1.7) and (1.8) are closely related to the hyperbolic metric on $\mathbb{R}_+^{n+1}$, defined by the Riemannian metric

$$ds^2 = x_n^{-2}\left(dx_0^2 + dx_1^2 + \ldots + dx_n^2\right). \tag{1.9}$$

Indeed, the equation (1.7) is the Laplace-Beltrami equation associated with the hyperbolic metric (1.9). The other equation (1.8) represents the eigenfunctions of the Laplace-Beltrami operator corresponding to the eigenvalue $1 - n$.

2   Hypermonogenic functions

Hypermonogenic functions form a right $C\ell_{n-1}$-module.

Lemma 6. *Let* $\nu = \{\nu_1, \dots, \nu_k\}$ *be a subset of* $\{1, \dots, n-1\}$ *with* $\nu_1 < \dots < \nu_k$. *If* $f : \Omega \to C\ell_n$ *is a continuous differentiable mapping, then*

$$M(fe_\nu) = (Mf)e_\nu.$$

Moreover, the hypermonogenic functions in an open subset Ω *of* $\mathbb{R}^{n+1}$ *form a right* $C\ell_{n-1}$-*module.*

Proof. Using Lemma 2 and (1.5), we infer that

$$Q'(fe_\nu) = P'(f)Q'(e_\nu) + Q'(f)P(e_\nu) = Q'(f)e_\nu,$$

completing the proof.   □

It is easy to note that $\frac{\partial Mf}{\partial x_j} = M\frac{\partial f}{\partial x_j}$ for $j = 0, 1, \dots, n-1$. Hence, the following result holds.

Lemma 7. *Let* Ω *be an open subset of* $\mathbb{R}^{n+1}$ *and* $f : \Omega \to C\ell_n$ *be hypermonogenic. Then*

$$\frac{\partial f}{\partial x_j}$$

is hypermonogenic for $j = 0, \dots, n-1$.

The product of hypermonogenic functions is not hypermonogenic. However, we may prove the following result.

Theorem 1. *Let* Ω *be an open subset of* $\mathbb{R}^{n+1}$ *and* $f : \Omega \to C\ell_n$ *be hypermonogenic. Then, the product* $F(x) = f(x)x$ *is hypermonogenic if and only if* f *is an* H-*solution.*

Proof. Substituting

$$\frac{\partial F}{\partial x_j} = \frac{\partial f}{\partial x_j}x + fe_j,$$

we obtain

$$DF = (Df)x + \sum_{j=0}^{n} e_j f e_j.$$

Using Lemma 2, we infer

$$\begin{aligned}
QF &= (Pf)x_n + (Qf)(x - x_n e_n)' \\
&= (f - (Qf)e_n)x_n + (Qf)x' + (Qf)x_n e_n \\
&= fx_n + (Qf)x'.
\end{aligned}$$

Hence,

$$MF = (Mf)\,x + \sum_{j=0}^{n} e_j f e_j + (n-1)\,f'. \tag{2.1}$$

If f is an H-solution, then f is vector valued and by Lemma 1 $MF = 0$. Assume that $MF = 0$. Since by the assumption f is hypermonogenic, we infer from the equality (2.1) that

$$0 = \sum_{j=0}^{n} e_j f e_j + (n-1)\,f'.$$

Applying Lemma 1, we deduce that f is vector valued. Since f is also hypermonogenic, it is also an H-solution. $\qquad\square$

Corollary 2. *For any $m \in \mathbb{N}$, the mapping x^m with $x = \sum_{i=0}^{n} x_i e_i$ is an H-solution in $\mathbb{R}^{n+1}$.*

Proof. Using Proposition 1 we infer that x^m is vector valued. Hence, the preceding theorem implies the assertion. $\qquad\square$

Proposition 2. *Let Ω be an open subset of $\mathbb{R}^{n+1}$ and $F : \Omega \to C\ell_n$ be hypermonogenic. Define a mapping $f : \Omega\backslash\{0\} \to C\ell_n$ by $f(x) = F(x)\,x^{-1}$. Then f is hypermonogenic in $\Omega\backslash\{0\}$ if and only if f is vector valued.*

Proof. Assume that F is hypermonogenic. Since $F(x) = f(x)\,x$, the equality (2.1) implies

$$0 = MF = (Mf)\,x + \sum_{j=0}^{n} e_j f e_j + (n-1)\,f'. \tag{2.2}$$

If f is hypermonogenic, then Lemma 1 yields the property that f is vector valued. On the other hand, if f is vector valued, then applying Lemma 1 and equality (2.2) we obtain that f is also hypermonogenic. $\qquad\square$

Corollary 3. *The mapping x^{-m} is an H-solution in $\mathbb{R}^{n+1}\backslash\{0\}$ for any $m \in \mathbb{N}$.*

Proof. Since $x^{-1} = \frac{\bar{x}}{|x|^2}$, the mapping x^{-1} is vector valued. The preceding theorem implies that x^{-1} is an H-solution. Inductively, we infer that x^{-m} is an H-solution for all $m \in \mathbb{N}$. $\qquad\square$

Proposition 3. *Let Ω be an open subset of $\mathbb{R}^{n+1}$ and $f : \Omega \to C\ell_n$ be a mapping with continuous partial derivatives. The equation $Mf = 0$ is* *equivalent with the system of equations*

$$
\begin{cases}
x_n \left(D_{n-1} \left(P_{n-1} f \right) - \dfrac{\partial \left(Q'_{n-1} f \right)}{\partial x_n} \right) + (n-1) Q'_{n-1} f = 0, \\[4mm]
D_{n-1} \left(Q_{n-1} f \right) + \dfrac{\partial P'_{n-1} \left(f \right)}{\partial x_n} = 0.
\end{cases}
\tag{2.3}
$$

Proof. If $Mf = 0$, then the equations (2.3) follow from Lemma 3. Conversely, if the system (2.3) holds, then multiplying the second equation from the left by x_n and from the right by e_n we obtain

$$
\begin{aligned}
0 &= x_n \left(D_{n-1} \left(Q_{n-1} f e_n \right) + \frac{\partial P'_{n-1} \left(f \right) e_n}{\partial x_n} \right) \\
&= x_n \left(D_{n-1} \left(Q_{n-1} f e_n \right) + \frac{e_n \partial P_{n-1} \left(f \right)}{\partial x_n} \right).
\end{aligned}
\tag{2.4}
$$

Since

$$
-\frac{\partial Q'_{n-1} \left(f \right)}{\partial x_n} = e_n^2 \frac{\partial Q'_{n-1} \left(f \right)}{\partial x_n} = e_n \frac{\partial Q_{n-1} \left(f \right) e_n}{\partial x_n},
$$

adding the first equation and (2.4), we conclude $Mf = 0$. $\square$

In the case $n = 1$, the system (2.3) represents the classical Cauchy-Riemann system for the holomorphic functions of one complex variable. In the case $n = 2$, the system (2.3) is researched by the second author and W. Hengartner in [12].

It is possible to integrate the system (2.3) as stated next.

Theorem 2. *Let Ω be an open subset of $\mathbb{R}^{n+1}$ and set $x = (x_0, \dots, x_n)$ $= (\tilde{x}, x_n)$ for any $x \in \mathbb{R}^{n+1}$. Then $f : \Omega \to C\ell_n$ is hypermonogenic if and only if the components of $Q_{n-1} f$ satisfy the equation (1.8), and for any $a \in \Omega$ and any ball $B(a, r)$ with $B(a, r) \subset \Omega$ there exists a mapping h from $(B(a,r) \cap \{x \mid x_n = a_n\})^{\sim}$ into $C\ell_{n-1}$ satisfying the equations*

$$
P_{n-1} f \left(x \right) - \int_{a_n}^{x_n} \overline{D_{n-1}} Q'_{n-1} \left(f \right) \left(\tilde{x}, t \right) dt + h \left(\tilde{x} \right)
\tag{2.5}
$$

$$
D_{n-1} h \left(\tilde{x} \right) =
\begin{cases}
-\dfrac{\partial Q'_{n-1} \left(f \right)}{\partial x_n} \left(\tilde{x}, a_n \right) + \\
\qquad (n-1) \dfrac{Q'_{n-1} \left(f \right) \left(\tilde{x}, a_n \right)}{a_n} & \text{if } a_n \neq 0, \\[4mm]
(n-2) \dfrac{\partial Q'_{n-1} \left(f \right)}{\partial x_n} \left(\tilde{x}, 0 \right) & \text{if } a_n = 0.
\end{cases}
$$

Proof. Assume that f is hypermonogenic. Then, by Corollary 1, the components of $Q_{n-1}f$ satisfy the equation (1.8). The system (2.3) implies that P_{n-1} has the desired form (2.5). Substituting this in the first equation of (2.3), we obtain

$$0 = \int_{a_n}^{x_n} -\Delta_{n-1} Q'_{n-1}(f)(\tilde{x}, t)\, dt + D_{n-1}h$$
$$-\frac{\partial Q'_{n-1}(f)}{\partial x_n} + (n-1)\frac{Q'_{n-1}(f)}{x_n}.$$

The equation (1.8) implies that

$$\Delta_{n-1} Q'_{n-1}(f)(\tilde{x}, t) = (n-1)\frac{\partial \frac{Q'_{n-1}(f)}{x_n}}{\partial x_n} - \frac{\partial^2 Q'_{n-1}f(x)}{\partial x_n^2}.$$

Hence, we obtain the second equation of (2.5). The converse statement follows similarly. □

Corollary 4. *Let n be odd and assume that the function*

$$f(x) = \sum_\alpha c(\alpha) x^\alpha$$

is hypermonogenic in the neighborhood $B(0, r)$ of the origin. Then, the coefficients $Q_{n-1}(c(\alpha))$ are determined by $Q_{n-1}(c(\alpha))$ with $\alpha_n = 1$ or $\alpha_n = n-1$. Moreover, if f is vanishing on the plane $x_n = 0$, then the coefficients $c(\alpha)$ are determined by the coefficients $Q_{n-1}(c(\alpha))$ with $\alpha_n = n-1$.

Proof. Using the equation (1.8) and denoting $Q_{n-1}(c(\alpha)) = b_n(\alpha)$, we obtain $b_n(\alpha) = 0$ for even $\alpha_n \leq n-3$ and

$$(n-1-\alpha_n) b_n(\alpha) = \sum_{i=0}^{n-1} \frac{(\alpha_i + 1)(\alpha_i + 2)}{\alpha_n - 1} b_n(\alpha + 2\varepsilon_i - 2\varepsilon_n), \quad \alpha_n > 1,$$

completing the proof of the first statement. If f is vanishing on the plane $x_n = 0$, then $h = 0$ in (2.5). This implies that

$$0 = \frac{\partial Q'_{n-1}(f)}{\partial x_n}(\tilde{x}, 0) = b_n(\alpha)$$

for any α with $\alpha_n = 1$. □

It is well known that holomorphic functions are conjugate gradients of harmonic functions. Similar results hold for hypermonogenic functions with respect to hyperbolic harmonic functions.

Theorem 3. *Let Ω be an open subset of $\mathbb{R}^{n+1}$. Then $f : \Omega \to C\ell_n$ is hypermonogenic if and only if for any $a \in \Omega$ and any ball $B(a,r)$ with $B(a,r) \subset \Omega$ there exists a mapping H from $B(a,r)$ into $C\ell_{n-1}$ satisfying the equations*

$$f = \overline{D}H \tag{2.6}$$

and

$$x_n \,\triangle\, H - (n-1)\frac{\partial H}{\partial x_n} = 0 \tag{2.7}$$

on $B(a,r)$.

Proof. Assume that a mapping $H : B(x,r) \to C\ell_{n-1}$ satisfies (2.7) and let $f = \overline{D}H$. Then, we have by (1.5) and Lemma 3

$$Q_{n-1}f = Q_{n-1}\left(\overline{D}_n H\right) = Q_{n-1}\left((D_n H')'\right)$$

$$= -(Q_{n-1}D_n H')' = -\left(\frac{\partial H}{\partial x_n}\right)'.$$

Substituting this into (2.7), we infer that f is hypermonogenic.

Conversely, assume that $f : \Omega \to C\ell_n$ is hypermonogenic. Let $B_{n+1}(a,r)$ be a ball in $\mathbb{R}^{n+1}$ centered at $a = (a_0, \dots, a_n)$ satisfying $B(a,r) \subset \Omega$. Let $s_\nu : (B(a,r) \cap \{x \mid x_n = a_n\})^{\sim} \to \mathbb{R}$ be a twice continuously differentiable solution of the Poisson equation

$$\triangle s_\nu(\tilde{x}) = (P_{n-1}f)_\nu(\tilde{x}, a_n)$$

for a ball $B_n((a_0, \dots, a_{n-1}), r)$, which exists, for example, by [1, p.171]. Set $s = \sum_{\nu \subset \{1, \dots, n-1\}} s_\nu e_\nu$ and define a mapping $H : B_{n+1}(a,r) \to C\ell_{n-1}$ by

$$H(x) = -\int_{a_n}^{x_n} Q'_{n-1}f(\tilde{x}, t)\, dt + D_{n-1}s.$$

Then, we have by Proposition 3

$$\overline{D}_n H(x) = e_n Q'_{n-1}f(x) - \int_{a_n}^{x_n} \overline{D_{n-1}}Q'_{n-1}f(\tilde{x}, t)\, dt + \overline{D_{n-1}}D_{n-1}s$$

$$= Q_{n-1}f(x)\, e_n + \int_{a_n}^{x_n} \frac{\partial P_{n-1}f}{\partial x_n}(\tilde{x}, t)\, dt + (P_{n-1}f)(\tilde{x}, a_n)$$

$$= f(x).$$

We still have to prove that the mapping H satisfies (2.7). Since f is hypermonogenic, we obtain from (1.6)

$$0 = x_n D_n \overline{D}_n H + (n-1)Q'_{n-1}\overline{D}_n H = x_n \,\triangle\, H - (n-1)Q_{n-1}D_n H'.$$

Recalling that the image space of the mapping H is in $C\ell_{n-1}$, Lemma 3 yields

$$Q_{n-1}(D_n H') = D_{n-1}(Q_{n-1} H') + \frac{\partial (P_{n-1} H')'}{\partial x_n} = \frac{\partial H}{\partial x_n}.$$

Hence, the mapping H satisfies (2.7), completing the proof.   □

Hypermonogenic functions may be obtained from H-solutions as follows.

Theorem 4. *A mapping f is hypermonogenic on a ball $B(a,r) \subset \mathbb{R}^{n+1}$ if and only if there exist H-solutions g_α such that*

$$f = \sum_{\alpha \subset \{1,\ldots,n-1\}} g_\alpha e_\alpha.$$

Proof. Assume that f is hypermonogenic on a ball $B(a,r) \subset \mathbb{R}^{n+1}$. Applying Theorem 3, we find a mapping H from $B(a,r)$ into $C\ell_{n-1}$, satisfying the equations (2.7) and $f = \overline{D}_n H$. Denote $H = \sum_{\alpha \subset \{1,\ldots,n-1\}} h_\alpha e_\alpha$ for real functions h_α. Then the mapping $g_\alpha = \overline{D}_n h_\alpha$ is vector valued and, by Theorem 3, an H-solution. Clearly, we have $f = \sum_{\alpha \subset \{1,\ldots,n-1\}} g_\alpha e_\alpha$.   □

Corollary 5. *If f is hypermonogenic, then it is real-analytic.*

Proof. The H-solutions are real-analytic by [6, Theorem 4]. Hence, the preceding theorem implies the statement.   □

The fundamental homogeneous polynomial H-solutions are defined as follows.

Definition 1. *Let α be a multi-index $\alpha \in \mathbb{N}_0^{n-1}$ and m a non-negative integer. The homogeneous polynomial L_m^α is defined by*

$$L_m^\alpha = \frac{1}{\alpha!} \frac{\partial^{|\alpha|} x^{m+|\alpha|}}{\partial x^\alpha}.$$

If n is odd, the homogeneous polynomial solution

$$T_m^\alpha(x) = \sum_{|\beta|=m} c(\beta) x^\beta,$$

with $|\alpha| = m - n + 1$, is defined by (2.5) and Corollary 4 – with $h = 0$ and $Q_{n-1} c(\beta) = 0$ for β, with $\beta_n = 1$, and $Q_{n-1} c(\beta) = \delta_{(\alpha,n-1)}(\beta)$ for β, with $\beta_n = n - 1$.

Theorem 5. *The set $\{L_m^\alpha \mid |\alpha| \leq m\}$ is a basis of the right $C\ell_{n-1}$-module of hypermonogenic functions if n is even. If n is odd, then the set*

$$\{L_m^\alpha \mid |\alpha| \leq m\} \cup \{T_m^\alpha \mid |\alpha| = m - n + 1\}$$

is a basis of the right $C\ell_{n-1}$-module of hypermonogenic functions.

Proof. Assume that n is even. Using [5, Theorem 25] we obtain that the set $\{L_m^\alpha \mid |\alpha| \le m\}$ is a basis of the right $C\ell_{n-1}$-module generated by the homogeneous polynomial H-solutions of degree m. If f is a homogenous hypermonogenic polynomial of degree m, then by Theorem 4 there exist homogeneous polynomial H-solutions p_α of degree m satisfying $f = \sum p_\alpha e_\alpha$. Applying [5, Theorem 24] we know that the set $\{L_m^\alpha \mid |\alpha| \le m\}$ is a basis of the right $C\ell_{n-1}$-module generated by H-solutions. Hence, it is also a basis of the right $C\ell_{n-1}$-module of hypermonogenic functions. The odd case is proved similarly. □

Theorem 6. *Let f be hypermonogenic in a neighborhood of a point $x = (x_0, \dots, x_n)$ with $x_n = 0$. If n is even, there exist constants $b_k(\alpha) \in C\ell_{n-1}$ such that*

$$f = \sum_{k=0}^{\infty} \sum_{|\alpha|=0}^{k} L_k^\alpha b_k(\alpha)$$

in some neighborhood of x. If f is odd, there exist constants $b_k(\alpha), c(\alpha) \in C\ell_{n-1}$ such that

$$f = \sum_{k=0}^{\infty} \left(\sum_{|\alpha|=0}^{k} L_k^\alpha b_k(\alpha) + \sum_{|\alpha|=k-n+1} T_k^\alpha c_k(\alpha) \right).$$

Proof. Let n be even. Assume that f is hypermonogenic in a neighborhood of a point x. If $T(y) = y + a$ for $a \in \mathbb{R}^{n+1}$ with the last coordinate $a_n = 0$, then $f \circ T$ is also hypermonogenic. Hence, we may assume that $x = 0$. Since f is hypermonogenic, f is real-analytic and, therefore, admits the presentation

$$f(y) = \sum_{\beta \in \mathbb{N}_0^{n+1}} a(\beta) y^\beta$$

in some neighborhood $B_r(0)$. Applying M, we obtain

$$0 = Mf(y) = \sum_{k=0}^{\infty} M\left(\sum_{|\beta|=k} a(\beta) y^\beta \right).$$

Since $M\left(\sum_{|\beta|=k} a(\beta) y^\beta \right)$ is a homogeneous polynomial of degree k, we infer

$$M\left(\sum_{|\beta|=k} a(\beta) y^\beta \right) = 0.$$

This implies that $\sum_{|\beta|=k} a(\beta) y^\beta$ is hypermonogenic. Applying Theorem 5 we obtain the result. The odd case is proven similarly. □

REFERENCES

[1] S. Axler, *Harmonic Function Theory*, Springer, New York, 1992.

[2] F. Brackx, R. Delanghe, and F. Sommen, *Clifford Analysis*, Pitman, Boston, London, Melbourne, 1982.

[3] J. Gilbert and M. Murray, Clifford algebras and Dirac operators in harmonic analysis, *Cambridge Studies in Advanced Mathematics* **26**, Cambridge, New York, 1991.

[4] S. -L. Eriksson-Bique, Comparison of quaternionic analysis and its modification, to appear in *The Proceedings of Delaware*, 1–16.

[5] S. -L. Eriksson-Bique, On modified Clifford analysis, submitted for publication, 1–22.

[6] S. -L. Eriksson-Bique, Real analytic functions on modified Clifford analysis, submitted for publication, 1–12.

[7] S. -L. Eriksson-Bique, The binomial theorem for hypercomplex numbers, *Ann. Acad. Sci. Fenn.* **24** (1999), 225–229.

[8] S. -L. Eriksson-Bique and H. Leutwiler, On modified quaternionic analysis in $\mathbb{R}^3$, *Arch. Math.* **70** (1998), 228–234.

[9] Th. Hempfling, Beiträge zur modifizierten Clifford-analysis, Dissertation, Univ. Erlangen-Nürnberg, August, 1997.

[10] Th. Hempfling, Multinomials in modified Clifford analysis, *C. R. Math. Rep. Acad. Sci.* No. 18, Canada, 1996, 2,3, 99–102.

[11] Th. Hempfling and H. Leutwiler, Modified quaternionic analysis in $\mathbb{R}^4$, in *Proceedings of Clifford Algebras and Their Application in Mathematical Physics*, Aachen, 1996, *Fundamental Theories of Physics* **94**, Kluwer, Dordrecht, 1998, 227–237.

[12] W. Hengartner and H. Leutwiler, Hyperholomorphic functions in $\mathbb{R}^3$, to appear.

[13] H. Leutwiler, Modified Clifford analysis, *Complex Variables* **17** (1992), 153–171.

[14] H. Leutwiler, Modified quaternionic analysis in $\mathbb{R}^3$, *Complex Variables* **20** (1992), 19–51.

[15] H. Leutwiler, More on modified quaternionic analysis in $\mathbb{R}^3$, *Forum Math.* **7** (1995), 279–305.

[16] H. Leutwiler, Rudiments of a function theory in $\mathbb{R}^3$, *Expo. Math.* **14** (1996), 097–123.

[17] A. Sudbery, Quaternionic analysis, *Math. Proc. Cambridge Philos. Soc.* **85** (1979), 199–225.

Sirkka-Liisa Eriksson-Bique
Department of Mathematics
University of Joensuu
P.O. Box 111
Fin-80101-Joensuu, Finland
E-mail: Sirkka-Liisa.Eriksson-Bique@Joensuu.FI

Heinz Leutwiler
Department of Mathematics
University of Erlangen-Nürnberg
Bismarckstrasse 1 1/2
D-91054 Erlangen, Germany
E-mail: leutwil@mi.uni-erlangen.de

Received: September 15, 1999; Revised: November 23, 1999

Reproducing Kernels for Hyperbolic Spaces

Paula Cerejeiras

ABSTRACT In this article we discuss the existence of a Poisson-Szegö kernel for the Laplace-Beltrami equation associated to an n-dimensional hyperbolic space. This solution is obtained via the Dirac operator. We restrict the treatment to the class of spaces with the property that the conformal group preserving the metric also preserves a predefined circle. We shall use the above-mentioned kernels in order to solve boundary value problems.

Keywords: Möbius transformations, Poisson-Szegö kernel.

1 Introduction

In classical potential theory there exists a connection between the Dirichlet problem in a given domain and the boundary behaviour of holomorphic functions on that domain. Hence, this theory is independent of the domains we are considering. However, for general manifolds, the boundary of the domain plays an essential role. In this case, one has to look at the behaviour of the Laplace-Beltrami operator. In fact, in this operator the tangential directions which arise in higher dimensional manifolds are directly expressed. This was the motivation for the study of the solutions of the Laplace-Beltrami equation in the last years (see, for instance, Leutwiler, Cnops, or Ryan [C], [Le], [Ry], [TRy]). We should also mention the work of Stein and Weiss. They studied this subject, which they called *the invariant potential theory with respect to the Laplace-Beltrami operator*, by use of several complex variables techniques.

In this article, we propose the treatment of a particular type of Boundary Value Problems (BVP) generated by the Laplace-Beltrami equation on n-dimensional Riemannian manifolds in which the metrics $ds = \frac{\sqrt{dx_1^2 + \cdots + dx_n^2}}{\lambda(x)}$ are invariant under special conformal groups. This problem was already studied by Loo-Keng Hua in 1977 (see [Loo]). We use the conformal invariance of the Laplace-Beltrami operator under special groups in order to establish, in Section 6, a class of equivalent BVP's. A generalized solution

AMS Subject Classification: 30G35, 31C12, 58J32.

for these Dirichlet-like problems, corresponding to the Laplace-Beltrami equation, is obtained via the explicit calculation of the Poisson-Szegö formulas. Also, some properties of the corresponding Poisson-Szegö kernels are described.

2   Notation

In this section, we shall introduce some notations and general results concerning Clifford algebras.

Consider the vectorial space $\mathbb{R}^n$, together with its orthonormal basis $e_1, \ldots, e_n$. We denote by $\mathbb{R}_{0,n}$ the real Clifford algebra generated by $\mathbb{R}^n$ endowed with the multiplication rules $e_i e_j + e_j e_i = -2\delta_{i,j}$, $i,j = 1, \ldots, n$. Therefore, a 2^n-dimensional real associative algebra, where

$$\mathcal{B} = \{1, e_A = e_{h_1} \cdots e_{h_s}\}$$

with $A = \{h_1, \ldots, h_s\} \subset N = \{1, \ldots, n\}, 1 \le h_1 < \cdots < h_s \le n\}$ stands for its basis. Each $x \in \mathbb{R}_{0,n}$ can be written as a linear combination of the elements of the basis. The particular linear combination of basic elements with equal length k is designated a k−vector, and we shall denote it by $[x]_k$ the k-vector part of $x \in \mathbb{R}_{0,n}$. In particular, the product of two vectors $x, y \in \mathbb{R}^n$ can be decomposed in a symmetric and an anti-symmetric part:

$$xy = \frac{1}{2}(xy + yx) + \frac{1}{2}(xy - yx) = x \cdot y + x \wedge y$$

where $x \cdot y = -\sum_{i=1}^{n} x_i y_i$ is scalar-valued and corresponds to the (Clifford) inner product of x and y, while the anti-symmetric part $x \wedge y = \sum_{i<j} e_i e_j (x_i y_j - x_j y_i)$ is bi-vectorial and is called the *outer product* of x and y.

We define the involutory anti-automorphism, called *reversion* on the Clifford algebra and denoted as $x \to x^*$, by its action on the basic elements $1^* = 1$ and $(e_{h_1} \cdots e_{h_s})^* = (-1)^{[s/2]} e_{h_1} \cdots e_{h_s}$, for all $h_1, \ldots, h_s$ satisfying $1 \le h_1 < \cdots < h_s \le n$.

Consider now the group $O(0, n)$ of all orthogonal transformations on $\mathbb{R}^n$. Then the Clifford group

$$\Gamma(0, n) = \{s = \prod_{i=1}^{k} s_i, k \in \mathbb{N}, s_i \in \mathbb{R}^n, s_i^2 \ne 0\}$$

is known to define a two-fold covering for $O(0, n)$.

A Clifford-valued function in a given domain $\Omega \in \mathbb{R}^n$ is defined as

$$f(x) = \sum_{A \subset N} f_A(x) e_A, x \in \Omega,$$

where each $f_A : \Omega \to \mathbb{R}$. Hence, function spaces of Clifford-valued functions are established as modules over the Clifford algebra $\mathbb{R}_{0,n}$ by imposing the coefficient functions f_A to be in the corresponding real-valued function space. Therefore, we shall use for the Clifford-valued function spaces the same notations as in the real-valued case.

The first order differential operator, $\partial = \sum_{i=1}^{n} e_i \partial_i$, represents the Euclidean Dirac operator which factorizes the Laplace operator, $\Delta = \sum_{i=1}^{n} \partial_i^2$, by means of $\Delta = -\partial^2$.

A Clifford-valued function f is said to be (left) monogenic in $\Omega \subset \mathbb{R}^n$ if it satisfies $\partial f = 0$ in Ω.

3   Conformal mappings

A Möbius transformation is defined as a conformal bijective transformation which maps $\mathbb{R}^n \cup \{\infty\}$ onto itself and preserves generalized spheres. Arising as a natural consequence of Liouville's conformality theorem, for spaces of dimensions higher than two, only Möbius transformations are conformal.

3.1   Vahlen matrices

The large variety of problems demanding the use of these type of transformations justifies the use of conformal mappings. However, as already pointed out by Ahlfors in [A], the classical approaches to the treatment of such mappings lead in general to complicated formulas. Repeating an original approach of Vahlen from 1902, in which he successfully described these kind of transformations by means of 2×2 matrices with entries in Clifford algebras, in a quite similar way as done in the complex case, Ahlfors proposed, in the early 80's, the use of this matrix description in order to simplify the calculations.

Theorem 1 (Vahlen's Theorem). *The set of matrices* $A = \begin{pmatrix} a & b \\ c & d \end{pmatrix} \in$ $(\mathbb{R}_{0,n})^{2 \times 2}$ *satisfying*

1. $a, b, c, d \in \Gamma(0, n) \cup \{0\}$

2. $bd^*, ac^*, a^*b, c^*d \in \mathbb{R}^{0,n}$

3. $\Delta(A) = ad^* - bc^* \neq 0$

is a matrix representation of the Clifford group $\Gamma(1, n+1)$.

In the next decade, several mathematicians were applying this method to obtain new results regarding singular integral operators, orthogonal decompositions of $L_2(\mathbb{R}^n)$, and conformal invariance of special manifolds (see, for instance, [C], [TRy], [PeT]). One proof of the above theorem can be found in [C]. This proof relies on a projective construction elaborated by Fillmore and Springer (see [FiSp]) and on the use of Clifford algebra techniques. In 1989, in his doctoral thesis, Maks extended this theorem to the case of Möbius transformations acting on vector spaces with arbitrary signature. He showed that it is necessary to impose three more conditions in order to obtain Vahlen's theorem.

3.2   Möbius transformations

We shall denote the group of Möbius transformations in $\mathbb{R}^n$ by $\mathbb{M}(n)$. Since each element

$$g(x) = (ax + b)(cx + d)^{-1} \in \mathbb{M}(n) \qquad (3.1)$$

is related to the orthogonal group $O(1, n+1)$, each Möbius transformation is, therefore, linked to a matrix satisfying conditions $1-3$ of theorem 1. As in the complex case, the inverse of such a transformation can be obtained by means of the inverse of the corresponding matrix (in the Clifford sense). That is,

$$g^{-1}(y) = (d^*y - b^*)(-c^*y + a^*)^{-1} \qquad (3.2)$$

is again an element of $\mathbb{M}(n)$. Using the sphere-preserving property of these transformations, we can associate to each given sphere S in $\mathbb{R}^n$ the subgroup of $\mathbb{M}(n)$ which preserves it. We denote this group by *fix-group* G_S. These subgroups are of particular interest when reproduced properties have to be considered, as will be seen in Section 5. Regarding several invariant properties studied by Cnops in [C], we shall mention the following theorem.

Theorem 2. *Let $\mathbb{R}^n$ be a Riemannian manifold with a metric*

$$ds = \frac{\sqrt{dx_1^2 + \cdots + dx_n^2}}{\lambda(x)}$$

which is invariant under the fix-group G_S. Then, the second order differential operator

$$\Delta_{inv} = [\lambda(x)]^{1+\frac{n}{2}} \Delta [\lambda(x)]^{1-\frac{n}{2}} \qquad (3.3)$$

is also invariant under the action of G_S.

Also, geometric interpretation can be provided for Riemannian manifolds endowed with metrics invariant under the action of this type of subgroups. The scalar function $\mu(x) = |cx + d|^{-2}$ is the so-called *local contraction factor* which expresses the relation between the vectorial differentials dx and dy (see [A]). It is equal, up to a non-zero constant, to the quotient $\frac{\lambda_2(g(x))}{\lambda_1(x)}$, where $\frac{1}{\lambda_1(x)}$ and $\frac{1}{\lambda_2(g(x))}$ are the metric functions which are invariant, respectively, under the fix-groups G_{S_1} and G_{S_2}, with S_1 and S_2 being spheres of $\mathbb{R}^n$. Hence, the set G_{S_1,S_2} of Möbius transformations mapping the sphere S_1 onto S_2 can be viewed as a link between the n-dimensional Riemannian manifolds endowed with the weight metrics,

$$ds = \frac{\sqrt{dx_1^2 + \cdots + dx_n^2}}{\lambda_i(x)}, \quad i = 1, 2,$$

respectively.

In 1979, Sudbery in the article *Quaternionic Analysis*, introduced a quaternionic functional preserving monogenicity. Later generalized to Clifford algebras, the functional associated to (3.1)

$$\gamma_g[f](y) = \frac{yc - a}{|yc - a|^n} \; f \circ g^{-1}(y) \tag{3.4}$$

is also called *conformal weight of order* 1. It maps the right module of left monogenic functions in $\mathbb{R}^n \setminus \{-c^{-1}d\}$ into the right module of left monogenic functions in $\mathbb{R}^n \setminus \{ac^{-1}\}$. It provides a tool for studying invariant properties of solutions of ∂^k (see [TRy], [PeT]), and it was used by Zöll to prove several results involving Clifford-valued differential forms. By means of the above functional, an isometry between L_2 spaces is established.

Theorem 3. *Let S_1 and S_2 be two spheres. The functional γ_g defines an isometry between the correspondent Hilbert spaces of square-integrable functions $L_2(S_1)$ and $L_2(S_2)$.*

Proof. In fact, for $g \in G_{S_1,S_2}$ we obtain that

$$\begin{aligned}
(\gamma_g[f], \gamma_g[h])_{S_2} &= \int_{S_2} \overline{\gamma_g[f](\eta)} \gamma_g[h](\eta) dV_{S_2} \\
&= \int_{S_2} \overline{f \circ g^{-1}(\eta)} \; h \circ g^{-1}(\eta) |\eta c - a|^{2-2n} dV_{S_2}, \\
&= \int_{S_1} \overline{f(\xi)} \; h(\xi) dV_{S_1} \\
&= (f, h)_{S_1},
\end{aligned}$$

for all $f, h \in L_2(S_1)$. Notice that due to the local contraction factor (3.2), we have $\mu_{g^{-1}}(\eta) = |\eta c - a|^{-2}$ as the Jacobian of the transformation g^{-1}, and, therefore, the surface elements are transformed according to

$$\mu_{g^{-1}}^{n-1}(\eta) dV_{S_2}(\eta) = dV_{S_1}(\xi).$$

$\square$

4   Orientable Riemannian manifolds

One approach to the treatment of orientable Riemannian manifolds is its identification with the space of differential k-forms, using the so-called Hodge-de Rham theory. In a sequence of articles, Leutwiler proposed an original treatment for the Laplace-Beltrami equation by means of harmonic differential forms in the Hodge-de Rham sense. This produced a system *like* Cauchy-Riemann's which allows us to study the scalar solutions of the Laplace-Beltrami equation.

4.1   Laplace-Beltrami operator

Definition 1. *Let M be an arbitrary orientable n-dimensional Riemannian manifold with metric $ds = g(x)\sqrt{dx_1^2 + \cdots + dx_n^2}$. Furthermore, let $[g_{ij}]$ be the $n \times n$ (real) matrix associated to the metric function g. The Laplace-Beltrami operator in M is then defined as*

$$
\begin{aligned}
\Delta_{LB} &= \operatorname{div}\left[\operatorname{grad}(g)\right] \\[2mm]
&= \frac{1}{\sqrt{|g|}} \sum_{i=1}^{n} \frac{\partial}{\partial x_i}\left(\sqrt{|g|} \sum_{j=1}^{n} g^{ij} \frac{\partial}{\partial x_j}\right)
\end{aligned}
\tag{4.1}
$$

where $[g^{ij}]$ stands for the inverse matrix of $[g_{ij}]$ and $|g| = \det[g_{ij}]$.

In our case, where the metric is given by $ds = \frac{\sqrt{dx_1^2 + \cdots + dx_n^2}}{\lambda(x)}$, we can express the matrix associated to the metric at each point of the manifold as $g_{ij}(x) = g(x)\delta_{ij}$, so that we have

$$
\Delta_{LB} = [\lambda(x)]^2 \Delta + (n-2)\lambda(x)\partial\lambda(x) \cdot \partial
\tag{4.2}
$$

as the correspondent Laplace-Beltrami operator. Following Leutwiler's idea, we perform the decomposition of this operator in terms of the Hodge-de Rham system of an exterior differential d and its adjoint inner differential d^*, acting on the spaces of k-differential forms $\wedge_k(\mathbb{R}^n)$ according to

$$
\begin{cases}
d^* : \wedge_k(\mathbb{R}^n) \longrightarrow \wedge_{k-1}(\mathbb{R}^n), \quad k = 1, \ldots, n \\
\qquad d^*(\text{0-form}) = 0, \quad d^*d^* = 0 \\[3mm]
d : \wedge_k(\mathbb{R}^n) \longrightarrow \wedge_{k+1}(\mathbb{R}^n), \quad k = 0, \ldots, n-1 \\
\qquad d(\text{n-form}) = 0, \quad dd = 0
\end{cases}
$$

and expressing the Laplace-Beltrami operator as (see [Le])

$$
\Delta_{LB} = (d + d^*)^2 = d^*d + dd^*.
$$

Moreover, we shall identify its real-valued solutions with 0-forms.

Definition 2 (λ-harmonic function).   *Let* $(\mathbb{R}^n, \lambda)$ *be a Riemannian manifold. A real-valued function* f *is said to be a λ-harmonic function on* $\Omega \subset \mathbb{R}^n$ *if* $\Delta_{LB} f = 0$ *on* Ω*. We shall denote by* $\mathcal{H}_\lambda(\Omega)$ *the space of all λ-harmonics on* Ω*.*

We can relate the invariant operator (3.3) with the Laplace-Beltrami operator through the equation

$$\Delta_{LB} = \{\Delta_{inv} - [\Delta_{inv} 1]\} = \Delta_{inv} + \frac{n(n-2)}{4},$$

which implies the invariance of (4.1) under the action of the fix-group G_S.

4.2   Two models of hyperbolic spaces

Two important examples of Riemannian manifolds are the Poincaré models for hyperbolic spaces (i.e., spaces with constant negative curvature $\kappa = -1$.) These models are obtained by introducing in an n-dimensional vector space a metric which will induce the hyperbolic structure. The models will be classified as *planar* or *spherical* according to the associated invariance group. Hence, the so-called *planar model* corresponds to the upper half space

$$\mathbb{R}^n_+ = \{\, x = x_1 e_1 + \cdots + x_n e_n \in \mathbb{R}^n \; : \; x_n > 0 \,\}$$

endowed with the metric $ds = \frac{\sqrt{dx_1^2 + \cdots + dx_n^2}}{x_n}$ which has $\mathcal{M}(\mathbb{R}^{n-1})$, the group of Möbius transformations mapping the hyperplane $\mathbb{R}^{n-1}$ into itself, as the invariance group. Simple calculations show this manifold to have constant curvature $\kappa = -1$. The spherical model is obtained by introducing in the unit disk

$$\mathbb{D} = \{\, x \in \mathbb{R}^n \; : \; |x| < 1 \,\}$$

the metric $ds = \frac{\sqrt{dx_1^2 + \cdots + dx_n^2}}{1 - |x|^2}$. This metric is proved to be invariant under the fix-group for the unit sphere S^{n-1}. Also, one shows this manifold to be isometric to the first model, up to a non-zero constant, which proves this to be again a model for hyperbolic manifolds. For these models we have the Laplace-Beltrami operators $\Delta_{\mathbb{R}^n_+} = x_n^2 \Delta + (2-n)x_n \partial_{x_n}$ and $\Delta_{\mathbb{D}} = (1 - |x|^2)^2 \Delta + 2(n-2)(1 - |x|^2)\sum_{i=1}^n x_i \frac{\partial}{\partial x_i}$, respectively. These stand for particular cases of elliptical partial differential equations which have been studied by Huber, Leutwiler, and others (see [Hu], [Le]). The solutions for the second operator are a particular case of generalized axially symmetric potentials with special parameters and, therefore, analytic in every region which does not intersect the hyperplane $x_n = 0$.

5   BVP for the spherical model

In this section, we shall adapt the classical techniques used by Loo-Keng Hua in order to solve the boundary value problem raised by the Laplace-Beltrami equation associated to the manifold $\mathbb{D}$. We shall call it the *Dirichlet-like problem for the spherical case.*

Definition 3. *Find the $\lambda_{S^{n-1}}$-harmonic function u in $\mathbb{D}$ satisfying*

$$\begin{cases} \Delta_{\mathbb{D}} u = 0 & in \ \mathbb{D}, \\ u = \phi & in \ S^{n-1} \end{cases} \tag{5.1}$$

for a given $\phi \in C(S^{n-1})$.

5.1   Poisson-Szegö kernel for $\mathbb{D}$

The solution for problem (5.1) will be given in terms of the family

$$g_x(\xi) = (\xi - x)(x\xi + 1)^{-1}, \quad x \in \mathbb{D} \tag{5.2}$$

of special conformal transformations which apply x into 0 and satisfy

$$1 - |g_x(\xi)|^2 = 1 - \frac{|\xi - x|^2}{|x\xi + 1|^2} = \left(\frac{1 - |x|^2}{|x\xi + 1|^2} \right) (1 - |\xi|^2).$$

Therefore, these transformations belong to the fix-group of the unit sphere S^{n-1}. From the change of variable (5.2), we obtain a relation among the Lebesgue measures

$$dV_{S^{n-1}}(\zeta) = \left(\frac{1 - |x|^2}{|x\xi + 1|^2} \right)^{n-1} dV_{S^{n-1}}(\xi).$$

Based on this, we establish the Poisson-Szegö kernel for the operator $\Delta_{\mathbb{D}}$ on the unit ball.

Theorem 4. *The Poisson-Szegö kernel*

$$\mathbb{P}_{S^{n-1}}(x, \xi) := \frac{1}{\omega_n} \left(\frac{1 - |x|^2}{|x\xi + 1|^2} \right)^{n-1}, \tag{5.3}$$

where $(x, \xi) \in \mathbb{D} \times S^{n-1}$ and ω_n stands for the area of the unit sphere, satisfies

i) $\mathbb{P}_{S^{n-1}}(x, \xi) > 0$, in $\mathbb{D} \times S^{n-1}$

ii) $\mathbb{P}_{S^{n-1}}(\cdot, \xi)$ is $\lambda_{S^{n-1}}$-harmonic in $\mathbb{D}$, for all fixed $\xi \in S^{n-1}$

iii) $\int_{S^{n-1}} \mathbb{P}_{S^{n-1}}(x, \xi) dV_{S^{n-1}}(\xi) = 1$ for all $x \in \mathbb{D}$.

We are now able to define the corresponding Poisson-Szegö formula and to introduce a solution for (5.1).

5.2   Decomposition of the Poisson-Szegö kernel

Definition 4. *For every $\phi \in C(S^{n-1})$, we shall denote by $\mathbf{P}_{S^{n-1}}[\phi]$ the Poisson-Szegö formula for ϕ,*

$$\mathbf{P}_{S^{n-1}}[\phi](x) = \int_{S^{n-1}} \mathbb{P}_{S^{n-1}}(x,\xi)\phi(\xi)dV_{S^{n-1}}(\xi) \tag{5.4}$$

for all $x \in \mathbb{D}$.

From the above definition, we obtain the following theorem.

Theorem 5. *The Poisson-Szegö formula satisfies*

$$f(x) = \int_{S^{n-1}} \mathbb{P}_{S^{n-1}}(x,\xi)f(\xi)dV_{S^{n-1}}(\xi)$$

for all $f \in C(\overline{\mathbb{D}})$ and $\Delta_{\mathbb{D}} f = 0$ on $\mathbb{D}$.

Then the extension of $\phi \in C(S^{n-1})$

$$\phi^{\star}(x) = \begin{cases} \mathbf{P}_{S^{n-1}}[\phi](x) & \text{if } x \in \mathbb{D}, \\ \phi(x) & \text{if } x \in S^{n-1} \end{cases} \tag{5.5}$$

stands for a solution of the problem (5.1). Moreover, it can be proven (see [Loo]) that the Poisson-Szegö kernel admits the decomposition in terms of Gegenbauer polynomials

$$\begin{aligned} \mathbb{P}_{S^{n-1}}(x,\xi) &= \frac{1}{\omega_n}\left(\frac{1-|x|^2}{|x\xi+1|^2}\right)^{n-1} \\ &= \sum_{l=0}^{+\infty} \frac{2l+n-2}{n-2}\tau_l(|x|)|x|^l P_l^{(\frac{n}{2}-1)}(x\cdot\xi) \end{aligned}$$

where each coefficient τ_l is given by

$$\tau_l(\rho) = \frac{\Gamma(\frac{n}{2})\Gamma(l+n-1)}{\Gamma(n-1)\Gamma(l+\frac{n}{2})}F(l,1-\frac{n}{2},l+\frac{n}{2};\rho^2)$$

with the extra property that $\lim_{\rho \to 1} \tau_l(\rho) = 1$.

6   Generalized BVP problem

In order to generalize the boundary value problem (5.1), we need the following theorem concerning the relationship between different manifolds (see [C], [Ce]).

Theorem 6. *Let M_i, $i = 1, 2$ be two orientable Riemannian n-dimensional manifolds endowed, respectively, with the metric assigned by the functions $\frac{1}{\lambda_i}$ which are invariant under the action of the fix group G_i associated to the sphere S_i. Then the corresponding Laplace-Beltrami operators are invariant under the action of G_{S_1, S_2}, that is,*

$$\Delta_{LB_1} f = \Delta_{LB_2} f \circ g^{-1} \tag{6.1}$$

for all $g \in G_{S_1, S_2}$ which satisfy $g(M_1) \subset M_2$.

This theorem states that, under the above conditions, $h \circ g^{-1}$ is λ_2-harmonic on $g(M_1)$ if and only if h is λ_1-harmonic on M_1. Therefore, we can reformulate the Dirichlet-like problem for an arbitrary disk in the following way: let $\mathbb{B}_r(a)$ be the disk centered at a and of radius $r > 0$. Consider the particular manifold $\mathbb{B}_r(a)$ endowed with the metric established by the positively defined function $\frac{1}{\lambda(x)}$, invariant under G_S, the fix-group for the boundary S of $\mathbb{B}_r(a)$. Also, consider its associated Laplace-Beltrami operator Δ_{LB}.

Definition 5. *Determine a λ_S-harmonic function v in $\mathbb{B}_r(a)$ which satisfies*

$$\begin{cases} \Delta_{LB} v = 0 & in \ \mathbb{B}_r(a), \\ v = \psi & in \ S \end{cases} \tag{6.2}$$

for a fixed $\psi \in C(S)$.

6.1   Generalization of the Poisson-Szegö kernel

Using an arbitrary Möbius transformation $g \in G_{S^{n-1}, S}$ such that $g(\mathbb{D}) = \mathbb{B}_r(a)$, we shall define $\phi \in C(S^{n-1})$ as $\phi(x) = \psi \circ g(x)$. Then Theorem 6 ensures that $\mathbf{P}_{S^{n-1}}[\phi]$ is $\lambda_{S^{n-1}}$-harmonic if and only if $\mathbf{P}_{S^{n-1}}[\phi] \circ g^{-1}$ is λ_S-harmonic. Hence, calculations lead to

$$\mathbf{P}_{S^{n-1}}[\phi] \circ g^{-1}(y)$$

$$= \int_{S^{n-1}} \mathbb{P}_{S^{n-1}}(g^{-1}(y), \xi) \phi(\xi) dV_{S^{n-1}}(\xi)$$

$$= \int_S \mathbb{P}_{S^{n-1}}\left(g^{-1}(y), g^{-1}(\eta)\right) \phi(g^{-1}(\eta))(c\eta + d)^{2-2n} dV_S(g^{-1}(\eta))$$

where $g(\xi) = \eta \in S$ for all $\xi \in S^{n-1}$, while the local contraction factor verifies $\mu_{g^{-1}}(\xi) = \mu_g^{-1}(\eta)$.

Theorem 7. *The function defined as*

$$\mathbb{P}_S(y, \eta) := \frac{(c\eta + d)^{2-2n}}{\omega_n} \left(\frac{1 - |g^{-1}(y)|^2}{|g^{-1}(y) g^{-1}(\eta) + 1|^2} \right)^{n-1}, \tag{6.3}$$

where $(y, \eta) \in \mathbb{B}_r(a) \times S$ and ω_n stands for the area of the unit sphere, satisfies

i) $\mathbb{P}_S(y, \eta) > 0$, *in* $\mathbb{B}_r(a) \times S$,

ii) $\mathbb{P}_S(\cdot, \eta)$ *is* λ_S*-harmonic in* $\mathbb{B}_r(a)$, *for all fixed* $\eta \in S$,

iii) $\int_S \mathbb{P}_S(y, \eta) dV_S(\eta) = 1$ *for all* $y \in \mathbb{B}_r(a)$,

and, therefore, it represents the Poisson-Szegö kernel for the generalized Dirichlet-like problem.

Definition 6. *For every* $\psi \in C(S)$, *we shall denote by* $\mathbf{P}_S[\psi]$ *the Poisson-Szegö formula for* ψ,

$$\mathbf{P}_S[\psi](y) = \int_S \mathbb{P}_S(y, \eta) \psi(\eta) dV_S(g^{-1}(\eta)) \tag{6.4}$$

for all $y \in \mathbb{B}_r(a)$.

Hence, the extension

$$\psi^\star(y) = \begin{cases} \mathbf{P}_S[\psi](y) & \text{if } y \in \mathbb{B}_r(a) \\ \psi(x) & \text{if } y \in S \end{cases} \tag{6.5}$$

stands for a solution for the generalized Dirichlet-like problem.

7 Conclusions

The procedure used in the previous section is a simple method to obtain the Poisson-Szegö kernels for a large family of orientable manifolds by using the initial kernel for the spherical model of hyperbolic spaces. However, one has to note a disadvantage of this method. We are, as expected, depending on the shape of the domain in which we work. This generates extra difficulties in the construction of Bergman and Poisson kernels for the above-mentioned manifolds.

REFERENCES

[A]   L. V. Ahlfors, Möbius transformations and Clifford numbers, *Differential Geometry and Complex Analysis*, Springer-Verlag, Berlin, Heidelberg, 1985.

[C]   J. Cnops, Hurwitz pairs and applications of Möbius transformations, Habilitation Thesis, Ghent, Belgium, 1994.

[Ce] P. Cerejeiras, O operador de Dirac em espaços hiperbólicos, Ph.D. Thesis, Aveiro, Portugal, 1997.

314   Paula Cerejeiras

[FiSp] J. Fillmore and A. Springer, Möbius groups over general fields using Clifford algebras associated with spheres, *Int. Jour. Theor. Phys.* **29** (1990), 225–246.

[Hu] A. Huber, On the uniqueness of generalized axially symmetric potentials, *Annals of Mathematics*, Vol. **60**, No. 2 (September, 1954), 351–358.

[Le] H. Leutwiler, Remarks on modified Clifford analysis, in *Potential Theory - ICPT 94*, Král, et al., eds., Walter de Gruyter & Co., Berlin, New York, 1996, 389–397.

[Loo] Hua Loo-Keng, *Starting with the Unit Circle*, Springer-Verlag, New York, 1981.

[PeT] Jaak Peetre and Tao Qian, Möbius covariance of iterated Dirac operators, *J. of Austral. Math. Soc.*, Vol. **56** (Series A) (1994), 403–414.

[Ry] John Ryan, Applications of conformal covariance in Clifford analysis, *Clifford Algebras in Analysis and Related Topics*, Stud. Adv. Math., CRC Press, 1995, 129–156.

[Sud] A. Sudbery, Quaternionic analysis, *Math. Proc. Camb. Phil. Soc.*, Vol. **85** (1979), 199–225.

[TRy] Tao Qian and John Ryan, Conformal transformations and Hardy spaces arising in Clifford analysis, *J. Operator Theory*, Vol. **35** (1996), 349–372.

[Z] G. Zöll, Regular n-Forms in Clifford analysis, their behavior under change of variables and their residues, *Complex Variables*, Vol. **11** (1989), 25–38.

Paula Cerejeiras
Departamento de Matemática
Universidade de Aveiro
P-3810-193 Aveiro, Portugal
E-mail: pceres@mat.ua.pt

Received: September 30, 1999; Revised March 17, 2000

Index

D.

E.

F.

318

T.

V.

W.

Z.